T0122348

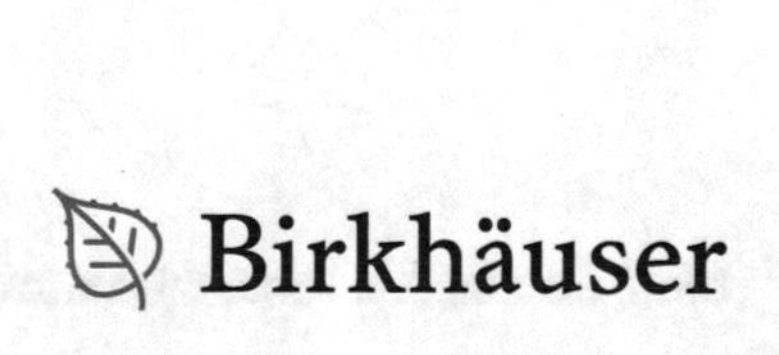
Birkhäuser

Progress in Nonlinear Differential Equations and Their Applications: Subseries in Control

Volume 90

More information about this series at https://www.springer.com/series/15137

Viorel Barbu

Controllability and Stabilization of Parabolic Equations

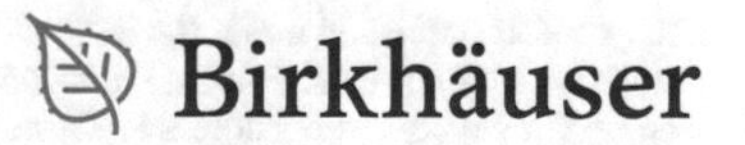

Viorel Barbu
A1. I CUZA UNIVERSITY
IASI, Romania

ISSN 1421-1750 ISSN 2374-0280 (electronic)
Progress in Nonlinear Differential Equations and Their Applications
ISBN 978-3-030-09550-5 ISBN 978-3-319-76666-9 (eBook)
https://doi.org/10.1007/978-3-319-76666-9

Mathematics Subject Classification (2010): 93D15, 93C10, 49N35, 35K55, 35K58

Printed on acid-free paper

This book is published under the trade name Birkhäuser, www.birkhauser-science.com by the registered company Springer International Publishing AG part of Springer Nature.
The registered company address is: Gewerbestrasse 11, 6330 Cham, Switzerland

Preface

This book is based on the author's works and lectures on controllability and stabilization of parabolic equations and part of it was used for a graduate course at the University of Iaşi, Romania.

In the inaugural lecture delivered at Warwick on the 7th of October 1970, Lawrence Markus has briefly defined the objectives of the mathematical control theory as:

> "... *the modification of differential equations, within prescribed limitations so that the solutions behave in some desired manner. The common feature of all these problems is that we prescribe the desired behaviour of the solutions and then we seek to modify the coefficients of the dynamical equations so as to induce this behaviour. In classical mathematical physics, we know all about the physical laws guiding the development of the phenomenon we may be studying – we do not know all the rules of the game – and we want to predict the outcome. In control theory, we know the rules, in fact we can change them within certain limitations, but we do know exactly how we want the game to end. Thus, the mathematical problems of control theory are inverse to the usual problems of mathematical physics.*"

The controllability and stabilization are without any doubt two fundamental problems of mathematical control theory which, for parabolic-like systems, have a special significance and difficulty due to the time irreversibility of dynamics generated by these systems.

The first part of the book is devoted to the internal null controllability of parabolic equations based essentially on the influential linear controllability result of G. Lebeau and L. Robbiano (1995), developed later on by A.V. Fursikov and O.Yu. Imanuvilov (1996), via a Carleman-type inequality for linear parabolic equations. Compared with controllability, internal and boundary stabilization are apparently weaker properties but, since they are usually realized by feedback controllers, they are structurally stable and so more convenient for applications and numerical simulations. The first stabilization method developed here is based on the spectral decomposition technique, which is used to represent the linear parabolic control systems in a convenient product space as a null controllable finite dimensional unstable system and a stable infinite dimensional one. Such a system is stabilizable by an open loop controller, and so a stabilizable linear feedback controller can be obtained by standard optimal control arguments from an algebraic Riccati equation. As a matter of fact, this stabilization method applies to general semilinear control systems of the form $\frac{dy}{dt} + Ay + Fy = Bu$ in a Hilbert space H, where the operator $-A$ is the infinitesimal generator of a C_0-analytic semigroup in H and has a compact resolvent. This is the class of so-called parabolic-like systems which, besides the standard heat and diffusion equations, includes Navier–Stokes and other related systems such as Boussinesq and magnetohydrodynamic equations.

The second method mostly applicable to boundary control systems consists in the explicit construction of a stabilizing feedback controller by using a finite number of unstable modes. Both methods provide stabilizable feedback controllers with finite dimensional structure and stabilization. The exact controllability of stochastic parabolic equations with linear multiplicative noise is also briefly treated in this book, though so far only partial results were obtained in this direction.

It should be mentioned, however, that none of the above topics was exhaustively treated in this book, which may be viewed only as a survey on controllability and stabilization techniques for parabolic boundary value problems. Many of these results, exact controllability in particular, keep their original strength and flavor, even though they were established at the end of the 1990s.

There are several important topics related to the content of this book which were not treated here and most notable is perhaps the controllability of Navier–Stokes equations extensively studied in the last two decades.

We wish to thank J.M. Coron for the invitation to write this book and publish it with the *Control/PNLDE* series.

Thanks are due also to my colleagues Gabriela Marinoschi, Cătălin-George Lefter and Ionuţ Munteanu who read and criticized various chapters.

Thanks are also due to the anonymous reviewers for their suggestions.

I am also indebted to Mrs. Elena Mocanu for typing and processing this book.

Iasi, Romania Viorel Barbu
January 1st, 2018

Contents

Acronyms

p_K	the Minkowski functional of the set K
$\mathrm{recc}(K)$	the recession cone of K
$\mathbb{C}$	the set of all complex numbers
$\mathbb{N}$	the set of all natural numbers
$\mathbb{R}$	the real line $(-\infty, \infty)$
$\mathbb{R}^d$	the d-dimensional Euclidean space
$\mathbb{R}^+$	$= (0, +\infty)$
$\mathbb{R}^d_+$	$= \{(x_1, \dots, x_d);\ x_d > 0\}$
$\mathcal{O}$	an open subset of $\mathbb{R}^d$
$\partial\mathcal{O}$	the boundary of $\mathcal{O}$
Q	$= \mathcal{O} \times (0, T)$
Σ	$= \partial\mathcal{O} \times (0, T)$, where $0 < T < \infty$
$\\|\cdot\\|_X$	the norm of the linear normed space X
X^*	the dual of the space X
$(\cdot, \cdot)_H$	the scalar product of the Hilbert space H
$x \cdot y$	the scalar product of the vectors $x, y \in \mathbb{R}^d$
$L(X, Y)$	the space of linear continuous operators from X to Y
∇f	the gradient of the function f
∂f	the subdifferential of the function f
B^*	the adjoint of the operator B
$\overline{C}$	the closure of the set C
$\mathrm{int}\, C$	the interior of the set C
$\mathrm{conv}\, C$	the convex hull of the set C
$D(A)$	the domain of the operator A
$R(A)$	the range of the operator A
I_C	the indicator function of the set C
sign	the signum function on X : $\mathrm{sign}\, x = x/\\|x\\|_X$ if $x \neq 0$, $\mathrm{sign}\, 0 = \{x;\ \\|x\\|_X \leq 1\}$
$C^k(\mathcal{O})$	the space of real-valued functions on $\mathcal{O}$ that are continuously differentiable up to order k, $k \leq \infty$

$C_0^k(\mathscr{O})$	the subspace of functions in $C^k(\mathscr{O})$ with compact support in $\mathscr{O}$
$\mathscr{D}(\mathscr{O})$	the space $C_0^\infty(\mathscr{O})$
$\frac{d^k u}{dt^k}, u^{(k)}$	the derivative of order k of the function $u : [a, b] \to X$
$\mathscr{D}'(\mathscr{O})$	the dual of $\mathscr{D}(\mathscr{O})$ (i.e., the space of distributions on $\mathscr{O}$)
$C(\overline{\mathscr{O}})$	the space of continuous functions on $\overline{\mathscr{O}}$
$L^p(\mathscr{O})$	the space of p-summable functions $u : \mathscr{O} \to \mathbb{R}$ endowed with the norm $\lvert u\rvert_p = (\int_{\mathscr{O}} \lvert u(x)\rvert^p dx)^{\frac{1}{p}}$, $1 \leq p < \infty$
$L_m^p(\mathscr{O})$	the space of p-summable functions $u : \mathscr{O} \to \mathbb{R}^m$
$W^{m,p}(\mathscr{O})$	the Sobolev space $\{u \in L^p(\mathscr{O});\ D^\alpha u \in L^p(\mathscr{O}),$ $\lvert\alpha\rvert \leq m,\ 1 \leq p \leq \infty\}$
$W_0^{m,p}(\mathscr{O})$	the closure of $C_0^\infty(\mathscr{O})$ in the norm of $W^{m,p}(\mathscr{O})$
$W^{-m,q}(\mathscr{O})$	the dual of $W_0^{m,p}(\mathscr{O})$; $\frac{1}{p} + \frac{1}{q} = 1$, $p < \infty$, $q > 1$
$H^k(\mathscr{O}), H_0^k(\mathscr{O})$	the spaces $W^{k,2}(\mathscr{O})$ and $W_0^{k,2}(\mathscr{O})$, respectively
$L^p(a, b; X)$	the space of p-summable functions from (a, b) to X, $1 \leq p \leq \infty$, $-\infty \leq a < b \leq \infty$
$C([a, b]; X)$	the space of X-valued continuous functions on $[a, b]$
$AC([a, b]; X)$	the space of absolutely continuous functions from $[a, b]$ to X
$W^{1,p}([a, b]; X)$	the Sobolev space $\left\{u \in AC([a, b]; X);\ \frac{du}{dt} \in L^p((a, b); X)\right\}$
ν, n	the outward normal to $\mathscr{O}$
$\frac{\partial u}{\partial \nu}, \frac{\partial u}{\partial n}$	the normal derivative of the function $u : \mathscr{O} \to \mathbb{R}$

Chapter 1
Preliminaries

Here we survey for later use some basic existence results for the infinite dimensional Cauchy problem, semilinear parabolic-like boundary value problems, and infinite dimensional control systems.

1.1 Notations

Given an open subset $\mathscr{O}$ of $\mathbb{R}^d$, $d \in \mathbb{N}$, we denote by $\overline{\mathscr{O}}$ its closure and $\partial\mathscr{O}$ its boundary. By $L^p(\mathscr{O})$, $1 \le p \le \infty$, we denote the standard space of L^p-Lebesgue integrable functions on $\mathscr{O}$ with the norm denoted $|\cdot|_p$ or $\|\cdot\|_{L^p(\mathscr{O})}$. By $\mathscr{D}'(\mathscr{O})$, we denote the space of distributions on $\mathscr{O}$, that is, the dual of $C_0^\infty(\mathscr{O})$. Denote by $W^{1,p}(\mathscr{O})$, $1 \le p \le \infty$, the Sobolev space

$$\left\{u \in L^p(\mathscr{O}),\ \frac{\partial u}{\partial x_i} \in L^p(\mathscr{O}),\ \ i = 1, \dots, d\right\},$$

where $\frac{\partial}{\partial x_i}$ is taken in $\mathscr{D}'(\mathscr{O})$. We denote by $C_0^\infty(\mathscr{O}) = \mathscr{D}(\mathscr{O})$ the space of infinitely differentiable functions with compact support in $\mathscr{O}$ and by $W_0^{1,p}(\mathscr{O})$ the closure of $C_0^\infty(\mathscr{O})$ in $W^{1,p}(\mathscr{O})$. For $p = 2$, we set

$$W^{1,2}(\mathscr{O}) = H^1(\mathscr{O}),\ \ W_0^{1,2}(\mathscr{O}) = H_0^1(\mathscr{O}),$$
$$H^2(\mathscr{O}) = \left\{u \in L^2(\mathscr{O});\ \frac{\partial^2 u}{\partial x_i \partial x_j} \in L^2(\mathscr{O}),\ i, j = 1, \dots, d\right\}.$$

The dual space of $W_0^{1,p}(\mathscr{O})$ is denoted by $W^{-1,p'}(\mathscr{O})$, $\frac{1}{p} + \frac{1}{p'} = 1$, and set $W^{-1,2}(\mathscr{O}) = H^{-1}(\mathscr{O})$. Given a Banach space X, denote by $C([0, T]; X)$ the

V. Barbu, *Controllability and Stabilization of Parabolic Equations*,
Progress in Nonlinear Differential Equations and Their Applications 90,
https://doi.org/10.1007/978-3-319-76666-9_1

space of X-valued continuous functions on $[0, T] \subset [0, \infty)$ and by $L^p(0, T; X)$ the space of X-valued L^p-Bochner integrable functions on $(0, T)$. We denote by $C_w([0, T]; X)$ the space of weakly continuous X-valued functions on $[0, T]$. By $W^{1,p}([0, T]; X)$, we denote the infinite dimensional Sobolev space $\left\{u \in L^p(0, T; X);\ \frac{du}{dt} \in L^p(0, T; X)\right\}$, where $\frac{du}{dt}$ is taken in sense of X-valued vectorial distributions on $(0, T)$. If X, Y are Banach spaces, $L(X, Y)$ is the space of linear continuous operators from X to Y with the operatorial norm $\|\cdot\|_{L(X,Y)}$. The norm of X will be denoted $\|\cdot\|_X$ or simply $\|\cdot\|$.

1.2 The Nonlinear Cauchy Problem in Banach Spaces

Let X be a real Banach space with the dual space X^* and let $A : D(A) \subset X \to 2^X$ be a nonlinear multivalued operator. Here $D(A)$ is the domain of A, that is,

$$D(A) = \{u \in X;\ Au \neq \emptyset\}.$$

The multivalued mapping A is very often identified with a subset of $X \times X$ by

$$A = \{(v, u) \in X \times X;\ v \in Au\}.$$

The operator A is said to be *accretive* if

$${}_X(v_1 - v_2, \eta)_{X^*} \geq 0 \ \text{ for } \eta \in J(u_1 - u_2),\ \forall\, v_i \in Au_i,\ i = 1, 2,$$

where $J : X \to X^*$ is the duality map of the space X and $\langle \cdot, \cdot \rangle_{X^*}$ is the duality function on $X \times X^*$. Equivalently,

$$\|(I + \lambda A)^{-1} f - (I + \lambda A)^{-1} g\|_X \leq \|f - g\|_X,$$

for all $f, g \in R(I + \lambda A)$ and $\lambda > 0$. Here $R(I + \lambda A)$ is the range of the operator $I + \lambda A$. The operator A is said to be *m-accretive* if A is accretive and $R(I + \lambda A) = X$ for all $\lambda > 0$ (equivalently, for some $\lambda > 0$). The operator is said to be *quasi-m-accretive* if $A + \alpha I$ is m-accretive for some $\alpha > 0$ and *quasi-accretive* if $A + \alpha I$ is accretive.

If X is a real Hilbert space with the scalar product $(\cdot, \cdot)$, then A is m-accretive if and only if it is maximal monotone in $X \times X$, that is, $(v_1 - v_2, u_1 - u_2) \geq 0$ for all $v_i \in Au_i$, $i = 1, 2$, and A has no proper monotone extension on $X \times X$ (see [17], p. 39, [18, 26]).

If X is a Hilbert space and $\varphi : X \to]-\infty, +\infty]$ is a convex and lower semicontinuous function, then its subdifferential $\partial\varphi : X \to 2^X$ is defined by $\partial\varphi(x) = \{z \in X;\ \varphi(x) \leq \varphi(y) + (z, x - y),\ \forall\, y \in X\}$. The operator $A = \partial\varphi$ is m-accretive (equivalently, maximal monotone) (see, [26, 54], p.47).

Consider the Cauchy problem

$$\begin{aligned}&\frac{dy}{dt} + Ay \ni 0 \text{ in } (0, T),\\ &y(0) = y_0.\end{aligned} \tag{1.1}$$

The function $y : [0, T] \to X$ is said to be a *strong solution* to (1.1) if it is absolutely continuous on $[0, T]$ and there is $\eta \in L^1(0, T; X)$ such that $\eta(t) \in Ay(t)$, $t \in (0, T)$, and

$$\begin{aligned}&\frac{dy}{dt}(t) + \eta(t) = 0, \quad \text{a.e. } t \in (0, T),\\ &y(0) = y_0.\end{aligned} \tag{1.2}$$

The function $y \in C([0, T]; X)$ is said to be a *mild* solution to (1.2) if

$$y(t) = \lim_{h\to 0} y_h(t), \ \forall t \in [0, T], \tag{1.3}$$

where $y_h : [0, T] \to H$ is the step function,

$$y_h(t) = y_i^h \text{ for } t \in [ih, (i + 1)h), \ i = 0, 1, \dots, N - 1, \tag{1.4}$$

$$y_{i+1}^h + hAy_{i+1}^h \ni y_i^h. \tag{1.5}$$

$y_0^h = y_0$ and $Nh = T$.

In other words, a *mild solution* is a continuous X-valued function on $[0, T]$ which is the limit of the solution y_h to the finite difference scheme (1.4)–(1.5) associated with problem (1.1). Equivalently,

$$y(t) = \lim_{h\to\infty} \left(I + \frac{t}{n} A\right)^{-n} y_0, \ \forall t \in [0, T]. \tag{1.6}$$

We have the following fundamental result (see [26], p. 138).

Theorem 1.1 (Crandall & Liggett). *Let A be a quasi-m-accretive operator in a real Banach space X. Then, for each $y_0 \in \overline{D(A)}$, there is a unique mild solution y to the Cauchy problem* (1.1). *If the space X is reflexive and $y_0 \in D(A)$, then the mild solution y is strong solution and $\frac{dy}{dt} \in L^\infty(0, T; X)$ (that is, $y \in W^{1,\infty}([0, T]; X)$). If X is a Hilbert space and $A = \partial\varphi$, then, for $y_0 \in D(\varphi) = \{\varphi;\ \varphi(y) < \infty\}$, $y \in W^{1,2}([0, T]; X)$, $\varphi(y) \in L^\infty(0, T; X)$, and $\sqrt{t}\,\frac{dy}{dt} \in L^2(0, T; X)$, $\varphi(y) \in L^1(0, T; X)$ for $y_0 \in \overline{D(A)} = \overline{D(\varphi)}$.*

Here $\overline{D(A)}$ is the closure of $D(A)$ in X. We note also that the map $y_0 \to y(t)$ defines a continuous semigroup of quasi-contractions on $\overline{D(A)}$, $S(t) : \overline{D(A)} \to \overline{D(A)}$ also denoted by e^{-tA}.

We have also the following more general result (see [26], p. 130).

Theorem 1.2. *Let A be quasi-accretive in a Banach space X and let C be a closed convex cone of X such that*

$$D(A) \subset C \subset \bigcap_{0<\lambda<\lambda_0} \mathbb{R}(1+\lambda A)$$

for some $\lambda_0 > 0$. Then, for each $y_0 \in \overline{D(A)}$, problem (1.1) *has a unique mild solution which is a strong solution if X is reflexive and $y_0 \in D(A)$.*

The standard way to prove these theorems is to approximate equation (1.1) by

$$\begin{aligned} &\frac{dy}{dt} + A_\lambda y = 0, \ t \in [0,T], \\ &y(0) = y_0, \end{aligned} \tag{1.7}$$

where $A_\lambda = \lambda^{-1}(I-(I+\lambda A)^{-1}),\ \lambda > 0$ is the *Yosida approximation* of A.

Example 1.1. Let $X = L^1(\mathcal{O})$, where $\mathcal{O}$ is a bounded and open domain of $\mathbb{R}^d$ and let $A : D(A) \subset L^1(\mathcal{O}) \to L^1(\mathcal{O})$ be the operator

$$\begin{aligned} Ay &= \{-\Delta\eta;\ \eta \in \beta(y),\ \text{a.e. in } \mathcal{O}\},\ \forall\, y \in D(A), \\ D(A) &= \{y \in L^1(\mathcal{O});\ \exists\, \eta \in W_0^{1,1}(\mathcal{O}),\ \eta \in \beta(y),\ \text{a.e. in } \mathcal{O}\}, \end{aligned}$$

where $\beta : D(\beta) \subset \mathbb{R} \to 2^{\mathbb{R}},\ \beta(0) \ni 0$, is a maximal monotone graph in $\mathbb{R} \times \mathbb{R}$, that is, β is monotone and $\mathbb{R}(I+\beta) = \mathbb{R}$. Then (see [26], p. 117) the operator A is m-accretive in $L^1(\mathcal{O})$ and, by Theorem 1.1, we have

Theorem 1.3. *For each $y_0 \in L^1(\mathcal{O})$, such that $y_0(x) \in \overline{D(\beta)}$, a.e., $x \in \mathcal{O}$, there is a unique mild solution to the equation*

$$\begin{aligned} &\frac{\partial y}{\partial t} - \Delta\beta(y) \ni 0 \ \textit{in}\ (0,T) \times \mathcal{O}, \\ &y(0,x) = y_0(x),\ x \in \mathcal{O}, \\ &\beta(y) = 0 \qquad\qquad \textit{on}\ (0,T) \times \partial\mathcal{O}. \end{aligned} \tag{1.8}$$

It turns out that the mild solution y to (1.8) is just a solution in sense of distributions on $(0,T) \times \mathcal{O}$.

Equation (1.8) can be also studied in a different functional setting, namely in the Sobolev space $H^{-1}(\mathcal{O})$. In fact, the operator

$$\widetilde{A}y = \{-\Delta\eta;\ \eta \in \beta(y),\ \eta \in H_0^1(\mathcal{O})\}$$

is m-accretive in $H^{-1}(\mathcal{O})$ (see, e.g., [26], p. 237) and so, again by Theorem 1.1, equation (1.8) has, for $y_0 \in \overline{D(\widetilde{A})}$, a unique strong solution $y \in C([0,T];\ H^{-1}(\mathcal{O}))$ such that $\frac{dy}{dt} \in L^\infty(0,T;H^{-1}(\mathcal{O})),\ \beta(y) \in L^\infty(0,T;H_0^1(\mathcal{O}))$.

1.3 Semilinear Parabolic Equations

We shall review here the main existence result for the parabolic boundary value

$$
\begin{aligned}
&\frac{\partial y}{\partial t}(x,t)-\sum_{i,j=1}^{d}\frac{\partial}{\partial x_i}\left(a_{ij}(x)\frac{\partial y}{\partial x_j}(x,t)\right)+f(x,y(x,t))=f_0(x,t) \\
&\qquad\qquad\qquad\qquad\qquad\qquad\qquad\qquad \text{in } Q=\mathcal{O}\times(0,T), \\
&\alpha_1\frac{\partial y}{\partial \nu}+\alpha_2 y=0 \qquad\qquad\qquad\qquad\quad \text{on } \Sigma=\partial\mathcal{O}\times(0,T), \\
&y(x,0)=y_0(x) \qquad\qquad\qquad\qquad\qquad\;\; \text{in } \mathcal{O}.
\end{aligned}
\tag{1.9}
$$

Here $\mathcal{O}$ is an open and bounded set of $\mathbb{R}^d$ with the boundary $\partial\mathcal{O}$ of class C^2 and $f_0 \in L^2(Q),\ y_0 \in L^2(\mathcal{O})$ are given functions. We assume that

$$\alpha_i \geq 0,\ \alpha_1+\alpha_2>0,\ \alpha_i \in C(\partial\mathcal{O}),\ i=1,2,\dots,d, \tag{1.10}$$

$$a_{ij}\in C^1(\overline{\mathcal{O}}),\ a_{ij}=a_{ji},\ \sum_{i,j=1}^{d} a_{ij}\xi_i\xi_j \geq \omega|\xi|^2,\ \forall \xi \in \mathbb{R}^d, \tag{1.11}$$

where $\omega > 0,\ \xi=(\xi_1,\dots,\xi_d)$. Finally, $f:\overline{\mathcal{O}}\times\mathbb{R}\to\mathbb{R}$ is continuous function satisfying the following assumptions

$$f\in C^1(\overline{\mathcal{O}}\times\mathbb{R}),\ f_y(x,y)\geq -\gamma,\ \forall (x,y)\in\overline{\mathcal{O}}\times\mathbb{R}, \tag{1.12}$$

$$f_x \in L^\infty(Q), \tag{1.13}$$

where $f_y=\nabla_y f,\ f_x=\nabla_x f$. By $\frac{\partial}{\partial \nu_A}$ we have denoted the outward conormal derivative corresponding to the elliptic operator $\sum_{i,j=1}^{d}\frac{\partial}{\partial x_i}\left(a_{ij}\frac{\partial}{\partial x_j}\right)$, that is,

$$\frac{\partial y}{\partial \nu_A}=\sum_{i,j=1}^{d} a_{ij}\frac{\partial y}{\partial x_j}\nu_j,$$

where $\nu=(\nu_1,\dots,\nu_d)$ is the outward normal to $\mathcal{O}$.

In the special case $a_{ij}=\delta_{ij}$ (Kronecker's symbol),

$$\frac{\partial y}{\partial \nu_A}=\frac{\partial y}{\partial \nu}=\nabla y\cdot\nu \ \text{ on } \partial\mathcal{O}$$

is the outward normal derivative of y. We have

Theorem 1.4. *Assume that*

$$y_0 \in H^2(\mathscr{O}),\ \alpha_1 \frac{\partial y_0}{\partial \nu} + \alpha_2 y_0 = 0 \ \text{ on } \partial\mathscr{O}, \tag{1.14}$$

$$f_0 \in W^{1,1}([0,T]; L^2(\mathscr{O})). \tag{1.15}$$

Then there is a unique strong solution y to (1.9) *satisfying*

$$y \in C([0,T]; H^1(\mathscr{O})) \cap L^\infty(0,T; H^2(\mathscr{O})) \cap W^{1,\infty}([0,T]; L^2(\mathscr{O})). \tag{1.16}$$

If one merely assumes that

$$y_0 \in H^1(\mathscr{O})\ (y_0 \in H_0^1(\mathscr{O}) \text{ if } \alpha_1 = 0),\ \ f_0 \in L^2(Q),\ \ j(x, y_0) \in L^1(\mathscr{O}), \tag{1.17}$$

then

$$y \in W^{1,2}([0,T]; L^2(\mathscr{O})) \cap L^2(0,T; H^2(\mathscr{O})) \cap C_w([0,T]; H^1(\mathscr{O})). \tag{1.18}$$

Finally, if $y_0 \in L^2(\mathscr{O})$ and $f_0 \in L^2(Q)$, then

$$y \in C([0,T]; L^2(\mathscr{O})) \cap L^2(0,T; H^1(\mathscr{O})),\ \sqrt{t}\, y \in L^2(0,T; H^2(\mathscr{O})), \tag{1.19}$$

$$\sqrt{t}\, y_t \in L^2(Q). \tag{1.20}$$

Here $j(x,r) \equiv \int_0^r f(x,s)ds$.

By strong solution to (1.9) we mean a continuous function $y : [0,T] \to L^2(\mathscr{O})$ which is a.e. differentiable and satisfies a.e. the equation on Q. By y_t we mean of course $\frac{dy}{dt}$, the strong derivative of $y \in W^{1,2}([0,T]; L^2(\mathscr{O}))$.

A convenient way to treat problem (1.9) is to write it as a Cauchy problem of the form (1.1) in the space $H = L^2(\mathscr{O})$. Namely, we view $y = y(x,t)$ as an H-valued function $y : [0,T] \to H$ and rewrite (1.9) as

$$\begin{aligned} &\frac{dy}{dt}(t) + Ay(t) = f_0(t), \quad t \in (0,T), \\ &y(0) = y_0, \end{aligned} \tag{1.21}$$

where $A : H \to H$ is the operator

$$Ay = -\sum_{i,j=1}^{d} \frac{\partial}{\partial x_i}\left(a_{ij}\frac{\partial y}{\partial x_j}\right) + f(x,y),\ \forall\, y \in D(A),$$

with the domain

$$D(A) = \left\{ y \in H^2(\mathscr{O});\ \alpha_1 \frac{\partial y}{\partial \nu} + \alpha_2 y = 0 \text{ on } \partial\mathscr{O};\ f(x,y) \in L^2(\mathscr{O}) \right\}. \tag{1.22}$$

We have

Lemma 1.1. *The operator A is quasi-m-accretive in $H \times H$.*

Proof. As seen earlier, this means that

$$(\lambda(y-\bar{y}) + Ay - A\bar{y}, y - \bar{y}) \geq 0,\ \forall\, y, \bar{y} \in D(A),\ \lambda > \lambda_0, \tag{1.23}$$

and

$$R(\lambda I + A) = L^2(\mathscr{O}),\ \forall\, \lambda > \lambda_0, \tag{1.24}$$

where $R(\lambda I + A)$ is the range of $\lambda I + A$ and $(\cdot, \cdot)$ is the scalar product in $L^2(\mathscr{O})$. Inequality (1.23) is an immediate consequence of (1.10)–(1.12). To prove (1.24), we fix $g \in L^2(\mathscr{O})$ and consider the equation

$$\lambda y + Ay = g. \tag{1.25}$$

We set

$$\begin{aligned}
f_\varepsilon(x,r) &= \frac{1}{\varepsilon}(r - (1 + \varepsilon f(x,\cdot))^{-1} r)\ \forall\, r \in \mathbb{R},\ \varepsilon > 0,\\
A_0 y &= -\sum_{i=1}^{d} \frac{\partial}{\partial x_i}\left(a_{ij}\frac{\partial y}{\partial x_j}\right),\\
D(A_0) &= \left\{ y \in H^2(\mathscr{O});\ \alpha_1 \frac{\partial y}{\partial \nu_A} + \alpha_2 y = 0 \text{ on } \partial\mathscr{O} \right\}.
\end{aligned}$$

This is the Yosida approximation of $f(x,\cdot)$ and it is easily seen that for each $x \in \mathscr{O}$, $f_\varepsilon(x,\cdot)$ is Lipschitzian, monotonically increasing and

$$\begin{gathered}
f_\varepsilon(x,r) = f(x, J_\varepsilon(r)) = \frac{1}{\varepsilon}(r - J_\varepsilon(r)),\ \forall\, r \in \mathbb{R},\\
J_\varepsilon(r) = (1 + \varepsilon f(x,\cdot))^{-1}(r),\ \lim_{\varepsilon \downarrow 0} f_\varepsilon(x,r) = f(x,r),\ \forall\, x \in \mathscr{O},\ r \in \mathbb{R},\\
|J_\varepsilon(r) - J_\varepsilon(\bar{r})| \leq (1 - \varepsilon\gamma)^{-1}|r - \bar{r}|,\ \forall\, r, \bar{r} \in \mathbb{R}.
\end{gathered} \tag{1.26}$$

Consider the approximating equation

$$\lambda y + A_0 y + f_\varepsilon(x,y) = g,\ x \in \mathscr{O},\ \varepsilon > 0, \tag{1.27}$$

and note that it has a unique solution $y_\varepsilon \in H^2(\mathscr{O})$.

Indeed, for each $z \in L^2(\mathscr{O})$, arbitrary but fixed, the equation

$$\begin{aligned}
&\left(\lambda + \frac{1}{\varepsilon}\right) y - \sum_{i,j=1}^{d} \frac{\partial}{\partial x_i}\left(a_{ij}\frac{\partial y}{\partial x_j}\right) = g + \frac{1}{\varepsilon} J_\varepsilon(z) \text{ in } \mathscr{O},\\
&\alpha_1 \frac{\partial y}{\partial \nu} + \alpha_2 y = 0 \qquad \text{on } \partial\mathscr{O},
\end{aligned}$$

has, by the Lax-Milgram lemma and elliptic regularity theorem, a unique solution $y = F(z) \in H^2(\mathcal{O})$ (see, e.g., [55]). Moreover, by an elementary calculation involving Green's formula, we see by (1.26) that

$$\|Fz - F\bar{z}\|_{L^2(\mathcal{O})} \leq \rho\|z - \bar{z}\|_{L^2(\mathcal{O})}, \ 0 < \rho < 1, \ \text{ for } \lambda \geq \lambda_0.$$

Then, by Banach's fixed point theorem, there is a solution $y_\varepsilon \in H^2(\mathcal{O})$.

A Priori Estimates

Assume here $\alpha_1 > 0$, but the case $\alpha_1 = 0$ can be similarly treated. Multiplying (1.27) by y_ε and taking into account that, by (1.26) and (1.12),

$$\begin{aligned} f_\varepsilon(x, r)r = f(x, J_\varepsilon(r)J_\varepsilon(r)) + \varepsilon f_\varepsilon^2(x, r) &\geq -\gamma|y_\varepsilon(r)|^2 + f(x, 0)y_\varepsilon(r) + \varepsilon f_\varepsilon^2(r) \\ &\geq -2\gamma|r^2|, \forall r \in \mathbb{R}, \end{aligned}$$

we obtain via Green's formula that

$$\lambda\|y_\varepsilon\|^2_{L^2(\mathcal{O})} + \omega\|\nabla y_\varepsilon\|^2_{L^2(\mathcal{O})} \leq C(\|y_\varepsilon\|^2_{L^2(\mathcal{O})} + 1).$$

Hence, for $\lambda \geq \lambda_0$ large enough,

$$\|y_\varepsilon\|_{H^1(\mathcal{O})} \leq C, \ \forall \varepsilon > 0. \tag{1.28}$$

(We shall use the same symbol C to denote several positive constants independent of ε.) Next, multiply (1.27) by $f_\varepsilon(x, y_\varepsilon)$ and integrate on $\mathcal{O}$. Noticing that

$$\begin{aligned} -\sum_{i,j=1}^{d} \int_{\mathcal{O}} \frac{\partial}{\partial x_i}\left(a_{ij}\,\frac{\partial y_\varepsilon}{\partial x_j}\right) f_\varepsilon(x, y_\varepsilon)dx = &\sum_{i,j=1}^{d} \int_{\mathcal{O}} a_{ij}\,\frac{\partial y_\varepsilon}{\partial x_j}\left(\frac{\partial y_\varepsilon}{\partial x_i}\,\frac{\partial f_\varepsilon}{\partial y} + \frac{\partial f_\varepsilon}{\partial x_i}\right) dx \\ &+ \int_{\partial\mathcal{O}} \frac{\alpha_2}{\alpha_1}\, f_\varepsilon(x, y_\varepsilon)y_\varepsilon\, d\sigma, \end{aligned}$$

we obtain, via the trace theorem, the estimate

$$\sum_{i,j=1}^{d} \int_{\mathcal{O}} \left|\frac{\partial}{\partial x_i}\left(a_{ij}\,\frac{\partial y_\varepsilon}{\partial x_j}\right)\right|^2 dx + \int_{\mathcal{O}} f_\varepsilon^2(x, y_\varepsilon)dx \leq C, \ \forall \varepsilon > 0,$$

and, therefore, we have

$$\|y_\varepsilon\|_{H^2(\mathcal{O})} + \|f_\varepsilon(\cdot, y_\varepsilon)\|_{L^2(\mathcal{O})} \leq C, \ \forall \varepsilon > 0. \tag{1.29}$$

By (1.11) and (1.27), we obtain also

$$\begin{aligned} \lambda\|y_\varepsilon - y_\mu\|^2_{L^2(\mathcal{O})} &+ \omega\|\nabla(y_\varepsilon - y_\mu)\|^2_{L^2(\mathcal{O})} \\ &+ \int_{\mathcal{O}} (f_\varepsilon(x, y_\varepsilon(x)) - f_\mu(x, y_\varepsilon(x)))(y_\varepsilon(x) - y_\mu(x))dx \leq 0. \end{aligned}$$

Taking into account that, by (1.26),

$$\begin{aligned}(f_\varepsilon(\cdot, y_\varepsilon) - f_\mu(\cdot, y_\mu))(y_\varepsilon - y_\mu) \\ = (f(\cdot, J_\varepsilon(y_\varepsilon)) - f(\cdot, J_\mu(y_\mu)))(J_\varepsilon(y_\varepsilon) - J_\mu(y_\mu)) \\ +(f_\varepsilon(x, y_\varepsilon) - f_\mu(x, y_\mu))(\varepsilon f_\varepsilon(x, y_\varepsilon) - \mu f_\mu(x, y_\mu))\end{aligned} \tag{1.30}$$

and that $y \to f(x, y) + \gamma y$ is monotonically increasing, we obtain by (1.29) and (1.30) that

$$\lambda\|y_\varepsilon - y_\mu\|^2_{L^2(\mathcal{O})} + \omega\|y_\varepsilon - y_\mu\|^2_{H^1(\mathcal{O})} \le C(\varepsilon + \lambda), \ \forall\, \varepsilon, \lambda > 0.$$

Hence $\{y_\varepsilon\}$ is strongly convergent in $H^1(\mathcal{O})$ for $\varepsilon \to 0$. We set

$$y = \lim_{\varepsilon \downarrow 0} y_\varepsilon \ \text{ in } H^1(\mathcal{O}). \tag{1.31}$$

Moreover, by (1.29) it follows that $\{A_0 y_\varepsilon\}$ and $\{f_\varepsilon(\cdot, y_\varepsilon)\}$ are weakly compact in $L^2(\mathcal{O})$. Hence, for $\varepsilon \to 0$, we have

$$A_0 y_\varepsilon \longrightarrow A_0 y \text{ weakly in } L^2(\mathcal{O}), \tag{1.32}$$

$$f_\varepsilon(\cdot, y_\varepsilon) \longrightarrow \eta \quad \text{weakly in } L^2(\mathcal{O}), \tag{1.33}$$

$$y_\varepsilon \longrightarrow y \quad \text{weakly in } H^2(\mathcal{O}). \tag{1.34}$$

It remains to be shown that

$$\eta(x) = f(x, y(x)), \quad \text{a.e. } x \in \mathcal{O}. \tag{1.35}$$

To this end, we note that since

$$|y_\varepsilon - J_\varepsilon(x, y_\varepsilon)| = \varepsilon|f_\varepsilon(x, y_\varepsilon)|,$$

we have

$$J_\varepsilon(x, y_\varepsilon) \longrightarrow y \text{ strongly in } L^2(\mathcal{O})$$

and, therefore,

$$f_\varepsilon(x, y_\varepsilon(x)) \longrightarrow f(x, y(x)), \text{ a.e. } x \in \mathcal{O}$$

because $f_\varepsilon(x, y_\varepsilon) = f(x, J_\varepsilon(y_\varepsilon))$ and $y \to f(x, y)$ is continuous. Then, by (1.33), we get (1.35), as claimed. This completes the proof of Lemma 1.1.

Proof of Theorem 1.4 (Continued). Assume first that (1.14) and (1.15) hold. Hence $y_0 \in D(A)$ and so, according to Theorem 1.1, there is a unique function

$$y \in W^{1,\infty}([0, T]; H) \cap L^\infty(0, T; D(A))$$

which satisfies a.e. on $(0, T)$ equation (1.21). This implies that y satisfies (1.16). Moreover, multiplying (1.9) by $\frac{dy}{dt}$ and integrating on $\mathcal{O}$, we obtain

$$\left\| \frac{dy}{dt}(t) \right\|^2_{L^2(\mathcal{O})} + \frac{1}{2}\,\omega\,\frac{d}{dt}\,\|\nabla y(t)\|^2_{L^2(\mathcal{O})} + \frac{d}{dt}\int_{\mathcal{O}} j(x, y(t))dx \le \int_{\mathcal{O}} f_0 \frac{dy}{dt}\,dx. \tag{1.36}$$

Integrating on $(0, t)$, we get the estimate

$$\begin{aligned}&\int_0^t \left\| \frac{dy}{dt}(s) \right\|^2_{L^2(\mathcal{O})} ds + \omega\|\nabla y(t)\|^2_{L^2(\mathcal{O})} + \int_{\mathcal{O}} j(x, y(x, t))dx \\ &\quad \le C\left(\int_{\mathcal{O}} |\nabla y_0(x)|^2 dx + \int_{\mathcal{O}} j(x, y_0(x))dx + \int_0^t \|f_0(s)\|^2_{L^2(\mathcal{O})} ds \right).\end{aligned} \tag{1.37}$$

If multiply (1.9) by $t\,\frac{dy}{dt}$ and integrate on $(0, t)$, we obtain

$$\begin{aligned}&\int_0^t s \left\| \frac{dy}{dt}(s) \right\|^2 ds + \omega t\|\nabla y(t)\|^2_{L^2(\mathcal{O})} + t\int_{\mathcal{O}} j(x, y(x, t))dx \\ &\le C\left(\int_0^t \|\nabla y(s)\|^2_{L^2(\mathcal{O})} ds + \int_0^t\int_{\mathcal{O}} f_0^2(x, s)dx\,ds + \int_0^t\int_{\mathcal{O}} j(x, y(x, s))dx\,ds \right) \\ &\le C\int_0^t \|f_0(s)\|^2_{L^2(\mathcal{O})} ds + \|y_0\|^2_{L^2(\mathcal{O})},\ \forall t \in (0, T)\end{aligned} \tag{1.38}$$

because, as easily seen by (1.8)–(1.10), we have

$$\begin{aligned}&\frac{1}{2}\,\|y(t)\|^2_{L^2(\mathcal{O})} + \omega\int_0^t \|\nabla y(s)\|^2_{L^2(\mathcal{O})} ds + \int_0^t\int_{\mathcal{O}} j(y(x, s))dx\,ds \\ &\quad \le \frac{1}{2}\,\|y_0\|^2_{L^2(\mathcal{O})} + \gamma_0\int_0^t \|y(s)\|^2_{L^2(\mathcal{O})} ds + \int_0^t \|f_0(s)\|_{L^2(\mathcal{O})}\|y(s)\|_{L^2(\mathcal{O})} ds\end{aligned}$$

and so, by Gronwall's lemma,

$$\begin{aligned}\|y(t)\|^2_{L^2(\mathcal{O})} + \omega\int_0^t \|\nabla y(s)\|^2_{L^2(\mathcal{O})} ds + \int_0^t\int_{\mathcal{O}} j(y(x, s))dx\,ds \\ \le C\left(\|y_0\|^2_{L^2(\mathcal{O})} + \int_0^t \|f_0(s)\|^2_{L^2(\mathcal{O})} ds \right).\end{aligned}$$

By (1.29) and (1.34), it follows that

$$\|y\|_{H^2(\mathcal{O})} \le C.$$

Then, by (1.37), we obtain

$$\begin{aligned}&\int_0^T \left(\left\| \frac{dy}{dt}(t) \right\|^2_{L^2(\mathcal{O})} + \|y(t)\|^2_{H^2(\mathcal{O})} + \|f(t, y(t))\|^2_{L^2(\mathcal{O})} \right) dt \\ &\quad + \|y(t)\|^2_{H^1(\mathcal{O})} \le C\left(\|y_0\|^2_{H^1(\mathcal{O})} + \int_0^T \|f_0\|^2_{L^2(\mathcal{O})} dt \right)\end{aligned} \tag{1.39}$$

while (1.38) yields

$$\int_0^T t\left(\left\|\frac{dy}{dt}(t)\right\|^2 + \|y(t)\|^2_{H^2(\mathcal{O})} + \|f(t,y(t))\|^2_{L^2(\mathcal{O})}\right)dt \\ \le C\left(\|y_0\|^2_{L^2(\mathcal{O})} + \int_0^T \|f_0(t)\|^2_{L^2(\mathcal{O})}dt\right). \tag{1.40}$$

Assume now that (1.17) holds. Then, choose the sequences $\{y_0^k\}\subset D(A)$, $\{f_0^k\}\subset W^{1,1}([0,T];L^2(\mathcal{O}))$ such that

$$\begin{aligned} y_0^k &\longrightarrow y_0 \text{ strongly in } H_0^1(\mathcal{O}),\\ f_0^k &\longrightarrow f_0 \text{ strongly in } L^2(0,T;H) = L^2(Q). \end{aligned}$$

The corresponding solution y_k to (1.9) satisfies (1.16) and estimate (1.39). Moreover, subtracting equation for y_k and y_m and multiplying by $y_k - y_m$, we see that

$$\|y_k(t) - y_m(t)\|^2_{L^2(\mathcal{O})} + \omega\int_0^t \|\nabla(y_n - y_m)(s)\|^2_{L^2(\mathcal{O})}ds \\ \le C\left(\|y_0^k - y_0^m\|^2_{L^2(\mathcal{O})} + \int_0^t \|f_0^k(s) - f_0^m(s)\|^2_{L^2(\mathcal{O})}ds\right).$$

Hence there is $y \in C([0,T];H) \cap L^2(0,T;H^2(\mathcal{O}))$ such that for $k \to \infty$

$$y_k \longrightarrow y \text{ strongly in } C(0,T];H) \cap L^2(0,T;H^1(\mathcal{O}))$$

and, by estimate (1.37), it follows that

$$\begin{aligned} \frac{dy_k}{dt} &\longrightarrow \frac{dy}{dt} && \text{weakly in } L^2(0,T;H),\\ A_0 y_k &\longrightarrow A_0 y && \text{weakly in } L^2(0,T;H),\\ f(t,y_k) &\longrightarrow f(t,y) && \text{weakly in } L^2(0,T;H). \end{aligned}$$

Hence $y \in W^{1,2}([0,T];H) \cap L^2(0,T;D(A))$ is a solution to (1.9) (obviously unique) and satisfies (1.18). By a similar argument, it follows by estimate (1.40) that, if $y_0 \in L^2(\mathcal{O})$, $f_0 \in L^2(Q)$, there is a solution y which satisfies (1.19), (1.20). This completes the proof of Theorem 1.4. ■

A similar existence result follows for the equation

$$\begin{aligned} &\frac{\partial y}{\partial t} - \Delta y + f(x,y) + F(y) = 0 && \text{in } Q,\\ &y(0) = y_0 && \text{in } \mathcal{O},\\ &y = 0 && \text{on } \Sigma, \end{aligned} \tag{1.41}$$

where f satisfies conditions (1.12)–(1.13), and the operator $F : L^2(0, T; H_0^1(\mathcal{O})) \to L^2(0, T; L^2(\mathcal{O}))$ is Lipschitzian. Namely, we have

Theorem 1.5. *Let $y_0 \in L^2(\mathcal{O})$. Then equation* (1.41) *has a unique solution*

$$\begin{aligned} &y \in C([0, T]; L^2(\mathcal{O})) \cap L^2(0, T; H_0^1(\mathcal{O})),\\ &\frac{dy}{dt} \in L^2(0, T; H^{-1}(\mathcal{O})). \end{aligned} \tag{1.42}$$

Moreover,

$$\sqrt{t}\, y \in L^2(0, T; H^2(\mathcal{O})),\ \sqrt{t}\,\frac{dy}{dt} \in L^2(0, T; L^2(\mathcal{O})) \tag{1.43}$$

and, if $y_0 \in H_0^1(\mathcal{O})$, then

$$y \in L^2(0, T; H^2(\mathcal{O})) \cap L^\infty(0, T; H_0^1(\mathcal{O})),\ \frac{dy}{dt} \in L^2(0, T; L^2(\mathcal{O})). \tag{1.44}$$

Proof. It is easily seen via fixed point arguments that the operator

$$Ay = -\Delta y + F(y),\ D(A) = H_0^1(\mathcal{O}) \cap H^2(\mathcal{O})$$

is quasi-m-accretive in $H = L^2(\mathcal{O})$. Then, by Theorem 1.1, for $y_0 \in D(A)$, equation (1.41) has a unique solution $y \in L^\infty(0, T; H_0^1(\mathcal{O}) \cap H^2(\mathcal{O}))$ with $\frac{dy}{dt} \in L^\infty(0, T; L^2(\mathcal{O}))$. Moreover, arguing as in the proof of Theorem 1.4, one finds (1.42)–(1.44).

It is useful to note that the solution y to (1.41) is given by

$$y = \lim_{\varepsilon \to 0} y_\varepsilon \ \text{ in } C([0, T]; H), \tag{1.45}$$

where $y_\varepsilon \in L^2(0, T; H(\mathcal{O})) \cap C([0, T]; H)$ is the solution to the approximating equation

$$\begin{aligned} &\frac{\partial y_\varepsilon}{\partial t} - \Delta y_\varepsilon + f_\varepsilon(x, y_\varepsilon) + F(y_\varepsilon) = 0,\\ &y_\varepsilon(x, 0) = y_0(x) \text{ in } \mathcal{O},\\ &y_\varepsilon = 0 \text{ on } \Sigma. \end{aligned} \tag{1.46}$$

Indeed, arguing as above, it follows that, for each $g \in H$ and $\lambda \geq \lambda_0$ sufficiently large, the solution z_ε to the equation

$$\begin{aligned} &\lambda z_\varepsilon - \Delta z_\varepsilon + f_\varepsilon(x, z_\varepsilon) + F(z_\varepsilon) = g \ \text{ in } \mathcal{O},\\ &z_\varepsilon \in H_0^1(\mathcal{O}) \cap H^2(\mathcal{O}), \end{aligned}$$

is strongly convergent to the solution z to the elliptic boundary value problem

$$\lambda z - \Delta z + f(x, z) + F(z) = g \quad \text{in } \mathcal{O},$$
$$z \in H_0^1(\mathcal{O}) \cap H^2(\mathcal{O}).$$

(This simply follows by subtracting the equations, multiplying by $z_\varepsilon - z_{\varepsilon'}$, where $\varepsilon, \varepsilon' > 0$, and integrating on $\mathcal{O}$.) Then, by the Trotter–Kato theorem for nonlinear semigroups (see [26], p. 168), it follows (1.45), as claimed. On the other hand, by multiplying (1.46) by y_ε, Δy_ε and $t\Delta y_\varepsilon$, respectively, and integrating on $(0, t) \times \mathcal{O}$, we get as above the estimates (see (1.38)–(1.40))

$$|y_\varepsilon(t)|_2^2 + \int_0^t |\nabla y_\varepsilon(s)|_2^2 ds \le C(|y_0|_2^2 + 1), \ \forall t \in [0, T],$$
$$\|y_\varepsilon(t)\|^2_{H_0^1(\mathcal{O})} + \int_0^t |\Delta y_\varepsilon(s)|_2^2 ds \le C(\|y_0\|^2_{H_0^1(\mathcal{O})} + 1), \ t \in [0, T),$$
$$t\|y_\varepsilon\|^2_{H_0^1(\mathcal{O})} + \int_0^t s|\Delta y_\varepsilon|_2^2 ds \le C(|y_0|_2^2 + 1).$$

Hence, for $y_0 \in L^2(\mathcal{O})$, we have also, for $\varepsilon \to 0$,

$$\begin{aligned} y_\varepsilon &\longrightarrow y && \text{weakly in } L^2(0, T; H_0^1(\mathcal{O})), \text{ strongly in } C([0, T]; L^2(\mathcal{O})), \\ \sqrt{t}\, y_\varepsilon &\longrightarrow \sqrt{t}\, y && \text{weakly in } L^2(0, T; H^2(\mathcal{O})) \cap L^\infty(0, T; H_0^1(\mathcal{O})). \end{aligned} \tag{1.47}$$

1.4 Navier–Stokes Equations

Let $\mathcal{O}$ be an open subset of $\mathbb{R}^d$ with a smooth boundary $\partial\mathcal{O}$. The Navier–Stokes system

$$\begin{aligned} & y_t(x, t) = \nu_0 \Delta y(x, t) - (y \cdot \nabla) y(x, t) + f(x, t) + \nabla p(x, t), \ x \in \mathcal{O}, \ t \in (0, T), \\ & (\nabla \cdot y)(x, t) = 0, \ \forall (x, t) \in \mathcal{O} \times (0, T), \\ & y = 0 \text{ on } \partial\mathcal{O} \in (0, T), \\ & y(x, 0) = y_0(x), \ x \in \mathcal{O}, \end{aligned} \tag{1.48}$$

describes the nonslip motion of a viscous, incompressible, Newtonian fluid in an open domain $\mathcal{O} \subset \mathbb{R}^d$, $d = 2, 3$. Here $y = (y_1, y_2, \dots, y_d)$ is the fluid velocity field, $p = p(t, x)$ is the pressure, f is the density of an external force, and $\nu_0 > 0$ is the kinematic viscosity. Equations (1.48) are obtained by Newton's second law, while the condition $\nabla \cdot u = 0$ represents the incompressibility constraint. We have used the following notation:

$$\nabla\cdot y = \operatorname{div} y = \sum_{i=1}^{d} D_i y_i,\ \ D_i = \frac{\partial}{\partial x_i},\ i = 1,\dots,d,\ \ (y\cdot\nabla)y = \left\{\sum_{i=1}^{d} y_i D_i y_j\right\}_{j=1}^{d}.$$

If the force $f = f_e$ is independent of t, then the motion of the fluid is governed by the *stationary* Navier–Stokes equation

$$\begin{aligned} &-\nu_0\Delta y(x) + (y\cdot\nabla)y(x) = f_e(x) + \nabla p(x),\ x\in\mathcal{O},\\ &\nabla\cdot y = 0 \qquad \text{in } \mathcal{O},\\ &y = 0 \qquad \text{on } \partial\mathcal{O}. \end{aligned} \tag{1.49}$$

The linear equation

$$\begin{aligned} &y_t - \nu_0\Delta y = f + \nabla p \quad \text{in } \mathcal{O}\times(0,T),\\ &\nabla\cdot y = 0 \quad \text{in } \mathcal{O}\times(0,T),\\ &y = 0 \quad \text{in } \mathcal{O},\\ &y(x,0) = y_0(x) \quad \text{in } \mathcal{O}, \end{aligned} \tag{1.50}$$

is called the *Stokes equation*.

A convenient and standard way to treat the boundary value problem (1.48) is to represent it as an infinite dimensional Cauchy problem in an appropriate function space on $\mathcal{O}$. To this end, we shall introduce the following spaces

$$H = \{y\in(L^2(\mathcal{O}))^d;\ \nabla\cdot y = 0,\ y\cdot\nu = 0 \text{ on } \partial\mathcal{O}\}, \tag{1.51}$$

$$V = \{y\in(H_0^1(\mathcal{O}))^d;\ \nabla\cdot y = 0\}. \tag{1.52}$$

Here ν is the outward normal to $\partial\mathcal{O}$.

The space H is a closed subspace of $(L^2(\mathcal{O}))^d$ and it is a Hilbert space with the scalar product

$$(y,z) = \int_{\mathcal{O}} y\cdot z\,dx$$

and the norm

$$|y| = \left(\int_{\mathcal{O}} |y|^2 dx\right)^{1/2}. \tag{1.53}$$

(We shall denote by the same symbol $|\cdot|$ the norm in $\mathbb{R}^d$, $(L^2(\mathcal{O}))^d$ and H, respectively.)

The norm of the space V will be denoted by $\|\cdot\|$, i.e.,

$$\|y\| = \left(\int_{\mathcal{O}} |\nabla y(x)|^2 dx\right)^{1/2}.$$

We shall denote by $\Pi : (L^2(\mathcal{O}))^d \to H$ the orthogonal projection of $(L^2(\mathcal{O}))^d$ onto H (the Leray projector) and set

$$a(y,z) = \int_{\mathcal{O}} \nabla y \cdot \nabla z \, dx, \ \forall y, z \in V.$$

$$A = -\Pi \Delta, \ D(A) = (H^2(\mathcal{O}))^d \cap V.$$

Equivalently,

$$(Ay, z) = a(y, z), \ \forall y, z \in V.$$

The *Stokes operator* A is self-adjoint in H, $A \in L(V, V')$ (V' is the dual of V) and

$$(Ay, y) = \|y\|^2, \ \forall y \in V.$$

Finally, consider the trilinear functional

$$b(y, z, w) = \int_{\mathcal{O}} \sum_{i,j=1}^{d} y_i D_i z_j w_j \, dx, \ \forall y, z, w \in V,$$

and denote by $B : V \to V'$ the operator defined by

$$By = \Pi((y \cdot \nabla)y)$$

or, equivalently,

$$(By, w) = b(y, y, w), \ \forall w \in V.$$

Then, taking in account that $\Pi(\nabla p) = 0$, problem (1.48) can be written as

$$\begin{aligned} &\frac{dy}{dt}(t) + \nu_0 Ay(t) + By(t) = \Pi f(t), \ t \in (0, T), \\ &y(0) = y_0. \end{aligned} \tag{1.54}$$

(We have assumed of course that $y_0 \in H$.)

Similarly, equation (1.49) can be rewritten as

$$\nu_0 Ay + By = \Pi f_e.$$

Let $f \in L^2(0, T; V')$ and $y_0 \in H$. The function $y : [0, T] \to H$ is said to be a weak solution to equation (1.48) if

$$y \in L^2(0, T; V') \cap C_w([0, T]; H) \cap W^{1,1}([0, T]; V'),$$

$$\frac{d}{dt}(y(t), \psi) + \nu_0 a(y(t), \psi) + b(y(t), y(t), \psi) = (f(t), \psi), \text{ a.e. } t \in (0, T),$$
$$y(0) = y_0, \ \forall \psi \in V.$$

(Here $(\cdot, \cdot)$ is, as usual, the pairing between V, V' and the scalar product of H.) This equation can be, equivalently, written as

$$\begin{aligned} &\frac{dy}{dt}(t) + \nu_0 Ay(t) + By(t) = f(t), \text{ a.e. } t \in (0, T), \\ &y(0) = y_0, \end{aligned} \tag{1.55}$$

where $\frac{dy}{dt}$ is the strong derivative of function $y : [0, T] \to V'$.

The function y is said to be *strong solution* to (1.9) if $y \in W^{1,1}([0, T]; H) \cap L^2(0, T; D(A))$ and (1.55) holds with $\frac{dy}{dt} \in L^1(0, T; H)$ the strong derivative of function $y : [0, T] \to H$.

The existence theory for the Navier–Stokes equations is based on the following estimate on the trilinear functional b,

$$|b(y, z, w)| \leq C\|y\|_{m_1}\|z\|_{m_2+1}\|w\|_{m_3}, \tag{1.56}$$

where $\|y\|_{m_1} = \|y\|_{(H^{m_1}(\mathcal{O}))^d} = |A^{\frac{m_1}{2}} y|^2$ and (see, [113], [26], p. 252)

$$m_1 + m_2 + m_3 \geq \frac{d}{2} \ \text{ if } m_i \neq \frac{d}{2}, \ \forall i = 1, 2, 3,$$

$$m_1 + m_2 + m_3 > \frac{d}{2} \ \text{ if } m_1 = \frac{d}{2} \text{ for some } i.$$

A fundamental question so far only partially solved is whether problem (1.48) (equivalently, (1.55)) is well posed, that is, if the solutions to the Navier–Stokes equations exist for all time $t > 0$ and are sufficiently smooth to imply uniqueness. The answer is positive if $d = 2$ or if the initial data y_0 is sufficiently small in the space V, but the problem is open for $d = 3$.

We pause briefly to discuss this problem via the abstract existence theory presented in Section 1.3 (see [26]).

Theorem 1.6. *Let $d = 2, 3$ and $f \in W^{1,1}([0, T]; H)$, $y_0 \in D(A)$ where $0 < T < \infty$. Then there is a unique function*

$$y \in W^{1,\infty}([0, T^*]; H) \cap L^\infty(0, T^*; D(A)) \cap C([0, T^*]; V)$$

such that

$$\begin{aligned}&\frac{dy(t)}{dt} + \nu_0 Ay(t) + By(t) = f(t), \quad a.e.\ t \in (0, T^*),\\&y(0) = y_0,\end{aligned} \tag{1.57}$$

for some $T^* = T^*(\|y_0\|, f) \le T$. *If* $d = 2$, *then* $T^* = T$.
Moreover, y is right differentiable and

$$\frac{d^+}{dt} y(t) + \nu_0 Ay(t) + By(t) = f(t), \ \forall t \in [0, T^*). \tag{1.58}$$

Finally, if $d = 3$, $f \equiv 0$, *and* $\|y_0\| \le C^*(\nu_0)^{\frac{1}{2}}$, *where* C^* *is independent of* y_0, *then* $T^* = T$.

Proof. One considers the operator $B_N : V \to V'$, defined by

$$B_N y = \begin{cases} By & \text{if } \|y\| \le N, \\ \dfrac{N^2}{\|y\|^2} By & \text{if } \|y\| > N, \end{cases}$$

and the equation

$$\begin{aligned}&\frac{dy_N}{dt} + \nu_0 Ay_N + B_N y_N = 0, \ t \in (0, T),\\&y_N(0) = y_0.\end{aligned} \tag{1.59}$$

Taking into account that $\|By - Bz\|_{V'} \le C\|y - z\|(\|y\| + \|z\|)$, $\forall y, z \in V$, it follows that $\nu_0 A + B_N$ is quasi-m-accretive in $H \times H$ and so, by Theorem 1.1, the Cauchy problem (1.59) has a unique strong solution $y_N = e^{-t(\nu_0 A + B_N)} y_0$.
By (1.57) and (1.56), one gets for y_N the Leray energetical estimate

$$|y_N(t)|^2 + \nu_0 \int_0^t \|y_N(t)\|^2 \le |x|^2 + \int_0^t |f(t)|^2, \ \forall t \in (0, T). \tag{1.60}$$

Moreover, for $d = 2$, we get the estimate

$$\begin{aligned}&\|y_N(t)\|^2 + \nu_0 \int_0^t |Ay_N(s)|^2 ds \le C\left(\|y_0\|^2 + \int_0^t f(s)|^2 ds\right),\\&\qquad\qquad \forall t \in [0, T], \ y_0 \in V,\end{aligned} \tag{1.61}$$

and

$$\frac{1}{2}\frac{d}{dt}\|y_N(t)\|^2 + \nu_0 |Ay_N(t)|^2 \le C(\|y_N(t)\|^6 + f_T^2), \tag{1.62}$$

for $d = 3$. (Here C is a positive constant independent of N and $f_T^2 = \int_0^T |f(t)|^2 dt$.)

By (1.61) we see that, if $d = 2$, then, for $N \geq \sqrt{C}\|y_0\|$, $B_N(y_N) \equiv By_N$, and so for such an N, y_N is the solution to (1.57).

If $d = 3$, it follows by (1.62) that $\|y_N(t)\|^2 \leq \varphi(t)$, $\forall t \in [0, T]$, where φ is the solution to the equation ($\alpha > 0$)

$$\begin{aligned} &\varphi'(t) + \alpha\nu_0\varphi(t) = C(\varphi^3(t) + f_T^2),\ t \in [0, T],\\ &\varphi(0) = \varphi_0 = \|y_0\|^2. \end{aligned} \tag{1.63}$$

Since equation (1.63) has a local solution only, there is $T^* = T^*(\|y_0\|, f)$ such that

$$\|y_N(t)\| \leq \varphi^{\frac{1}{2}}(t) \leq C,\ \forall t \in [0, T^*],$$

and so, $\|y_N(t)\| \leq N$, $\forall t \in [0, T^*]$ for N sufficiently large.

Assume now that $f \equiv 0$. Solving equation (1.63) yields

$$\|y_N(t)\|^2 \leq \varphi(t) \leq e^{-\alpha\nu_0 t}\left(\frac{1}{\varphi_0^2} + \alpha\frac{C}{\nu_0}(e^{-2\alpha\nu_0 t} - 1)\right)^{-\frac{1}{2}},\ \forall t \in [0, T],$$

where $C > 0$ is independent of N and φ_0. If $C\varphi_0^2 \leq \alpha\nu_0$, that is, if

$$\|y_0\| \leq \left(\frac{\alpha\nu_0}{C}\right)^{\frac{1}{2}}, \tag{1.64}$$

then $\|y_N\|^2 \leq C_1$, $\forall t \in [0, T]$, and so, for N large enough, $y_N = y$ is a global solution to (1.57). If condition (1.64) does not hold, then as seen earlier $\|y_N(t)\| \leq C_1$ on some interval $[0, T]$ and, for N large, $y_N = y$ satisfies (1.55).

This completes the proof.

We have also (see [26], p. 260)

Theorem 1.7. *Let $y_0 \in H$, $f \in L^2(0, T; H)$, $T > 0$ and $d = 2$. Then there is a unique solution*

$$\begin{aligned} &y \in C([0, T]; V) \cap C_w([0, T]; H) \cap L^2(0, T; V),\\ &t^{1/2}y \in L^2(0, T; D(A)) \cap L^\infty(0, T; V),\\ &t^{1/2}\frac{dy}{dt} \in L^2(0, T; H),\ \frac{dy}{dt} \in L^{\frac{2}{1+\varepsilon}}(0, T; V') \end{aligned}$$

to equation (1.57), *i.e.,*

$$\begin{aligned} &\frac{dy}{dt}(t) + \nu_0 Ay(t) + By(t) = f(t),\ a.e.\ t \in (0, T)\\ &y(0) = y_0. \end{aligned} \tag{1.65}$$

If $y_0 \in V$, then $y \in L^\infty(0, T; V) \cap L^2(0, T; D(A))$.

For $N \to \infty$, it turns out that $\{y_N\}$ is weakly convergent in $L^2(0,T;V) \cap W^{1,\frac{4}{3}}(0,T;V')$ to a weak solution y to equation (1.55) (see [26], p. 264). Due to its low regularity, one cannot prove its uniqueness except on the interval $[0,T^*]$ mentioned above, or if $\|y_0\|$ is sufficiently small.

From the above construction of the solution y, it follows, however, that it coincides globally if $d=2$ or for $\|y_0\|$ small, and locally if $d=3$ with a smooth solution y_N to equation (1.65). Taking into account the Leray estimate (1.60), we have, for all $N \in \mathbb{N}$,

$$\nu_0\, m\{t \in [0,T];\ \|y_N(t)\| \geq N\} \leq N^{-2}\left(|x|^2 + \int_0^T |f(t)|^2 ds\right).$$

Hence in $3D$ the weak solution y satisfies everywhere equation (1.57), excepting a measurable subset $E_0 \subset (0,T)$ of an arbitrary small Lebesgue measure.

Remark 1.1. Theorem 1.6 extends to more general equations of the form

$$\begin{aligned}&\frac{\partial y}{\partial t} - \nu_0 \Delta y + (y\cdot\nabla)a_1 + (a_2\cdot\nabla)y + (y\cdot\nabla)y = \nabla p,\\ &\nabla\cdot y = 0,\ \ y = 0 \text{ on } [0,T]\times\partial\mathscr{O},\end{aligned}$$

where $a_i \in C^2(\overline{\mathscr{O}})$, $i=1,2$. The details are omitted.

1.5 Infinite Dimensional Linear Control Systems

Let H be a real Hilbert space with scalar product $(\cdot,\cdot)$ and norm denoted $|\cdot|_H$. Consider the Cauchy problem

$$\begin{aligned}&\frac{dy}{dt}(t) + A(t)y(t) = 0,\quad t\in(0,T),\\ &y(0) = y_0,\end{aligned} \tag{1.66}$$

where $\{A(t);\ t\in[0,T]\}$ is a family of closed and densely defined operators form H to itself.

We say that the Cauchy problem (1.66) is well posed if there exists a function $S(t,s) : \Delta = \{0 \leq s \leq t \leq T\} \to L(H,H)$ such that

1° For each $y_0 \in H$, the function $(t,s) \to S(t,s)y_0$ is continuous on Δ.
2° $S(s,s) = I,\ \forall s \in [0,T]$.
3° $S(t,s)S(s,\tau) = S(t,\tau)$ for all $0 \leq \tau \leq s \leq t \leq T$.
4° For each $s \in [0,T]$, there is a densely linear space $E(s) \subset H$ such that, for each $y_0 \in E(s)$, the function $t \to S(t,s)y_0$ is continuously differentiable on $[S,T]$ and

$$\frac{d}{dt}S(t,s)y_0 = A(t)S(t,s)y_0,\ \forall t \in [S,T], \tag{1.67}$$

$$\|S(t,s)\|_{L(H,H)} \le C,\ \forall (s,t) \in \Delta. \tag{1.68}$$

If conditions 1°–4° hold, we say that $\{A(t);\ t \in [0,T]\}$ *generates the evolution* $S(t,s)$.

If problem (1.66) is well posed and $f \in L^1(0,T;H)$, then by *mild solution* to the nonhomogeneous Cauchy problem

$$\begin{aligned} &\frac{dy}{dt}(t) + A(t)y(t) = f(t),\ \ t \in (0,T),\\ &y(0) = y_0, \end{aligned} \tag{1.69}$$

we mean the continuous function $y : [0,T] \to H$ given by

$$y(t) = S(t,0)y_0 + \int_0^t S(t,s)f(s)ds,\ \forall t \in [0,T]. \tag{1.70}$$

Examples

(i) If $A(t) \equiv A$ is independent of t, then problem (1.66) is well posed if and only if $-A$ is the *infinitesimal generator* of a C_0*-semigroup* $\{S(t) = e^{-tA}\}$ of linear continuous operators on H. By the Hille–Yosida theorem (see, e.g., [104]) this happens if there are $M,\ \omega \in \mathbb{R}$, such that

$$\|(\lambda I + A)^{-n}\|_{L(H,H)} \le M(\lambda - \omega)^{-n},\ \forall \lambda > \omega,\ n = 1, 2, \ldots \tag{1.71}$$

(ii) Let V be a Hilbert space such that $V \subset H$ with dense and continuous embedding and let V' be the dual space of V with the space pivot H, that is $V \subset H \subset V'$ with dense and continuous embeddings.

Let $A : [0,T] \to L(V,V')$ be such that, for each $y_0 \in V$, the function $t \to A(t)y_0$ is V'-valued measurable on $(0,T)$ and

$$\|A(t)y_0\|_{V'} \le C\|y_0\|_V,\ \forall t \in (0,T),\ y_0 \in V, \tag{1.72}$$

$${}_{V'}(A(t)y,y)_V + \alpha_1|y|_H^2 \ge \alpha_2\|y\|_V^2,\ \forall y \in V, \tag{1.73}$$

where $\alpha_2 > 0$ and $\alpha_1 \in \mathbb{R}$. Here ${}_{V'}(\cdot,\cdot)_V$ is the duality pairing between V and V' which coincides with the scalar product $(\cdot,\cdot)$ of H on $H \times H$.

Then $A_H(t)u = A(t)u \cap H,\ \forall u \in V$, generates an evolution on H. Moreover, we have (see, e.g., [26], p. 177)

Proposition 1.1. *For each $y_0 \in H$ and $f \in L^2(0,T;V')$, the Cauchy problem*

$$\begin{aligned} &\frac{dy}{dt} + A(t)y = f(t),\ t \in (0,T),\\ &y(0) = y_0, \end{aligned} \tag{1.74}$$

has a unique strong solution

$$y \in L^2(0,T;V) \cap C([0,T];H),\ \frac{dy}{dt} \in L^2(0,T;V'). \tag{1.75}$$

This means that $y : [0,T] \to V'$ is absolutely continuous and (1.74) holds a.e. on $(0,T)$. (See [26], p. 25.)

This result applies neatly to the liner parabolic equation

$$\begin{aligned} &y_t(x,t) - \Delta y(x,t) + b(x,t)\cdot\nabla y(x,t) + a(x,t)y(x,t) = f_0(x,t),\\ &\qquad\qquad\qquad\qquad\qquad\qquad \forall\,(x,t) \in Q = \mathscr{O} \times (0,T),\\ &\alpha_1 \frac{\partial y}{\partial \nu}(x,t) + \alpha_2 y(x,t) = 0,\ \forall\,(x,t) \in \Sigma = \partial\mathscr{O} \times (0,T),\\ &y(x,0) = y_0(x),\ x \in \mathscr{O}, \end{aligned} \tag{1.76}$$

where $y_t = \frac{\partial y}{\partial t}$, $\mathscr{O}$ is an open, bounded set of $\mathbb{R}^d$ with smooth boundary $\partial\mathscr{O}$, $a, b \in L^\infty(Q)$, $f_0 \in L^2(Q)$ and α_1, α_2 are as in (1.10). We take $H = L^2(\mathscr{O})$, $V = H^1(\mathscr{O})$ (or $V = H_0^1(\mathscr{O})$ if $\alpha_1 = 0$) and $A(t) : V \to V'$ defined by

$$\begin{aligned} {}_{V'}(A(t)y,z)_V = &\int_{\mathscr{O}} (\nabla y(x)\cdot\nabla z(x) + a(x,t)y(x)z(x) + b(x,t)\nabla y(x)\cdot z(x))dx\\ &+ \int_{\partial\mathscr{O}} \frac{\alpha_2}{\alpha_1}\, y(x)z(x)d\sigma. \end{aligned}$$

Then, by Proposition 1.1, we infer that for $y_0 \in L^2(\mathscr{O})$ there is a unique solution y to (1.76) satisfying

$$y \in L^2(0,T;H^1(\mathscr{O})) \cap C([0,T];L^2(\mathscr{O})),\ \frac{dy}{dt} \in L^2(0,T;(H^1(\mathscr{O}))'), \tag{1.77}$$

and, by Theorem 1.4, part (1.19)–(1.20),

$$\sqrt{t}\ y \in L^2(0,T;H^2(\mathscr{O})),\ \sqrt{t}\ y_t \in L^2(Q). \tag{1.78}$$

Consider now the linear control system

$$\begin{aligned} &\frac{dy}{dt} + A(t)y = B(t)u,\ t \in (0,T),\\ &y(0) = y_0, \end{aligned} \tag{1.79}$$

where $A(t)$ generates the evolution $S(t,s)$ and $B : [0,T] \to L(U,H)$ is strongly measurable and

$$\|B(t)\|_{L(U,H)} \le C < \infty,\ \text{ a.e. } t \in (0,T). \tag{1.80}$$

Here U is a real Hilbert space with the norm $\|\cdot\|_U$.

For each $u \in L^1(0, T; U)$, (1.79) has a unique mild solution $y \in C([0, T]; H)$ given by

$$y(t) = S(t, 0)y_0 + \int_0^t S(t, s)B(s)u(s)ds, \ t \in [0, T]. \tag{1.81}$$

The input function u is called the *controller* and the corresponding solution $y = y^u$ is called the *state* of the control system (1.79).

The control system (1.79) is said to be *exact null controllable* if, for each $y_0 \in H$, there is $u \in L^2(0, T; H)$ such that $y^u(T) = 0$ and $\|u\|_{L^2(0,T;U)} \le C_T |y_0|_H^2$.

We associate with (1.79) the *dual-backward* system

$$\frac{dp}{dt} - A^*(t)p = 0, \ t \in (0, T), \tag{1.82}$$

where $A^*(t)$ is the adjoint of $A(t)$, that is,

$$(A^*(t)y, z) = (y, A(t)z), \ \forall\, y \in D(A^*(t)), \ z \in D(A(t)).$$

We have

Theorem 1.8. *The dual system* (1.82) *is observable, that is, there is* $C_T > 0$ *such that*

$$|p(0)|_H^2 \le C_T \int_0^T \|B^*(t)p(t)\|_U^2 dt \tag{1.83}$$

for every mild solution p *to* (1.82)*, if and only if the control system* (1.79) *is exactly null controllable.*

Proof. Assume that (1.83) holds for each *mild* solution to (1.82) and consider for fixed $y_0 \in H$ and $\varepsilon > 0$ the minimization problem

$$\text{Minimize} \left\{ \int_0^T \|u(t)\|_U^2 dt + \frac{1}{2\varepsilon} |y(T)|_H^2 \right\} \text{ subject to (1.79).} \tag{1.84}$$

By a standard argument, it follows that problem (1.84) has a unique solution $(y_\varepsilon, u_\varepsilon)$ given by (the maximum principle)

$$\begin{aligned} &\frac{dy_\varepsilon}{dt} + A(t)y_\varepsilon = B(t)u_\varepsilon, \ t \in (0, T), \\ &y_\varepsilon(0) = y_0, \end{aligned} \tag{1.85}$$

$$\begin{aligned} &u_\varepsilon = B^*(t)p_\varepsilon, \ \text{a.e. } t \in (0, T), \\ &\frac{dp_\varepsilon}{dt} - A^*(t)p_\varepsilon = 0, \ t \in (0, T), \\ &p_\varepsilon(T) = -\frac{1}{\varepsilon}\, y_\varepsilon(T). \end{aligned} \tag{1.86}$$

By (1.85)–(1.86), we obtain that

$$\int_0^T \|B^*(t)p_\varepsilon(t)\|_U^2 dt + \frac{1}{\varepsilon}|y_\varepsilon(T)|_H^2 = (y_0, p_\varepsilon(0))_H$$

and so, by (1.83), it follows that

$$\int_0^T \|u_\varepsilon(t)\|_U^2 dt + \frac{1}{2\varepsilon}|y_\varepsilon(T)|_H^2 \le C_T|y_0|_H^2, \ \forall\, \varepsilon > 0. \tag{1.87}$$

Hence $\{u_\varepsilon\}$ is bounded in $L^2(0, T; U)$ and $|y_\varepsilon(T)|_H^2 \le 2\varepsilon C_T|y_0|_H^2$, $\forall\, \varepsilon > 0$.

On a subsequence, again denoted $\{\varepsilon\}$, we have

$$\begin{aligned} u_\varepsilon &\longrightarrow u^* && \text{weakly in } L^2(0, T; U),\\ y_\varepsilon(T) &\longrightarrow 0 && \text{strongly in } H,\\ y_\varepsilon(t) &\longrightarrow y^*(t) && \text{weakly in } H, \forall\, t \in (0, T). \end{aligned}$$

Clearly, (y^*, u^*) satisfies (1.79) and $y^*(T) = 0$, as desired. We note also that, by (1.87), we have

$$\|u^*\|_{L^2(0,T;U)} \le \sqrt{C_T}\,|y_0|_H,$$

and, therefore, the observability inequality (1.83) implies the uniform exact null controllability of system (1.79).

Conversely, if system (1.79) is exactly null controllable, then, for each y_0 with $|y_0|_H \le 1$, there is $\rho > 0$ such that

$$-S(T, 0)y_0 \in \mathscr{K} = \left\{\int_0^T S(T, s)B(s)u(s)ds;\ \|u\|_{L^2(0,T;U)} \le \rho\right\}. \tag{1.88}$$

Hence, for each $\xi \in H$,

$$-(S(T, 0)y_0, \xi) \le \sup_u \left\{\int_0^T (S(T, s)B(s)u(s), \xi)ds,\ \|u\|_{L^2(0,T;U)} \le \rho\right\}.$$

This yields

$$-(S(T, 0)y_0, \xi) \le \rho\left(\int_0^T |B^*(s)S^*(T, s)\xi|^2 ds\right)^{\frac{1}{2}},\ \forall\, \xi \in H,$$

and so

$$|S^*(T, 0)\xi|^2 \le \rho^2 \int_0^T |B^*(s)S^*(T, s)\xi|^2 ds,\ \forall\, \xi \in H,$$

which implies (1.83) with $p(0) = S^*(T, 0)\xi$.

We note that, if $(B^*(t))^{-1} \in L(H,U)$, $\forall t \in [0,T]$, and the function $t \to (B^*(t))^{-1}$ is in $L^2(0,T; L(H,H))$, then (1.83) trivially holds because

$$|p(0)|_H^2 \le C \int_0^T |p(t)|_H^2 dt.$$

Inequality (1.83) is also called the *observability* property of the dual backward system (1.79) and, in specific control systems governed by linear partial differential equations and, in particular, for linear parabolic systems of the form (1.9), it is implied by a Carleman-type inequality to be discussed later on. For instance, if $H = L^2(\mathcal{O})$, $U = L^2(\mathcal{O})$, $A = -\Delta$, $D(A) = H_0^1(\mathcal{O}) \cap H^2(\mathcal{O})$ and $Bu = \mathbf{1}_\omega u$, where ω is an open subset of $\mathcal{O}$ and $\mathbf{1}_\omega$ is its characteristic function, system (1.79) reduces to the controlled heat equation

$$\begin{aligned} &\frac{\partial y}{\partial t} - \Delta y = \mathbf{1}_\omega u && \text{in } (0,T)\times\mathcal{O},\\ &y = 0 && \text{on } (0,T)\times\partial\mathcal{O},\\ &y(x,0) = y_0(x),\ x \in \mathcal{O}, \end{aligned} \tag{1.89}$$

and the observability inequality (1.84) is

$$|p(0)|_2^2 \le C_T \int_0^T \int_\omega p^2(x,t)dx,dt,$$

for all the solutions p to the backward equation

$$\begin{aligned} &\frac{\partial p}{\partial t} + \Delta p = 0 && \text{in } (0,T)\times\mathcal{O},\\ &p = 0 && \text{on } (0,T)\times\partial\mathcal{O}. \end{aligned}$$

If $H = \mathbb{R}^n$, $U = \mathbb{R}^m$ and $A(t) \equiv A$, $B(t) \equiv B$, where A and B are time-independent $n \times n$ and, respectively, $n \times m$ matrices, then the observability condition (1.83) reduces to Kalman's rank condition

$$\text{rank}\|B, AB, \dots, A^{n-1}B\| = n, \tag{1.90}$$

which is equivalent with the exact null controllability of the finite-dimensional system (1.79). Indeed, in this case, the exact null controllability is equivalent with the observability condition

$$B^* e^{-tA^*} p_0 = 0,\ \ \forall t \in [0,T] \Longrightarrow p_0 = 0,$$

which, in turn, is equivalent with (1.90).

By (1.83) and (1.90), we see that the exact controllability property is sensitive (unstable) with respect to the structure of system (1.79) and this is the reason that it has a limited impact in automatic system theory. A simple example is just system (1.89), which can be approximated by the hyperbolic equation

$$\begin{aligned}&\frac{\partial y}{\partial t}-\varepsilon\frac{\partial^2 y}{\partial t^2}-\Delta y=\mathbf{1}_\omega u \text{ in } (0,T)\times\mathcal{O},\\&y=0 \text{ on } (0,T)\times\mathcal{O},\ \ y(x,0)=y_0(x),\ \ \frac{\partial y}{\partial t}(x,0)=y_1(x).\end{aligned}$$

However, due to the finite speed of propagation, the latter is not exactly controllable. From this perspective, a more convenient concept is that of stabilization. System (1.79) defined on $(0,\infty)$ is said to be *stabilizable* if there is a controller $u:[0,\infty)\to U$ such that the corresponding solution y satisfies

$$|y(t)|_H\le C\exp(-\gamma t)|y_0|_H,\ \forall t\ge 0,\ y_0\in H, \tag{1.91}$$

for some positive constants C and γ. Such a controller u is called a *stabilizing open loop controller*.

System (1.79) is said to be *feedback stabilizable* if there is a mapping Φ : $[0,\infty)\times H\to U$ called *feedback controller*, such that the solution y to the *closed loop system*

$$\begin{aligned}&\frac{dy}{dt}+A(t)y+B\Phi(t,y)=0,\ t\ge 0,\\&y(0)=y_0,\end{aligned}$$

exists and satisfies (1.91). These definitions extend mutatis-mutandis to nonlinear control systems of the form (1.79) for y_0 in a neighborhood of the origin.

It should be mentioned that, if $A,\ B$ are time-independent, then the exact controllability or, more generally, the existence of a stabilizing open loop controller $u\in L^2(0,T;U)$ implies the feedback stabilization by a special device involving algebraic infinite dimensional Riccati equations (see, e.g., [19], p. 210 and [51]). More precisely, the stabilizing feedback controller u in this case is given by

$$u=-B^*Ry,$$

where $R\in L(H,H)$, $R=R^*$, $R\ge 0$ is the solution to the algebraic Riccati equation

$$A^*R+RA+RBB^*R=C^*C, \tag{1.92}$$

where C is a linear densely defined closed operator in H such that

$$|Cy|_H^2\ge\gamma_0|y|_H^2,\ \forall\, y\in D(C).$$

The solution R to (1.92) is given by

$$(Ry_0, y_0) = \frac{1}{2} \inf \left\{ \int_0^\infty (|C(y(t)|_H^2 + \|u(t)\|_U^2)dt;\ \frac{dy}{dt} + Ay = Bu,\right.$$
$$\left. t \geq 0,\ y(0) = y_0 \right\}. \tag{1.93}$$

Remark 1.2. We refer to the survey [109] and to the books [51, 116] for an abstract theory of controllability of linear differential systems in Banach spaces and applications to partial differential equations of parabolic and hyperbolic type.

Chapter 2
The Carleman Inequality for Linear Parabolic Equations

This chapter is concerned with the Carleman estimates for the backward linear parabolic equations on smooth and bounded domains of $\mathbb{R}^d$ which implies observability that, as seen earlier, is the main tool to investigate the exact controllability of the forward parabolic controlled system.

2.1 The Carleman and Observability Inequality

Consider here the controlled linear parabolic equation

$$\begin{aligned} & y_t(x,t) - \Delta y(x,t) + b(x,t)\cdot\nabla y(x,t) + a(x,t)y(x,t) \\ & \qquad = m(x)u(x,t) + F(x,t),\ \forall\,(x,t)\in Q = \mathcal{O}\times(0,T), \\ & \alpha_1\frac{\partial y}{\partial \nu}(x,t) + \alpha_2 y(x,t) = 0,\ \forall\,(x,t)\in\Sigma = \partial\mathcal{O}\times(0,T), \\ & y(x,0) = y_0(x),\ x\in\mathcal{O}, \end{aligned} \tag{2.1}$$

where $\mathcal{O}$ is an open, bounded set of $\mathbb{R}^d$ with smooth boundary $\partial\mathcal{O}$ (of class C^2, for instance), $y_t = \frac{\partial y}{\partial t}$, and m is the characteristic function of an open set $\omega\subset\mathcal{O}$. Here $\frac{\partial}{\partial\nu}$ is the outward normal derivative, $\alpha_1\in C(\partial\mathcal{O}),\ \alpha_1\geq 0$ is constant, $\alpha_2\in C^1(\partial\mathcal{O}),\ \alpha_2\geq 0, \alpha_1+\alpha_2>0$.

Moreover, $a\in L^\infty(Q)$, $b\in C^1(\overline{Q};\mathbb{R}^d)$ and $F\in L^2(Q)$ are given functions while $u\in L^2(Q)$ is a control input. This is a parabolic equation of the form (1.76) which, as seen earlier, is well posed for $y_0\in L^2(\mathcal{O})$ and $u\in L^2(Q)$.

V. Barbu, *Controllability and Stabilization of Parabolic Equations*,
Progress in Nonlinear Differential Equations and Their Applications 90,
https://doi.org/10.1007/978-3-319-76666-9_2

We shall associate with (2.1) the backward dual parabolic problem

$$
\begin{aligned}
&p_t(x,t) + \Delta p(x,t) + \mathrm{div}_x(b(x,t)p(x,t)) - a(x,t)p(x,t) = g(x,t),\ (x,t) \in Q,\\
&\alpha_1 \frac{\partial p}{\partial \nu} + (\alpha_2 + \alpha_1 b \cdot \nu)p = 0 \ \text{ on } \Sigma,\ \ p(T) \in L^2(\mathscr{O}),
\end{aligned}
\tag{2.2}
$$

where $g \in L^2(Q)$ is a given function, $\frac{\partial p}{\partial \nu} = \nabla p \cdot \nu$ and ν is the outward normal to $\partial \mathscr{O}$. Recalling (1.77)–(1.78), it follows that, for $p(T) \in L^2(\mathscr{O})$, the solution p to (2.2) satisfies (see Section 1.3)

$$
\begin{gathered}
p \in C([0,T]; L^2(\mathscr{O})) \cap L^2(0,T; H^1(\mathscr{O})),\ (T-t)^{1/2} p_t \in L^2(Q)\\
\text{and } (T-t)^{1/2} p \in L^2(0,T; H^2(\mathscr{O})).
\end{gathered}
$$

If we denote by $A(t) : D(A(t)) \subset L^2(\mathscr{O}) \to L^2(\mathscr{O})$ the linear operator

$$
A(t)y = -\Delta y + b(t) \cdot \nabla y + a(t)y
$$

with the domain

$$
D(A(t)) = \left\{ y \in H^2(\mathscr{O});\ \alpha_1 \frac{\partial y}{\partial \nu} + \alpha_2 y = 0 \text{ on } \partial \mathscr{O} \right\},
$$

then equation (2.1) can be written as

$$
\begin{aligned}
&\frac{dy}{dt} + A(t)y = F,\ t \in (0,T),\\
&y(0) = y_0,
\end{aligned}
\tag{2.3}
$$

and (2.2) is the backward equation

$$
\frac{dp}{dt} - A^*(t)p = g,\ t \in (0,T),
\tag{2.4}
$$

where $A^*(t)$ is the dual operator

$$
A^*(t)p = -\Delta p - \mathrm{div}(b(t)p) + a(t)p,\ \ p \in D(A(t)),
$$

with the domain

$$
D(A^*(t)) = \left\{ p \in H^2(\mathscr{O}_0);\ \alpha_1 \frac{\partial p}{\partial \nu} + (\alpha_2 + \alpha_1 b(t).\nu)p = 0 \text{ on } \partial \mathscr{O}_0 \right\}.
$$

Taking into account the relevance of the observability of the dual system (2.4) for the exact null controllability of (2.3), this section is entirely devoted to the proof of the observability inequality (1.79) in this special case. This will be derived from a more precise estimate for solutions to equation (2.2), known in literature as the *Carleman inequality*.

Let ω be an open subset of $\mathcal{O}$ and let ω_0 be an open subset of ω such that $\overline{\omega}_0 \subset \omega$. An essential ingredient of the proof is the following technical lemma.

Lemma 2.1. *There is* $\psi \in C^2(\overline{\mathcal{O}})$ *such that*

$$\begin{aligned} &\psi(x) > 0, \ \forall\, x \in \mathcal{O}; \ \psi = 0 \ \text{ on } \partial\mathcal{O} \\ &|\nabla \psi(x)| > 0, \ \forall\, x \in \mathcal{O}_0 = \overline{\mathcal{O}} \backslash \omega_0. \end{aligned} \tag{2.5}$$

Proof. The existence of such a function ψ is easily seen in $1 - D$ or for rectangular domains $\mathcal{O}$, but we shall give below a simple proof in the case where $\mathcal{O}$ is an open convex set of $\mathbb{R}^d$ and refer to [76] and [116], p. 439, for the general case.

We consider the solution $\psi \in C^2(\overline{\mathcal{O}})$ to the Dirichlet problem

$$\begin{aligned} \Delta \psi &= -1 \text{ in } \mathcal{O}, \\ \psi &= 0 \qquad \text{on } \partial\mathcal{O}, \end{aligned}$$

By the maximum principle, we know that $\psi > 0$ on $\mathcal{O}$ and $\frac{\partial \psi}{\partial \nu} < 0$ on $\partial\mathcal{O}$. We recall (see [91]) that $u = \sqrt{\psi}$ is concave and the Hessian matrix of u is negative definite.

This means that every critical point $(x_0, y_0) \in \mathcal{O}$ is not degenerate and so, by the Morse theory, u has a unique local maximum (x_0, y_0) in $\mathcal{O}$. If $(x_0, y_0) \in \omega_0$, then clearly the function ψ satisfies conditions (2.5). Otherwise, we take

$$\widetilde{\psi}(x, y) = \psi(F^{-1}(x, y)), \ \forall\, (x, y) \in \overline{\mathcal{O}},$$

where F is a diffeomorphism of $\overline{\mathcal{O}}$ into itself such that $F(x_0, y_0) \in \omega_0$. It is clear that $\{(x, y) \in \mathcal{O};\ \nabla \widetilde{\psi}(x, y) = 0\} \subset \omega_0$, and so $\widetilde{\psi}$ satisfies (2.5). As regards the mapping F, it can be constructed as follows. We fix $(x_0^*, y_0^*) \in \omega_0$ and consider the operator $A_0 : I \to \mathbb{R}^d$

$$A_0(s(x_0^*, y_0^*) + (1 - s)(x_0, y_0)) = (x_0 - x_0^*, y_0 - y_0^*), \ \forall\, s \in (0, 1)$$

defined on the segment

$$I = [s(x_0^*, y_0^*) + (1 - s)(x_0, y_0);\ s \in [0, 1] \subset \mathcal{O}.$$

The operator A_0 has a maximal monotone extension $A : D(A) \subset \overline{\mathcal{O}} \to \mathbb{R}^d$, such that $A_0 \subset A$ and $\overline{D(A)} \subset \overline{\mathcal{O}}$ (see [17], p. 41, Theorem 1.4). Let $S(t) : \overline{D(A)} \to \overline{\mathcal{O}}$ be the semigroup generated by $-A$, that is,

$$\frac{d}{dt}(S(t)u_0) + A(S(t)u_0) \ni 0, \quad \text{a.e. } t > 0,$$

$$S(0)u_0 = u_0.$$

Since $A_0 \subset A$, we have

$$S(t)(x_0, y_0) = t(x_0^*, y_0^*) + (1-t)(x_0, y_0), \; \forall t \in [0, 1].$$

Now, we consider a smooth invertible approximation $A^\lambda \in C^1(\overline{\mathscr{O}}; \mathbb{R}^2)$ of A such that $(A^\lambda)^{-1} \in \text{Lip}(\mathbb{R}^2, \mathbb{R}^d)$ and, for each $\alpha > 0$, $(I + \alpha A^\lambda)^{-1} \to (I + \alpha A)^{-1}$ as $\lambda \to 0$. (Such an operator A^λ might be $\lambda I + A_\lambda * \rho_\lambda$, where A_λ is the Yosida approximation of A and ρ_λ is a mollifier.) Then, by the Kato-Trotter theorem (see [26], p. 168) for $\lambda \to 0$,

$$S_\lambda(t)u_0 \to S(t)u_0, \; \forall u_0 \in \overline{\mathscr{O}}, t \geq 0,$$

where $S_\lambda(t)$ is the semigroup generated by $-A^\lambda$. Clearly, $S_\lambda(1)$ is a diffeomorphism of $\overline{\mathscr{O}}$ into itself and, for λ small enough, $S_\lambda(1)(x_0, y_0) \in \omega_0$. Hence, $F = S_\lambda(1)$ is the desired mapping.

For $\lambda > 0$ and a function ψ which satisfies (2.5) we set

$$\alpha(x,t) = \frac{e^{\lambda\psi(x)} - e^{2\lambda\|\psi\|_{C(\overline{\mathscr{O}})}}}{(t(T-t))^k}, \quad \varphi(x,t) = \frac{e^{\lambda\psi(x)}}{(t(T-t))^k}, \quad \forall (x,t) \in Q, \tag{2.6}$$

where $k \geq 1$ is a natural number. However, for the purposes of this chapter one might choose $k = 1$.

Theorem 2.1 below (the Carleman inequality) is the main result of this section.

Theorem 2.1. *Let $a \in L^\infty(Q)$ and $b \in C^1(\overline{Q}; \mathbb{R}^d)$. Then there are $\lambda_0 > 0$ and $s_0 > 0$ such that, for $\lambda \geq \lambda_0$ and $s \geq s_0$,*

$$\begin{aligned} &\int_Q e^{2s\alpha}(s^3\lambda^4\varphi^3 p^2 + s\lambda^2\varphi|\nabla p|^2 + (s\varphi)^{-1}(p_t^2 + |\Delta p|^2))dx\,dt \\ &\quad \leq C_\lambda \left(s^3\lambda^4 \int_{Q_\omega} e^{2s\alpha}\varphi^3 p^2\,dx\,dt + \int_Q e^{2s\alpha} g^2 dx\,dt \right) \end{aligned} \tag{2.7}$$

for all solutions p to equation (2.3). *Here $Q_\omega = \omega \times (0,T)$ and C is a positive constant independent of s, λ and p.*

It should be observed that this inequality is symmetric with respect to 0 and T. Thus, by the change of variable $t \to T - t$, we see that (2.1) remains true and has the same form for the forward equation (2.1). As mentioned earlier, Theorem 2.1 is motivated by the exact controllability of equation (2.1) but it has, however, an intrinsic interest. We note also that Theorem 2.1 remains true if $-\Delta$ is replaced by the linear elliptic operator $L = -\frac{\partial}{\partial x_i}\left(a_{ij}(x,t)\frac{\partial}{\partial x_j}\right)$ with smooth coefficients.

We recall that the classical generic form for the Carleman inequality on a domain $Q \subset \mathbb{R}^{n+1}$ for a linear differential operator $P(D)$ is

$$\tau\|e^{\tau\theta}u\| \le C\|e^{\tau\theta}P(D)u\|,\ \forall\tau > \tau_0,\ u \in C_0^\infty(Q),\ \tau > \tau_0.$$

Here θ is a C^∞ function such that $|\nabla\theta| > 0$ in Q_0 and $\|\cdot\|$ is a Sobolev space norm. In the present situation $\theta = \alpha + \log\varphi$ and $Q_0 = Q_\omega$.

Theorem 2.1 implies the observability of system (2.1).

Corollary 2.1. *Under the assumptions of Theorem* 2.1, *there are* $\lambda_0>0$, s_0 *and* C, μ, *independent of* p *such that, for* $\lambda \ge \lambda_0$, $s \ge s_0$, *the following inequality holds*

$$\int_\Omega p^2(x,0)dx \le Ce^{\mu s}\left(\int_{Q_\omega} e^{2s\alpha}\varphi^3 p^2 dx\,dt + \int_Q g^2 dx\,dt\right) \tag{2.8}$$

for all solutions p *to equation* (2.3).

Proof. We multiply equation (2.3) by p and integrate on $\mathcal{O}$. After some calculation involving Green's formula, we get

$$\frac{1}{2}\left(\frac{d}{dt}|p(t)|_2^2 - \int_{\mathcal{O}}|\nabla p(x,t)|^2 dx\right) \ge -C(|p(t)|_2^2 + |g(t)|_2^2),\quad \text{a.e. } t\in(0,T),$$

where $|\cdot|_2$ is the $L^2(\mathcal{O})$ norm. Then, by the Gronwall lemma, we obtain

$$|p(\tau)|_2^2 \le C\left(|p(t)|_2^2 + \int_\tau^t |g(\theta)|_2^2 d\theta\right),\ 0 \le \tau \le t.$$

Hence

$$|p(0)|_2^2 \le C\left(\gamma(t)\int_{\mathcal{O}} e^{2s\alpha(x,t)}\varphi^3(x,t)p^2(x,t)dx + \int_0^T |g(t)|_2^2 dt\right),$$

where $\gamma(t) = \sup\{e^{-2s\alpha(x,t)}\varphi^{-3}(x,t);\ x \in \mathcal{O}\} \le Ce^{\frac{\mu s}{(t(T-t))^k}}$ and $\mu = 2e^{2\lambda\|\psi\|_{C(\overline{\mathcal{O}})}}$.

Now, we integrate the latter inequality on the interval (t_1, t_2) to get

$$|p(0)|_2^2 \int_{t_1}^{t_2} e^{-\frac{\mu s}{((T-t)t)^k}} dt \le C\left(\int_Q e^{2s\alpha}\varphi^3 p^2 dx\,dt + \int_Q g^2 dx\,dt\right),$$

for $0 < t_1 < t_2 < T$. Then, using estimate (2.7), we obtain (2.8), where C_λ is suitable chosen. This completes the proof.

Proof of Theorem 2.1. We set $z = e^{s\alpha} p$ and note that z satisfies the boundary value problem

$$\begin{aligned}
&z_t + \Delta z + (\lambda^2 s^2 \varphi^2 |\nabla\psi|^2 - \lambda^2 s\varphi|\nabla\psi|^2)z - (2s\lambda\varphi\nabla\psi - b)\cdot\nabla z\\
&\quad -(a - \operatorname{div} b + s\lambda\varphi(b\cdot\nabla\psi) + (s\alpha_t + \lambda s\varphi\Delta\psi)z = ge^{s\alpha} \text{ in } Q,\\
&\alpha_1 \frac{\partial z}{\partial \nu} + \left(\alpha_2 + \alpha_1 b\cdot\nu - \alpha_1 s\lambda\varphi\frac{\partial\psi}{\partial\nu}\right)z = 0 \text{ on } \Sigma,\\
&z(x,0) = z(x,T) = 0 \text{ in } \mathcal{O}.
\end{aligned} \tag{2.9}$$

Without no loss of generality, we may assume that $z \in H^{2,1}(Q)$. We set

$$\begin{aligned}
X(t)z &= -2(s\lambda^2\varphi|\nabla\psi|^2 z + s\lambda\varphi\nabla z\cdot\nabla\psi),\\
B(t)z &= -\Delta z - (\lambda^2 s^2\varphi^2|\nabla\psi|^2 + s\lambda^2\varphi|\nabla\psi|^2)z + s\alpha_t z,
\end{aligned}$$

and rewrite equation (2.9) as

$$z_t + X(t)z - B(t)z = e^{s\alpha} g + Z(t)z \text{ in } Q,$$

where $Z(t)z = -(-s\lambda\varphi\Delta\psi + \operatorname{div} b - s\lambda\varphi(b\cdot\nabla\psi) - a)z - b\cdot\nabla z$. This yields

$$\begin{aligned}
&\frac{d}{dt}\int_{\mathcal{O}} (B(t)z)(x,t)z(x,t)dx = \int_{\mathcal{O}} ((B(t)z)(x,t)z_t(x,t)\\
&\quad +(B(t)z_t)(x,t)z(x,t))dx + \int_{\mathcal{O}} (B_t(t)z)(x,t)z(x,t)dx\\
&\quad = 2\int_{\mathcal{O}} (B(t)z)(x,t)(B(t)z(x,t) - X(t)z(x,t) + Z(t)z(x,t) + e^{s\alpha} g)\,dx\\
&\quad + \int_{\mathcal{O}} (B_t z)(x,t)z(x,t)dx - \int_{\partial\mathcal{O}} \left(z\frac{\partial z_t}{\partial\nu} - z_t\frac{\partial z}{\partial\nu}\right) d\sigma dt.
\end{aligned}$$

Integrating by parts with respect to t on the surface integral on Σ, we get after some calculations that

$$\begin{aligned}
&2\int_Q (B(t)z(x,t))^2 dx\,dt + 2\int_Q (B(t)z)(x,t)(Z(t)z + e^{s\alpha} g)(x,t)dx\,dt + 2Y\\
&\quad \le -\int_Q (B_t(t)z)(x,t)z(x,t)dx\,dt - 2\int_\Sigma z_t\frac{\partial z}{\partial\nu} d\sigma dt,
\end{aligned} \tag{2.10}$$

where

$$\begin{aligned}
Y = -2\int_Q &(s\lambda^2\varphi|\nabla\psi|^2 z + s\lambda\varphi\nabla z\cdot\nabla\psi)\\
&(\Delta z + (\lambda^2 s^2\varphi^2|\nabla\psi|^2 + s\lambda^2\varphi|\nabla\psi|^2 - s\alpha_t)z dx\,dt.
\end{aligned} \tag{2.11}$$

We note that

$$\begin{aligned}
&\int_Q B_t(t)z\,z\,dx\,dt + 2\int_\Sigma z_t \frac{\partial z}{\partial \nu} d\sigma dt \\
&\quad = -\int_Q z^2(2\lambda^2 s^2 \varphi\varphi_t |\nabla\psi|^2 + s\lambda^2 \varphi_t |\nabla\psi|^2 - s\alpha_{tt})dx\,dt \\
&\quad\quad - 2\alpha_1^{-1}\int_\Sigma z_t z \left(\alpha_2 + \alpha_1 b\cdot\nu - \alpha_1 s\lambda\varphi \frac{\partial\psi}{\partial\nu}\right) d\sigma\,dt.
\end{aligned}$$

We set $\gamma(\lambda) = e^{2\lambda\|\psi\|_{C(\overline{\mathscr{O}})}}$. We have

$$\begin{aligned}
&|\alpha_t| \le \gamma(\lambda)|\varphi_t| \le C\gamma(\lambda)\varphi^{1+k^{-1}}, \\
&\qquad |\alpha_{tt}| \le \gamma(\lambda)|\varphi_{tt}| \le C\gamma(\lambda)\varphi^{1+2k^{-1}}.
\end{aligned}$$

Taking into account the boundary condition in (2.9), we get after some integration by parts that

$$\left|\int_\Sigma z_t \frac{\partial z}{\partial\nu} d\sigma dt\right| \le C\int_\Sigma (1 + s\lambda\varphi^2)z^2 d\sigma dt.$$

On the other hand, by the trace theorem and the interpolation inequality, we have

$$\int_\Sigma z^2 d\sigma dt \le C\|z\|^2_{L^2(0,T;H^{\frac{1}{2}+\varepsilon}(\mathscr{O}))} \le C\|z\|^{\frac{1}{2}-\varepsilon}_{L^2(0,T;L^2(\mathscr{O}))}\|z\|^{\frac{1}{2}+\varepsilon}_{L^2(0,T;H^1(\mathscr{O}))}.$$

This yields

$$\int_\Sigma \varphi^2 z^2 d\sigma dt \le C\left(\int_Q \varphi^3 z^2 dxdt + \int_Q \varphi|\nabla z|^2 dxdt\right) \tag{2.12}$$

and, therefore, for $s \ge 1$, we have

$$\begin{aligned}
&\left|\int_Q (B_t(t)z)z\,dx\,dt + 2\int_\Sigma z_t\frac{\partial z}{\partial\nu} d\sigma dt\right| \\
&\quad \le C(\lambda^2 s^2 + s\gamma(\lambda))\int_Q \varphi^3 z^2 dxdt \\
&\quad + Cs\lambda\int_Q \varphi|\nabla z|^2 dx\,dt.
\end{aligned}$$

(Here and everywhere in the sequel, C is a positive constant independent of s, λ, z and g.) Note also that

$$\begin{aligned}
&\left|2\int_Q (B(t)z)(Z(t)z + e^{s\alpha}g)\,dx\,dt\right| \\
&\quad \le \|B(t)z\|^2_{L^2(Q)} + \|e^{s\alpha}g\|^2_{L^2(Q)} \\
&\quad + C(s^2\lambda^2\|\varphi z\|^2_{L^2(Q)} + \|\nabla z\|^2_{L^2(Q)})
\end{aligned}$$

and so, by (2.10), we see that

$$Y \leq C(s^2\lambda^2 + s\gamma(\lambda))\int_Q \varphi^3 z^2 dx\, dt + C\int_Q (s\lambda\varphi|\nabla z|^2 + e^{2s\alpha}g^2)dx\, dt, \tag{2.13}$$

while, by (2.11), we obtain that

$$\begin{aligned} Y \geq -2s\int_Q &(\lambda^2\varphi|\nabla\psi|^2 z + \lambda\varphi\nabla\psi\cdot\nabla z)\\ &(\Delta z + (\lambda^2 s^2\varphi^2|\nabla\psi|^2 + s\lambda^2\varphi|\nabla\psi|^2)z)dx\, dt\\ &+ s^2\lambda\int_\Sigma z^2\varphi\alpha_t\nabla\psi\cdot\nu d\sigma dt - CD_1(s,\lambda,z) \end{aligned} \tag{2.14}$$

for $s, \lambda \geq \lambda_0$ sufficiently large, where

$$D_1(s,\lambda,z) = s^2\gamma(\lambda)\lambda^2\int_Q \varphi^3 z^2 dx\, dt + s\lambda\int_Q \varphi|\nabla z|^2 dxdt. \tag{2.15}$$

Since $\psi = 0$ on $\partial\Omega$ and $\psi \geq 0$ in Ω, it follows that $\frac{\partial\psi}{\partial\nu} \leq 0$ on Σ and so, by (2.9), we see that

$$\frac{\partial z}{\partial\nu}z \leq -b\cdot\nu z^2 \text{ on } \Sigma. \tag{2.16}$$

Then, by Green's formula, we obtain that

$$\begin{aligned} -\int_Q z\Delta z\varphi|\nabla\psi|^2 dx\, dt &= \int_Q \varphi|\nabla z|^2|\nabla\psi|^2 dx\, dt\\ &-\int_\Sigma \varphi|\nabla\psi|^2 z\frac{\partial z}{\partial\nu}d\sigma\, dt + \int_Q z\nabla z\cdot\nabla(\varphi|\nabla\psi|^2)dx\, dt\\ &\geq \frac{3}{4}\int_Q \varphi|\nabla\psi|^2|\nabla z|^2 dx\, dt - C(\lambda^2+1)\int_Q \varphi^2 z^2 dx\, dt. \end{aligned} \tag{2.17}$$

Here, to estimate the surface integral, we have used the boundary conditions in (2.9) along with the above interpolation inequality given before which, taking into account (2.12), (2.16), allows to estimate the above integral by

$$-\int_\Sigma \varphi|\nabla\psi|^2 z^2 d\sigma\, dt \geq -C\int_Q \varphi^2|z|^2 dxdt - \frac{1}{8}\int_Q \varphi|\nabla\psi|^2|\nabla z|^2 dt.$$

Note also that

$$
\begin{aligned}
&-2s\lambda\int_Q \varphi\nabla z\cdot\nabla\psi(\lambda^2 s^2\varphi^2|\nabla\psi|^2+s\lambda^2\varphi|\nabla\psi|^2)z\,dx\,dt\\
&\quad=-s\lambda\int_Q \nabla z^2\cdot\nabla\psi(\lambda^2 s^2\varphi^3+s\lambda^2\varphi^2)|\nabla\psi|^2 dx\,dt\\
&\quad=-s\lambda\int_Q \operatorname{div}(z^2\nabla\psi(\lambda^2 s^2\varphi^3+s\lambda^2\varphi^2)|\nabla\psi|^2)dx\,dt\\
&\qquad+s\lambda\int_Q z^2\operatorname{div}(\nabla\psi(\lambda^2 s^2\varphi^3+s\lambda^2\varphi^2)|\nabla\psi|^2)dx\,dt\\
&\quad\ge -s\lambda\int_\Sigma z^2(\lambda^2 s^2\varphi^3+s\lambda^2\varphi^2)|\nabla\psi|^2\frac{\partial\psi}{\partial\nu}\,d\sigma\,dt\\
&\qquad+\int_Q(3s^3\lambda^4\varphi^3+2s^2\lambda^4\varphi^2)|\nabla\psi|^4 z^2 dx\,dt\\
&\qquad-C\int_Q(\lambda^3 s^3\varphi^3+s^2\lambda^3\varphi^2)z^2 dx\,dt.
\end{aligned}
\tag{2.18}
$$

By (2.14), (2.15), (2.16), and (2.17), we get

$$
\begin{aligned}
Y\ge&\int_Q(s^3\lambda^4\varphi^3|\nabla\psi|^4 z^2+\frac{3}{2}s\lambda^2\varphi|\nabla\psi|^2|\nabla z|^2)dxdt-I_0(z)\\
&+\lambda s^2\int_\Sigma z^2\varphi\alpha_t\nabla\psi\cdot\nu d\sigma dt-2s\lambda\int_Q\varphi(\nabla z\cdot\nabla\psi)\Delta z\,dx\,dt-CD(s,\lambda,z),
\end{aligned}
\tag{2.19}
$$

where

$$
\begin{aligned}
I_0(z)&=s\lambda\int_\Sigma z^2(\lambda^2 s^2\varphi^3+s\lambda^2\varphi^2)|\nabla\psi|^2\frac{\partial\psi}{\partial\nu}d\sigma dt\le 0\\
D(s,\lambda,z)&=\int_Q((s^3\lambda^3\varphi^3+s^2\lambda^2\gamma(\lambda)\varphi^2)z^2+s\lambda\varphi|\nabla z|^2)dx\,dt.
\end{aligned}
$$

Next, by the Green formula, it follows after some calculation that

$$
\begin{aligned}
&-2s\lambda\int_Q\varphi(\nabla z\cdot\nabla\psi)\Delta z\,dx\,dt=-2s\lambda\int_\Sigma\varphi(\nabla\psi\cdot\nabla z)(\nabla z\cdot\nu)d\sigma\,dt\\
&+2s\lambda\int_Q\nabla z\cdot\nabla(\varphi\nabla z\cdot\nabla\psi)dx\,dt=-2s\lambda\int_\Sigma\varphi(\nabla\psi\cdot\nabla z)(\nabla z\cdot\nu)d\sigma\,dt\\
&+s\lambda\int_\Sigma\varphi|\nabla z|^2(\nabla\psi\cdot\nu)d\sigma\,dt-s\lambda^2\int_Q\varphi|\nabla\psi|^2|\nabla z|^2dx\,dt\\
&+\int_Q\left(2s\lambda^2\varphi(\nabla z\cdot\nabla\psi)^2-s\lambda\varphi\left(|\nabla z|^2\Delta\psi-\sum_{i,j=1}^d z_{x_i}z_{x_j}\psi_{x_ix_j}\right)\right)dx\,dt,
\end{aligned}
$$

where ν is the outward normal to $\partial\mathscr{O}$.

Since, as seen earlier, $\frac{\partial\psi}{\partial\nu}\le 0$ on $\partial\mathscr{O}$ and $\psi=0$ on $\partial\mathscr{O}$, we have

$$
\nu=-\frac{\nabla\psi}{|\nabla\psi|}\ \text{ on }\partial\mathscr{O},\ \ (\nabla\psi\cdot\nabla z)(\nabla z\cdot\nu)=-(\nabla\psi\cdot\nabla z)^2|\nabla\psi|^{-1}\text{ on }\partial\mathscr{O}.
$$

We have, therefore,

$$-2s\lambda \int_Q \varphi \nabla z \cdot \nabla \psi \, \Delta z \, dx \, dt \geq 2s\lambda \int_\Sigma \varphi (\nabla \psi \cdot \nabla z)^2 |\nabla \psi|^{-1} d\sigma dt$$
$$-s\lambda \int_\Sigma \varphi |\nabla z|^2 |\nabla \psi| d\sigma \, dt - s\lambda^2 \int_Q \varphi |\nabla \psi|^2 |\nabla z|^2 dx \, dt - C D(s, \lambda, z).$$

Then, by (2.13), (2.14) and (2.18), we obtain that

$$\begin{aligned} &s^3\lambda^4 \int_Q \varphi^3 |\nabla \psi|^4 z^2 dx \, dt + s\lambda^2 \int_Q \varphi |\nabla \psi|^2 |\nabla z|^2 dx \, dt \\ &\leq C \left(D(s, \lambda, z) + \int_Q g^2 e^{2s\alpha} dx \, dt \right) - I(z) + I_0(z), \end{aligned} \tag{2.20}$$

where

$$I(z) = s\lambda \int_\Sigma (2(\nabla \psi \cdot \nabla z)^2 - |\nabla z|^2 |\nabla \psi|^2) \varphi |\nabla \psi|^{-1} + s z^2 \varphi \alpha_t \nabla \psi \cdot \nu d\sigma \, dt.$$

We note that, if $\alpha_1 = 0$, then $\nu = \pm \frac{\nabla z}{|\nabla z|}$ on Σ and so $|\nabla z \cdot \nabla \psi| = |\nabla z| |\nabla \psi|$ on Σ.
This implies that $I(z) \geq 0$ and, since $I_0(z) \leq 0$, we infer that

$$\begin{aligned} &s^3\lambda^4 \int_Q \varphi^3 |\nabla \psi|^4 z^2 dx \, dt + s\lambda^2 \int_Q \varphi |\nabla \psi|^2 |\nabla z|^2 dx \, dt \\ &\leq C \left(D(s, \lambda, z) + \int_Q e^{2s\alpha} g^2 dx \, dt \right). \end{aligned} \tag{2.21}$$

We shall prove now that the latter inequality still remains true if $\alpha_1 > 0$. To this purpose, we set $\bar{z} = e^{s\bar{\alpha}} p$, where

$$\bar{\alpha} = \frac{e^{-\lambda\psi} - e^{2\lambda \|\psi\|_{C(\overline{\Omega})}}}{(t(T-t))^k}, \quad \overline{\varphi} = \frac{e^{-\lambda\psi}}{(t(T-t))^k}.$$

Clearly, $\bar{\alpha} \leq \alpha$, $\overline{\varphi} \leq \varphi$ in Q and $\bar{\alpha} = \alpha$, $\overline{\varphi} = \varphi$ on Σ. We note that $\bar{z}$ satisfies equation (2.9), where φ is replaced by $\overline{\varphi}$ and λ by $-\lambda$. We have, therefore,

$$\bar{z}_t + \overline{X}(t)\bar{z} - \overline{B}(t)\bar{z} = e^{s\overline{\alpha}} g + \overline{Z}(t)\bar{z} \text{ in } Q,$$

where $\overline{X}(t)\bar{z} = -2(s\lambda^2\overline{\varphi}|\nabla\psi|^2 - s\lambda\overline{\varphi}\nabla\bar{z}\cdot\nabla\psi)$, $\overline{B}(t)\bar{z} = -\Delta\bar{z} - (\lambda^2 s^2 \overline{\varphi}^2 |\nabla\psi|^2 + s\lambda^2\overline{\varphi}|\nabla\psi|^2)\bar{z} + s\overline{\alpha}_t\bar{z}$, $\overline{Z}(t)\bar{z} = (a - s\lambda\overline{\varphi}\Delta\psi - \operatorname{div} b + s\lambda\varphi(b\cdot\nabla\psi)\bar{z} - b\cdot\nabla\bar{z}$.
We obtain as above (see (2.13), (2.14))

$$-2s \int_Q (\lambda^2 \overline{\varphi} |\nabla \psi|^2 \bar{z} - \lambda \overline{\varphi} \nabla \psi \cdot \nabla \bar{z})(\Delta \bar{z} + (\lambda^2 s^2 \overline{\varphi}^2 |\nabla \psi|^2 + s\lambda^2 \overline{\varphi} |\nabla \psi|^2)\bar{z}) dx \, dt$$
$$\leq C \left(D(s, \lambda, \bar{z}) + \int_Q e^{2s\bar{\alpha}} g^2 dx \, dt + s^2\lambda \int_\Sigma \bar{z}^2 \overline{\varphi} \overline{\alpha}_t \nabla \psi \cdot \nu d\sigma dt \right)$$

because $\overline{\varphi} \le \varphi$ and $\overline{\alpha} \le \alpha$. Arguing as above, we obtain also that (see (2.16))

$$-\int_Q \bar{z}\Delta\bar{z}\overline{\varphi}|\nabla\psi|^2 dx\, dt \ge \frac{3}{4}\int_Q \overline{\varphi}|\nabla\psi|^2|\nabla\bar{z}|^2 dx\, dt - C(\lambda^2+1)\int_Q \overline{\varphi}^2\bar{z}^2 dx\, dt \tag{2.22}$$

and (see (2.17))

$$\begin{aligned} 2s\lambda\int_Q \overline{\varphi}\nabla\bar{z}\cdot\nabla\psi(\lambda^2 s^2\overline{\varphi}^2|\nabla\psi|^2 + s\lambda^2\overline{\varphi}|\nabla\psi|^2)\bar{z}dx\, dt \\ \ge \int_Q (3s^3\lambda^4\overline{\varphi}^3 + 2s^2\lambda^4\overline{\varphi}^2)\bar{z}^2 dx\, dt + I_0(\bar{z}) - CD(s,\lambda,\bar{z}). \end{aligned} \tag{2.23}$$

(Here we must take into account that $\nabla\overline{\varphi} = -\lambda\nabla\psi\cdot\overline{\varphi}$.)

Arguing as in the previous case, we obtain also that

$$\begin{aligned} 2s\lambda\int_Q \overline{\varphi}(\nabla\bar{z}\cdot\nabla\psi)\Delta\bar{z}dx\, dt \ge -2s\lambda\int_\Sigma \overline{\varphi}(\nabla\psi\cdot\nabla\bar{z})^2|\nabla\psi|^{-1}d\sigma\, dt \\ +s\lambda\int_\Sigma \overline{\varphi}|\nabla\bar{z}|^2|\nabla\psi|d\sigma\, dt + s\lambda^2\int_Q \overline{\varphi}|\nabla\psi|^2|\nabla z|^2 dx\, dt - CD(s,\lambda,\bar{z}). \end{aligned}$$

Substituting the latter inequality along with (2.22), (2.23) into (2.20), we obtain

$$\begin{aligned} \int_Q (s^3\lambda^4\overline{\varphi}^3|\nabla\psi|^4\bar{z}^2 + s\lambda^2\overline{\varphi}|\nabla\psi|^2|\nabla\bar{z}|^2)dx\, dt \\ \le C\left(D(s,\lambda,\bar{z}) + \int_Q e^{2s\alpha}g^2 dx\, dt + I(\bar{z}) - I_0(\bar{z})\right). \end{aligned} \tag{2.24}$$

By (2.19) and (2.24), we see that

$$\begin{aligned} \int_Q (s^3\lambda^4\varphi^3|\nabla\psi|^4 z^2 + s\lambda^2\varphi|\nabla\psi|^2|\nabla z|^2)dx\, dt \\ +\int_Q (s^3\lambda^4\overline{\varphi}^3|\nabla\psi|^4\bar{z}^2 + s\lambda^2\overline{\varphi}|\nabla\psi|^2|\nabla\bar{z}|^2)dx\, dt \\ \le C\left(D(s,\lambda,z) + \int_Q e^{2s\alpha}g^2 dx\, dt\right), \end{aligned} \tag{2.25}$$

because $\overline{\varphi} \le \varphi$ and $\bar{z} = e^{s(\bar{\alpha}-\alpha)}z$, $\nabla\bar{z} = \nabla z e^{s(\bar{\alpha}-\alpha)} + s\nabla(\bar{\alpha}-\alpha)z$.

Indeed, the latter imply that $z = \bar{z}$, $\nabla z = \nabla\bar{z}$ on Σ

$$|\nabla\bar{z}| \le |\nabla z| + s\lambda|\nabla\psi|\varphi|z| \text{ in } Q$$

and we have, therefore,

$$I_0(z) = I_0(\bar{z}),\ I(z) = I(\bar{z}),\ D(s,\lambda,\bar{z}) \le CD(s,\lambda,z).$$

Thus, by (2.25), we get the desired estimate (2.21), that is,

$$\begin{aligned}&\int_Q (s^3\lambda^4\varphi^3|\nabla\psi|^4 z^2 + s\lambda^2\varphi|\nabla\psi|^2|\nabla z|^2)dx\,dt\\&\qquad\le C\left(D(s,\lambda,z)+\int_Q e^{2s\alpha}g^2dx\,dt\right).\end{aligned}\tag{2.26}$$

On the other hand, by (2.9), we see that

$$(s\lambda)^{-1}\int_Q (z_t+\Delta z)^2\varphi^{-1}dx\,dt \le C\left(D(s,\lambda,z)+\int_Q g^2e^{2s\alpha}dx\,dt\right).$$

This yields

$$\begin{aligned}&(s\lambda)^{-1}\int_Q (z_t+|\Delta z|^2)\varphi^{-1}dx\,dt\\&\qquad\le C\left(D(s,\lambda,z)+\int_Q g^2e^{2s\alpha}dx\,dt\right)+\frac{1}{2}(s\lambda)^{-1}\int_Q z_t^2\varphi^{-1}dx\,dt\end{aligned}$$

because $z(0)=z(T)=0$,

$$\int_Q z_t\Delta z\varphi^{-1}dxdt = -\int_Q \nabla z\cdot\nabla(z_t\varphi^{-1})dx\,dt + \int_\Sigma \varphi^{-1}\frac{\partial z}{\partial\nu}z_t d\sigma\,dt$$

while, by the boundary conditions in (2.9),

$$\begin{aligned}&\int_\Sigma \varphi^{-1}\frac{\partial z}{\partial\nu}z_t d\sigma dt = \frac{1}{2}\int_\Sigma\left(s\lambda\frac{\partial\psi}{\partial\nu}-\left(\frac{\alpha_2}{\alpha_1}+b\cdot\nu\right)\varphi^{-1}\right)(z)_t^2d\sigma\,dt\\&=\frac{\alpha_2}{2\alpha_1}\int_\Sigma z^2\left(\left(1+\frac{\alpha_1}{\alpha_2}b\cdot\nu\right)\varphi^{-1}\right)_t d\sigma\,dt \le C\int_Q (z^2+|\nabla z|^2)dx\,dt\end{aligned}\tag{2.27}$$

by the trace theorem. We obtain, therefore, that for $s,\lambda\ge\lambda_0$

$$(s\lambda)^{-1}\int_Q (z_t^2+|\Delta z|^2)\varphi^{-1}dx\,dt \le C\left(D(s,\lambda,z)+\int_Q g^2e^{2s\alpha}dx\,dt\right)\tag{2.28}$$

and recall also that, by (2.26), we have that

$$\begin{aligned}&s^3\lambda^4\int_Q \varphi^3|\nabla\psi|^4z^2dx\,dt + s\lambda^2\int_Q\varphi|\nabla\psi|^2|\nabla z|^2dx\,dt\\&\qquad\le C\left(D(s,\lambda,z)+\int_Q e^{2s\alpha}g^2dx\,dt\right).\end{aligned}\tag{2.29}$$

Recalling the definition of $D(s, \lambda, z)$ and the fact that $|\nabla\psi(x)| \geq \rho > 0, \forall x \in \mathscr{O}_0 = \overline{\mathscr{O}}\backslash\omega_0,\ \omega_0 \subset\subset \omega$, it follows by (2.29) that

$$s^3\lambda^4\rho^4 \int_{Q_0} \varphi^3 z^2 dx\, dt + s\lambda^2\rho^2 \int_{Q_0} \varphi|\nabla z|^2 dx\, dt$$
$$\leq C\left((s^2\lambda^2\gamma(\lambda) + s^3\lambda^3)\int_Q \varphi^3 z^2 dx\, dt + s\lambda\int_Q \varphi|\nabla z|^2 dx\, dt + \int_Q e^{2s\alpha} g^2 dx\, dt\right),$$

where $Q_0 = \mathscr{O}_0\times(0, T)$. Hence there are $\lambda_0 > 0$ and $s_0 = s_0(\lambda)$ such that, for $\lambda \geq \lambda_0$ and $s \geq s_0(\lambda)$, we have

$$\int_{Q_0} (s^3\lambda^4\varphi^3 z^2 + s\lambda^2\varphi|\nabla z|^2) dx\, dt$$
$$\leq C\left(\int_{Q_{\omega_0}} (s^3\lambda^4\varphi^3 z^2 + s\lambda\varphi|\nabla z|^2) dx\, dt + \int_Q e^{2s\alpha} g^2 dx\, dt\right).$$

Finally,

$$\int_Q (s^3\lambda^4\varphi^3 z^2 + s\lambda^2\varphi|\nabla z|^2) dx\, dt$$
$$\leq C_\lambda\left(\int_{Q_{\omega_0}} (s^3\lambda^4\varphi^3 z^2 + s\lambda\varphi|\nabla z|^2) dx\, dt + \int_Q e^{2s\alpha} g^2 dx\, dt\right). \tag{2.30}$$

Coming back to p, we get

$$\int_Q e^{2s\alpha}(s^3\lambda^4\varphi^3 p^2 + s\lambda^2\varphi|s\lambda\varphi p\nabla\psi + \nabla p|^2) dx\, dt$$
$$\leq C_\lambda\left(\int_{Q_{\omega_0}} e^{2s\alpha}\left(s^3\lambda^4\varphi^3 p^2 + s\lambda^2\varphi|s\lambda\varphi p\nabla\psi + \nabla p|^2\right) dx\, dt + \int_Q e^{2s\alpha} g^2 dx\, dt\right). \tag{2.31}$$

We note that, for any $\delta \in (0, 1)$, we have

$$2s^2\lambda^3 \int_Q e^{2s\alpha}\varphi^2 p\nabla\psi \cdot \nabla p dx\, dt$$
$$\geq -\delta^{-1}s^3\lambda^4 \int_Q e^{2s\alpha}\varphi^3|\nabla\psi|^2 p^2 dxdt - \delta s\lambda^2 \int_Q e^{2s\alpha}\varphi|\nabla p|^2 dxdt. \tag{2.32}$$

Then, by (2.31), it follows that, for $\delta^{-1}\|\nabla\psi\|^2_{C(\overline{\mathscr{O}})} < 1$ and for $\lambda \geq \lambda_0$ large enough, $s \geq s_0(\lambda)$, we have

$$\int_Q e^{2s\alpha}(s^3\lambda^4\varphi^3 p^2 + s\lambda^2\varphi|\nabla p|^2) dxdt$$
$$\leq C_\lambda\left(\int_{Q_{\omega_0}} e^{2s\alpha}(s^3\lambda^4\varphi^3 p^2 + s\lambda^2\varphi|\nabla p|^2) dxdt + \int_Q e^{2s\alpha} g^2 dxdt\right).$$

Next, we choose $\chi \in C_0^\infty(\mathscr{O})$ such that $\chi = 1$ in $\overline{\omega}_0$ and $\chi = 0$ in $\mathscr{O}\backslash\omega$. If multiply equation (2.3) by $\chi\varphi e^{2s\alpha}p$ and integrate on Q, we get after some calculation involving Green's formula that

$$\int_Q e^{2s\alpha}\varphi\chi|\nabla p|^2 dx\,dt \le C_\lambda s^2\lambda^2 \int_{Q_\omega} e^{2s\alpha}\varphi^3 p^2 dx\,dt + C_\lambda \int_Q e^{2s\alpha} g^2 dx\,dt.$$

Substituting into (2.32), we get

$$\begin{aligned}&\int_Q e^{2s\alpha}(s^3\lambda^4\varphi^3 p^2 + s\lambda^2\varphi|\nabla p|^2)dx\,dt\\ &\quad \le C_\lambda\left(s^3\lambda^4\int_{Q_\omega} e^{2s\alpha}\varphi^3 p^2 dx\,dt + \int_Q e^{2s\alpha}g^2 dx\,dt\right).\end{aligned}$$

Hence

$$\begin{aligned}&\int_Q e^{2s\alpha}(s^3\varphi^3 p^2 + s\varphi|\nabla p|^2)dx\,dt\\ &\quad \le C_\lambda\left(s^3\int_{Q_\omega} e^{2s\alpha}\varphi^3 p^2 dx\,dt + \int_Q e^{2s\alpha}g^2 dx\,dt\right)\end{aligned} \tag{2.33}$$

for all $\lambda \ge \lambda_0$ sufficiently large and $s \ge s_0(\lambda)$.

Finally, by (2.28) and (2.33), we see that

$$\begin{aligned}s^{-1}&\int_Q e^{2s\alpha}(p_t^2 + |\Delta p|^2)\varphi^{-1} dx\,dt\\ &\le C_\lambda\left(\int_Q e^{2s\alpha}(s^3\varphi^3 p^2 + s\varphi|\nabla p|^2)dx\,dt + \int_Q g^2 e^{2s\alpha} dx\,dt\right)\\ &\le C_\lambda\left(s^3\int_Q e^{2s\alpha}\varphi^3 p^2 dx\,dt + \int_Q e^{2s\alpha}g^2 dx\,dt\right)\end{aligned}$$

for $\lambda \ge \lambda_0$, $s \ge s_0(\lambda)$. Along with (2.33), the latter implies (2.3), thereby completing the proof. ■

Remark 2.1. The constant C_λ arising in Carleman's inequality depends on $|a|_\infty$ and $|\text{div}\, b|_\infty$, $|b|_\infty$ only. As a function of λ, $C_\lambda \le C\gamma(\lambda)$ for $\lambda \ge \lambda_0$. Notice also that Theorem 2.1 extends to general second order parabolic equations of the form

$$p_t + \sum_{i,j=1}^n (a_{ij}p_{x_i})_{x_j} + \text{div}\,(bp) - ap = g$$

with smooth coefficients a_{ij}, b. It turns out, however, that a similar result remains true if $a_{i,j} \in W^{1,\infty}(Q)$, $b \in L^\infty(0,T;L^r(\mathscr{O}))$, $r > 2n$ and $a \in L^\infty(0,T;W_{r_1}^{-\mu}(\Omega)$ for $0 < \mu < \frac{1}{2}$ and r_1 sufficiently large. (See also [21].)

Theorem 2.1 remains true for the solutions $p \in H^2(\mathcal{O})$ to the elliptic boundary-value problem

$$\begin{aligned} &\Delta p(x,t) + \mathrm{div}_x(b(x,t)p(x,t)) + c(x,t) \cdot \nabla p(x,t) \\ &\qquad -a(x,t)p(x,t) = g(x,t), \ \forall (x,t) \in Q, \\ &\alpha_1 \frac{\partial p}{\partial \nu} + (\alpha_2 + \alpha_1 b \cdot \nu)p = 0 \text{ on } \Sigma, \end{aligned} \tag{2.34}$$

where $t \in [0,T]$ is a parameter. Namely, one has

Theorem 2.2. *Under the assumptions of Theorem 2.1, there are $\lambda_0 > 0$ and s_0 such that, for $\lambda \geq \lambda_0$ and $s \geq s_0$,*

$$\begin{aligned} &\int_{\mathcal{O}} e^{2s\alpha}(s^3\varphi^3 p^2 + s\varphi|\nabla p|^2 + (s\varphi)^{-1}|\Delta p|^2)dx \\ &\quad \leq C_\lambda \left(s^3 \int_\omega e^{2s\alpha}\varphi^3 p^2 dx + \int_{\mathcal{O}} e^{2s\alpha} g^2 dx \right), \quad \text{a.e. } t \in (0,T) \end{aligned} \tag{2.35}$$

Proof. We shall argue as in the proof of Theorem 2.1. As a matter of fact, the proof is identic, but we shall *forget* the terms p_t, z_t and α_t, φ_t from the previous calculation. Namely, we set $z = e^{s\alpha}p$ and obtain the elliptic equation (see (2.9))

$$\begin{aligned} &\Delta z + (\lambda^2 s^2 \varphi^2 |\nabla\psi|^2 - \lambda^2 s\varphi|\nabla\psi|^2)z \\ &-2s\lambda\varphi\nabla\psi \cdot \nabla z + (b+c) \cdot \nabla z - (a + s\lambda\varphi\nabla\psi \cdot (b+c) \\ &-\mathrm{div}\, b - \lambda s\varphi\Delta\psi)z = g e^{s\alpha}, \end{aligned} \tag{2.36}$$

that is,

$$X(t)z - B_0(t)z = e^{s\alpha}g + Z(t)z, \tag{2.37}$$

where

$$\begin{aligned} X(t)z &= -2(s\lambda^2\varphi^2|\nabla\psi|^2 z + s\lambda\varphi\nabla z \cdot \nabla\psi), \\ B_0(t)z &= -\Delta z - (\lambda^2 s^2\varphi^2|\nabla\psi|^2 + s\lambda^2\varphi|\nabla\psi|^2)z, \\ Z(t)z &= (a - \mathrm{div}\, b - s\lambda\varphi\Delta\psi + s\lambda\varphi(b+c) \cdot \nabla\psi)z - (b+c) \cdot \nabla z. \end{aligned}$$

This yields

$$\int_{\mathcal{O}} B_0(t)z(B_0(t)z - X(t)z + Z(t)z + e^{s\alpha}g)dx = 0.$$

For simplicity, we shall assume that $\alpha_1 = 0$. Then the same calculation as in the proof of Theorem 2.1 leads us to (see (2.20))

$$\begin{aligned} &s^3\lambda^4 \int_{\mathcal{O}} \varphi^3|\nabla\psi|^4 z^2 dx + s\lambda^2 \int_{\mathcal{O}} \varphi|\nabla\psi|^2|\nabla z|^2 dx \\ &\qquad \leq C\left(D(s,\lambda,z)(t) + \int_{\mathcal{O}} e^{2s\alpha} g^2 dx \right), \end{aligned} \tag{2.38}$$

where

$$D(s,\lambda,z)(t)=\int_{\mathscr{O}}((s^3\lambda^3\varphi^3+s^2\lambda^4\varphi^2)z^2+s\lambda\varphi|\nabla z|^2)dx,\ t\in(0,T).$$

On the other hand, by (2.36), we see that

$$(s\lambda)^{-1}\int_{\mathscr{O}}|\Delta z|^2\varphi^{-1}dx\le C\left(D(s,\lambda,z)(t)+\int_{\mathscr{O}}e^{2s\alpha}g^2dx\right).$$

Then, by (2.38), we obtain that

$$\begin{aligned}s^3\lambda^4\int_{\mathscr{O}}\varphi^3|\nabla\psi|^4z^2dx+s\lambda^2\int_{\mathscr{O}}\varphi|\nabla\psi|^2|\nabla z|^2dx+(s\lambda)^{-1}\int_{\mathscr{O}}|\Delta z|^2\varphi^{-1}dx\\ \le C\left(D(s,\lambda,z)(t)+\int_{\mathscr{O}}e^{2s\alpha}g^2dx\right).\end{aligned}$$

Then, arguing as in the proof of Theorem 2.1 (see (2.10), (2.29)), one obtains inequality (2.30) on $\mathscr{O}$ and, consequently, (2.31), which leads as before to the desired inequality (2.35). This completes the proof.

2.2 Notes on Chapter 2

Theorem 2.1 was established by A.V. Fursikov and O.Yu. Imanuvilov [76]. The proof given here closely follows [20, 22]. As shown in [81], it extends to more general linear second order parabolic equations with nonregular lower order coefficients and the right-hand side in a Sobolev space of negative order. Sharper Carleman estimates for linear parabolic equations were recently obtained by J. Le Rousseau and G. Lebeau [87] and J. Le Rousseau and L. Robbiano [88]. More precisely, the results of [88] refer to a Carleman inequality of the form (2.7) for solutions of the equation $y_t-\operatorname{div}(a(t,x)\nabla y)=f$ in $(0,T)\times\mathscr{O}$, where a is piecewise smooth in a spatial variable and discontinuous across a smooth interface.

Chapter 3
Exact Controllability of Parabolic Equations

This chapter is concerned with the presentation of some basic results on the exact internal and boundary controllability of linear and semilinear parabolic equations on smooth domains of $\mathbb{R}^d$. The exact controllability of linear stochastic parabolic equations with linear multiplicative Gaussian noise is also briefly studied. The main ingredient to exact controllability is the observability inequality for the dual parabolic equations established in Chapter 2.

3.1 Exact Controllability of Linear Parabolic Equations

We come back to the controlled linear parabolic equation (2.1), that is,

$$\begin{aligned}
&y_t(x,t) - \Delta y(x,t) + b(x,t)\cdot\nabla y(x,t) + a(x,t)y(x,t),\\
&\qquad\qquad = m(x)u(x,t) + F(x,t),\ \forall\,(x,t)\in Q = \mathcal{O}\times(0,T),\\
&\alpha_1\frac{\partial y}{\partial\nu}(x,t) + \alpha_2 y(x,t) = 0,\ \forall\,(x,t)\in\Sigma = \partial\mathcal{O}\times(0,T),\\
&y(x,0) = y_0(x),\ x\in\mathcal{O},
\end{aligned}\tag{3.1}$$

where $\mathcal{O}$ is an open, bounded, and smooth set of $\mathbb{R}^d$ (of class C^2 for instance) and $m = \mathbf{1}_\omega$ is the characteristic function of an open set $\omega\subset\mathcal{O}$.

Let $y_0\in L^2(\mathcal{O})$ be arbitrary but fixed. If $F(x,T)\equiv 0$, then system (3.1) is said to be *exactly null controllable* if there is $u\in L^2(Q)$ such that $y^u(T)\equiv 0$.

We have denoted by $y^u\in C([0,T];L^2(\mathcal{O}))\cap L^2(0,T;H^1(\mathcal{O}))$ the solution to (3.1) and recall (see (1.76)–(1.77)) that

$$\sqrt{t}\,y_t^u\in L^2(Q),\ \sqrt{t}\,y^u\in L^2(0,T;H^2(\mathcal{O}))$$

V. Barbu, *Controllability and Stabilization of Parabolic Equations*,
Progress in Nonlinear Differential Equations and Their Applications 90,
https://doi.org/10.1007/978-3-319-76666-9_3

and if $y_0 \in H^1(\mathscr{O})$ ($y_0 \in H_0^1(\mathscr{O})$ if $\alpha_1 = 0$) then

$$y^u \in C_w([0,T]; H^1(\mathscr{O})) \cap L^2(0,T; H^2(\mathscr{O})), \ y_t^u \in L^2(Q).$$

Let $a \in L^\infty(\mathscr{O})$, $b \in C^1(\overline{\mathscr{O}}; \mathbb{R}^d)$, $\alpha_2 \in C^1(\partial\mathscr{O})$, $F \in L^2(\mathscr{O})$ and let $y_e \in H^2(\mathscr{O})$ be a steady-state (equilibrium) solution to system (3.1), that is, a solution to elliptic boundary value problem

$$\begin{aligned} -\Delta y_e + b \cdot \nabla y_e + a y_e &= F \ \text{ in } \mathscr{O}, \\ \alpha_1 \frac{\partial y_e}{\partial \nu} + \alpha_2 y_e &= 0 \ \text{ on } \partial\mathscr{O}. \end{aligned} \tag{3.2}$$

The steady-state solution y_e is said to be exactly controllable if there is $u \in L^2(Q)$ such that $y^u(T) \equiv y_e$. Subtracting equations (3.1) and (3.2), we may reduce the exact controllability of the steady-state solution y_e to the exact null controllability of system (3.1) with modified coefficients a, b.

We may view $v = mu$ as an *internal controller*, i.e., as a distributed controller with the support in $\omega \times (0,T)$ and for this reason we also refer to this property as the *internal exact controllability*.

As it is well known, (3.1) describes the heat propagation and also the density fluctuation $y = y(x,t)$ of a diffusing material at $x \in \mathscr{O}$ at time t. The term $F + mu - ay$ in equation (3.1) represents sources of substance, composed of a fixed source and another one which is proportional with the concentration y. If the concentration of the substance y is specified on Σ, one has $\alpha_1 = 0$, while, if $\alpha_2 = 0$, one assumes to have a zero flow of substance through boundary (impermeable boundary). The exact null controllability problem formulated above is whether one can steers in time T the concentration y to equilibrium state y_e by applying a source mu of material active on some subdomain ω of $\mathscr{O}$. A similar problem if the source u is applied on the boundary $\partial\mathscr{O}$ or on some parts of it.

In the absence of exterior sources, the solution of diffusion systems naturally decreases to an equilibrium state, but the role of the controller u introduced here is to achieve this goal in a finite specified time T. For practical purposes, it is desirable to find such a control u in an arbitrary small domain $\omega \in \Omega$ or on the boundary. If equation (3.1) reduces to the heat equation

$$\begin{aligned} &\frac{\partial y}{\partial t} - \Delta y = mu, && \text{in } (0, T \times \mathscr{O}, \\ &y = 0 && \text{on } (0,T) \times \partial\mathscr{O}, \\ &y(x,0) = y_0(x) && \text{on } \mathscr{O} \subset \mathbb{R}^3, \end{aligned} \tag{3.3}$$

the exact null controllability in time T means that one wants to drive the temperature y in zero by applying a heating (cooling) source $u(t,x)$ on a subdomain ω of conductor Ω.

Let us illustrate the problem with the following simple example pertaining the exact null controllability of the $1-D$ heat equation (3.3) on a finite interval of $\mathbb{R}$.

Problem 3.1. Given $y_0 \in L^2(0,\pi)$ and an interval $(a,b) \in (0,\pi)$, find a controller $v \in L^2(0,T)$ such that the solution y to the equation

$$\begin{aligned}&\frac{\partial y}{\partial t}-\frac{\partial^2 y}{\partial x^2}=v(t)m \text{ on } (0,T)\times(0,\pi),\\&y(x,0)=y_0(x),\ y(0,t)=y(\pi,t)=0,\ \forall t\in(0,T),\end{aligned}$$

satisfies the condition

$$y(x,T)=0,\ \forall x\in(0,\pi).$$

Here, m is the characteristic function $\mathbf{1}_{[a,b]}$ of the interval $[a,b]\subset(0,\pi)$.

This problem can be solved directly by invoking Theorem 1.8 and Corollary 2.1, that is, showing that the solution p to the backward dual equation

$$\begin{aligned}&\frac{\partial p}{\partial t}+\frac{\partial^2 p}{\partial x^2}=0 \qquad \text{in } (0,T)\times(0,\pi),\\&p(0,t)=p(\pi,t)=0\ \forall t\in(0,T),\end{aligned}\tag{3.4}$$

satisfies the observability inequality

$$\int_0^T p^2(x,0)dx \le C\int_0^T\left(\int_a^b p(x,t)dx\right)^2 dt. \tag{3.5}$$

(We note that, in this case, $H=L^2(0,\pi)$, $U=\mathbb{R}$, $B:\mathbb{R}\to L^2(0,\pi)$ is given by $Bu=mu$ and $B^*p=\int_0^\pi mp\,dx=\int_a^b p(x)dx$.) To prove (3.5), we express the solution p to equation (3.4) with the final value $p(x,T)=p_T(x)$ as the Fourier series

$$p(x,t)=\sum_{j=1}^\infty p_j e^{-j^2(T-t)}\sin(jx),$$

where $p_j=\frac{\sqrt{\pi}}{\sqrt{2}}\int_0^\pi p_T\sin(jx)dx$. Then, by Parseval's identity, inequality (3.5) reduces to

$$\sum_{j=1}^\infty p_j^2 e^{-2j^2T}\le C\int_0^T\left(\sum_{j=1}^\infty p_j e^{-j^2t}h_j\right)^2 dt, \tag{3.6}$$

where $h_j=\int_a^b \sin(jx)dx$.

Given the system $\{e^{-j^2t}\}_{j=0}^{\infty} \subset L^2(0,T)$, consider $\{\varphi_j\}_{j=1}^{\infty} \subset L^2(0,T)$ such that

$$\int_0^T \varphi_j(t)e^{-k^2t}dt = \delta_{jk},\ \forall\, j,k. \tag{3.7}$$

Such a system $\{\varphi_j\}_{j=1}^{\infty}$ is given by

$$\varphi_j(t) = (e^{-j^2t} - q_j(t))\|e^{-j^2t} - q_j\|_{L^2(0,T)}^{-2},$$

where

$$q_j = \arg\ \min\left\{\int_0^T |e^{-j^2t} - q(t)|^2 dt;\ q \in \Lambda_j\right\}$$

and Λ_j is the linear closed space of $L^2(0,T)$ spanned by $\{e^{-k^2t};\ k \neq j\}$.

By (3.7), we have

$$\begin{aligned}
e^{-j^2T}p_j &= \frac{e^{-j^2T}}{h_j}\int_0^T \varphi_j(t)\left(\sum_{k=1}^{\infty} h_j e^{-k^2t}p_k\right)dt \\
&= \frac{e^{-j^2T}}{h_j}\int_0^T \varphi_j(t)\left(\sum_{k=1}^{\infty} p_k e^{-k^2t}h_k\right)dt \\
&\leq \frac{e^{-j^2T}}{|h_j|}\left(\int_0^T \left(\sum_{k=1}^{\infty} p_k e^{-k^2t}h_k\right)^2 ds\right)^{\frac{1}{2}} \|\varphi_j\|_{L^2(0,T)}.
\end{aligned}$$

This yields

$$\sum_{j=1}^{\infty} p_j^2 e^{-2j^2T} \leq \sum_{j=1}^{\infty} \frac{e^{-2j^2T}}{|h_j|^2}\ \|\varphi_j\|_{L^2(0,T)}^2 \int_0^T \left(\sum_{k=1}^{\infty} p_k e^{-k^2t}h_k\right)^2 dt.$$

We see, therefore, that a sufficient condition to have (3.6) is that $|h_j| \geq M$, for all j. We note that this happens and so the exact controllability problem has a solution if $b-a$ and $b+a$ are not rational numbers. Of course, even if this happens, we can find a smaller interval $(a',b') \subset (a,b)$ with this property. Hence, for each subinterval ω, there is a controller $\widetilde{u}(t,x) \equiv u(t)m(x)$ with support in ω such that $y^u(T) = 0$. This approach, which was extended to more general linear parabolic equations (see D. Russell [108, 109]) based on sharp analysis of Fourier series corresponding to eigenfunctions of the Laplace operator, was the first successful attempt to solve the controllability problem for parabolic equations. (We note here also the pioneering works of V. Mizel and T. Seidman [96], H. Fattorini and D. Russell [69] and refer to the monograph [116] by M. Tucsnack and G. Weiss for other recent results in this area.) However, it is limited to time independent parabolic operators and in more dimensions it involves sophisticated results of harmonic analysis.

Here we shall derive an exact controllability result for equation (3.1) via Carleman's inequality (2.8). To this end, we shall associate with (3.1) the dual backward system (2.3), that is,

$$\begin{aligned} &p_t + \Delta p + \mathrm{div}_x(bp) - ap = 0 \quad \text{in } Q \\ &\alpha_1 \frac{\partial p}{\partial \nu} + (\alpha_2 + \alpha_1 b \cdot \nu)p = 0 \quad \text{on } \Sigma \end{aligned} \tag{3.8}$$

already considered in Section 2.1.

Keeping in mind Theorem 1.1 and the observability inequality (2.8), we have at this stage all ingredients for studying the exact null controllability of equation (3.1). Throughout in the sequel, the functions α and φ are defined by equations (2.6), where $k = 1$. The main result is Theorem 3.1 below.

Theorem 3.1. *Let $\mathcal{O}$ be a C^2-open and bounded domain of $\mathbb{R}^d$ and $a \in L^\infty(Q)$, $b \in C^1(\overline{Q}; \mathbb{R}^d)$. Let $F_0 \in L^2(Q)$ be such that*

$$|F(x,t)| \leq |F_0(x,t)| e^{s\alpha(x,t)} \varphi^{3/2}(x,t), \ \ a.e. \ (x,t) \in Q \tag{3.9}$$

where $s \geq s_0$, $\lambda \geq \lambda_0$, are as in Theorem 2.1. Then, for each $y_0 \in L^2(\mathcal{O})$, there is $u \in L^2(Q)$ such that $y^u(x,T) \equiv 0$ and

$$\|u\|_{L^2(Q)} \leq C(|y_0|_2 + \|F_0\|_{L^2(Q)}).$$

Here $y^u \in C([0,T]; L^2(\mathcal{O})) \cap L^2(0,T; H^1(\mathcal{O}))$ is the solution to (3.1) and $C = C(\|a\|_\infty, \|\nabla b\|_\infty)$. By $|\cdot|_p$, $1 \leq p \leq \infty$, we denote as usually the norm of the space $L^p(\mathcal{O})$. Theorem 3.1 amounts to saying that system (3.1) is exactly null controllable uniformly with respect to y_0 and F.

Proof. We note that, in the special case $F \equiv 0$, Theorem 3.1 is a direct consequence of Theorem 1.8. For the more general case considered here, we shall apply the argument used in the proof of Theorem 1.8. Namely, consider the optimal control problem:

$$\text{Minimize} \int_Q u^2 dx\, dt + \frac{1}{\varepsilon} \int_{\mathcal{O}} y^2(x,T) dx \ \ \text{subject to (3.1).} \tag{3.10}$$

By a standard argument, it follows that, for each $\varepsilon > 0$, problem (3.10) has a unique solution $(y_\varepsilon, u_\varepsilon)$. Moreover, by the maximum principle (see (1.85)–(1.86)), we have

$$u_\varepsilon = m p_\varepsilon, \ \text{a.e. in } Q \tag{3.11}$$

where $p_\varepsilon \in C([0,T]; L^2(\mathcal{O})) \cap L^2(0,T; H_0^1(\mathcal{O}))$ is the solution to the backward dual system (3.8), i.e.,

$$\begin{aligned} &(p_\varepsilon)_t + \Delta p_\varepsilon + \mathrm{div}_x(bp_\varepsilon) - ap_\varepsilon = 0 && \text{in } Q \\ &\alpha_1 \frac{\partial p_\varepsilon}{\partial \nu} + (\alpha_2 + \alpha_1 b \cdot \nu) p_\varepsilon = 0 && \text{on } \Sigma \\ &p_\varepsilon(T) = -\frac{1}{\varepsilon} y_\varepsilon(T) && \text{in } \mathcal{O}. \end{aligned} \tag{3.12}$$

Now, we multiply (3.12) by y_ε, equation (3.1) (where $y = y_\varepsilon$) by p_ε and integrate on Q. We get, after summing the two identities, that

$$\int_{Q_\omega} p_\varepsilon^2 dx\, dt + \frac{1}{\varepsilon}\int_{\mathcal{O}} y_\varepsilon^2(x,T)dx = -\int_{\mathcal{O}} y_0(x) p_\varepsilon(x,0)dx - \int_Q F p_\varepsilon dx\, dt. \tag{3.13}$$

By Corollary 2.1, we have

$$\left| \int_{\mathcal{O}} y_0(x) p_\varepsilon(x,0)dx \right| \le C \left(\int_{Q_\omega} p_\varepsilon^2 dx\, dt \right)^{1/2} |y_0|_2$$

whilst the Carleman estimate (2.7) and condition (3.9) imply that

$$\begin{aligned} \left| \int_Q F p_\varepsilon dx\, dt \right| &\le C \left(\int_Q e^{2s\alpha} \varphi^3 p_\varepsilon^2 dx\, dt \right)^{1/2} \left(\int_Q F_0^2(x,t) dx\, dt \right)^{1/2} \\ &\le C \left(\int_{Q_\omega} e^{2s\alpha} \varphi^3 p_\varepsilon^2 dx\, dt \right)^{\frac{1}{2}} \|F_0\|_{L^2(Q)} \end{aligned}$$

for $s \ge s_0$ and $\lambda \ge \lambda_0$. (Here and everywhere in the sequel, we shall denote by the same symbol C a positive constant independent of ε and y_0.)

Putting the latter inequality into (3.13), we get the estimate

$$\int_Q u_\varepsilon^2 dx\, dt + \frac{1}{\varepsilon}\int_{\mathcal{O}} y_\varepsilon^2(x,T)dx \le C(|y_0|_2^2 + \|F_0\|_{L^2(Q)}^2), \ \forall\, \varepsilon > 0. \tag{3.14}$$

This means that on a subsequence, again denoted ε, we have

$$u_\varepsilon \to u^* \text{ weakly in } L^2(Q)$$

and $|y_\varepsilon(T)|_2^2 \to 0$ as $\varepsilon \to 0$. By (3.1) it follows that

$$\begin{aligned} \{y_\varepsilon\} &\text{ is bounded in } L^2(0,T;H^1(\mathcal{O})) \cap C([0,T];L^2(\mathcal{O})), \\ \{\sqrt{t}(y_\varepsilon)_t\} &\text{ is bounded in } L^2(Q), \\ \{\sqrt{t}\, y_\varepsilon\} &\text{ is bounded in } L^2(0,T;H^2(\mathcal{O})) \cap L^\infty(0,T;H^1(\mathcal{O})). \end{aligned}$$

Hence, selecting further subsequence, if necessary, we may assume that

$$\begin{aligned}
y_\varepsilon &\longrightarrow y^* \quad \text{weakly in } L^2(0,T;H^1(\mathcal{O})) \text{ and weak–star in } L^2(Q),\\
\sqrt{t}(y_\varepsilon)_t &\longrightarrow \sqrt{t}y^*_t \text{ weakly in } L^2(Q),\\
y_\varepsilon(t) &\longrightarrow y^*(t) \text{ strongly in } L^2(\mathcal{O}) \text{ and uniformly on each } [\delta,T],
\end{aligned}$$

where $0 < \delta < T$. Clearly, $y^* = y^{u^*}$ and $y^*(T) = 0$. Moreover, letting ε tend to zero into (3.14), we see that u^* satisfies the desired estimate. This completes the proof.

Theorem 3.1 has several important consequences. In particular, it implies the exact controllability of the steady-state solutions y_e to equation (3.1).

Corollary 3.1. *Under assumptions of Theorem* 3.1, *let b, a, F be independent of t and let y_e be a steady state solution to* (3.1). *Then for each $y_0 \in L^2(\mathcal{O})$ there is $u \in L^2(Q)$ such that $y^u(T) \equiv y_e$.*

Proof. One applies Theorem 3.1 to $\bar{y} = y - y_e$ which satisfies the homogeneous equation (3.1) with the initial value $\bar{y}_0 = y_0 - y_e$.

Note also that Theorem 3.1 implies the exact boundary controllability of the boundary control system

$$\begin{aligned}
&y_t(x,t) - \Delta y(x,t) + b(x,t)\cdot\nabla y(x,t) + a(x,t) = F(x,t) \ \text{ in } Q,\\
&\alpha_1\frac{\partial y}{\partial\nu} + \alpha_2 y = u \ \text{ on } \Gamma_1\times(0,T) = \Sigma_1,\\
&\alpha_1\frac{\partial y}{\partial\nu} + \alpha_2 y = 0 \ \text{ on } \Gamma_2\times(0,T) = \Sigma_2,\\
&y(x,0) = y_0(x),
\end{aligned}\tag{3.15}$$

where $\partial\mathcal{O} = \Gamma_1 \cup \Gamma_2$ and $\Gamma_1 \cap \Gamma_2 = \emptyset$.

Corollary 3.2. *Under assumptions of Theorem* 3.1, *for each $y_0 \in H^1(\mathcal{O})$, there are $u \in L^2(\Sigma_1)$ and $y \in C([0,T]; H^1(\mathcal{O})) \cap L^2(0,T; H^2(\mathcal{O}))$, $y_t \in L^2(Q)$, which satisfy* (3.15) *and $y(T) \equiv 0$.*

Proof. The idea is to "inflate" a little bit the domain $\mathcal{O}$ and apply Theorem 3.1 on this new domain. Namely, let $\widetilde{\mathcal{O}}$ be an open bounded subset such that $\widetilde{\mathcal{O}} \supset \mathcal{O}$ and $\partial\widetilde{\mathcal{O}} = \Gamma_2 \cup \Gamma_3$.

We set $\omega = \widetilde{\mathcal{O}}\backslash\overline{\mathcal{O}}$ extend a, b to smooth functions on $\overline{\mathcal{O}} \times (0,T)$ and apply Theorem 3.1 to equation (3.1) on $\widetilde{\mathcal{O}}$ with the Dirichlet homogeneous boundary conditions and the initial value conditions

$$y(x,0) = \widetilde{y}_0(x) \text{ on } \widetilde{\mathcal{O}}$$

where $\widetilde{y}_0$ is an H^1_0–extension of y_0 to $\widetilde{\mathcal{O}}$.

In fact, as the boundary is of class C^1 the function y_0 can be extended by the usual method to a function $y^0 \in H^1(\mathbb{R}^d)$ (i.e., by symmetry if $\mathcal{O}$ is flat and by a similar argument using a partition of unity on a finite covering of $\overline{\mathcal{O}}$ in the general case.) We set $\widetilde{y}_0 = \rho y^0$ where $\rho \in C_0^\infty(\mathbb{R}^d)$ and $\rho = 1$ in a neighborhood of $\mathcal{O}$. We take $F = 0$ on $\widetilde{\mathcal{O}}\backslash\mathcal{O}$). Consequently, by virtue of Theorem 3.1 there are $\widetilde{y} \in C([0,T]; H_0^1(\widetilde{\mathcal{O}})) \cap L^2(0,T; H^2(\widetilde{\mathcal{O}}))$ and $\widetilde{y} \in L^2(\widetilde{\mathcal{O}} \times (0,T))$ satisfying (3.1) on $\widetilde{\mathcal{O}} \times (0,T)$ and such that $\widetilde{y}(T) = 0$.

Let u be the trace of $\alpha_1 \widetilde{y} + \alpha_2 \frac{\partial \widetilde{y}}{\partial \nu}$ to $\Sigma_1 = \Gamma_1 \times (0,T)$. (It should be recalled that by the trace theorem u is well defined and belongs to $L^2(\Sigma_1)$.) Clearly, the restriction y of $\widetilde{y}$ to $\mathcal{O} \times (0,T)$ satisfies all requirements of Corollary 3.2.

We note also that, if $y_0 \in L^2(\mathcal{O})$, then by the regularity theory of parabolic equations and by the trace theorem it follows that Corollary 3.2 remains true with a controller $u \in L^2_{\rm loc}(0,T; L^2(\Gamma_1))$, $\sqrt{t}u \in L^2(\Sigma_1)$.

Coming back to Theorem 3.1 it should be observed that the Carleman inequality (2.7) leads to a sharper controllability result than that expressed in Theorem 3.1.

Theorem 3.2. *Under the assumptions of Theorem* 3.1, *there are* $u \in L^2(Q)$ *and* $y \in C([0,T]; L^2(\mathcal{O})) \cap L^2(0,T; H^1(\mathcal{O}))$, *which satisfy* (3.1) *and such that*

$$\int_Q e^{-2s\alpha}\varphi^{-3}u^2 dx\,dt + s^3 \int_Q e^{-2s\alpha} y^2 dx\,dt \le C \int_Q e^{-2s\alpha}\varphi^{-3} F^2 dx\,dt. \tag{3.16}$$

Proof. Let

$$\alpha_\varepsilon = \frac{e^{\lambda\psi} - e^{2\lambda\|\psi\|_{C(\overline{\mathcal{O}})}}}{(t+\varepsilon)(T+\varepsilon-t)},$$

where ε is positive and sufficiently small. Consider the optimal control problem

$$\text{Minimize} \int_Q e^{-2s\alpha}\varphi^{-3}u^2 dx\,dt + s^3 \int_Q e^{-2s\alpha_\varepsilon} y^2 dx\,dt \text{ subject to (3.1).}$$

Let $(y_\varepsilon, u_\varepsilon)$ be an optimal pair. Then, by the maximum principle, we have

$$u_\varepsilon = m e^{2s\alpha}\varphi^3 p_\varepsilon, \text{ a.e. in } Q,$$

where p_ε is the solution to

$$\begin{aligned}
&(p_\varepsilon)_t + \Delta p_\varepsilon + \operatorname{div}(b p_\varepsilon) - a p_\varepsilon = s^3 y_\varepsilon e^{-2s\alpha_\varepsilon} \text{ in } Q,\\
&\alpha_1 \frac{\partial p_\varepsilon}{\partial \nu} + (\alpha_2 + b\cdot\nu) p_\varepsilon = 0 \text{ on } \Sigma,\\
&p_\varepsilon(T) = 0 \text{ in } \mathcal{O}.
\end{aligned}$$

By the transversality condition in the maximum principle in the above optimal control problem, we have also $p_\varepsilon(0) = 0$. Then, multiplying the latter by y_ε and (3.1) by p_ε, we get as above (see (3.13))

$$\int_{Q_\omega} e^{2s\alpha}\varphi^2 p_\varepsilon^2 dx\,dt + s^3 \int_Q e^{-2s\alpha_\varepsilon} y_\varepsilon^2 dx\,dt = -\int_Q p_\varepsilon F dx\,dt.$$

By (3.9), this yields

$$\int_{Q_\omega} e^{2s\alpha}\varphi^3 p_\varepsilon^2 dx\,dt + s^3\int_Q e^{-2s\alpha_\varepsilon} y_\varepsilon^2 dx\,dt \le C\int_Q e^{-2s\alpha}\varphi^{-3}F^2 dx\,dt$$

and, therefore,

$$\int_{Q_\omega} e^{-2s\alpha}\varphi^{-3}u_\varepsilon^2 dx\,dt + \frac{1}{2}s^3\int_Q e^{-2s\alpha_\varepsilon} y_\varepsilon^2 dx\,dt \le C\int_Q e^{-2s\alpha}\varphi^{-3}F^2 dx\,dt.$$

Letting ε tend to zero, we get

$$\begin{aligned} u_\varepsilon &\longrightarrow u \text{ weakly in } L^2(Q)\\ y_\varepsilon &\longrightarrow y \text{ weakly in } L^2(Q)\end{aligned}$$

and conclude as above that (y, u) satisfy system (3.1) along with estimate (3.16).

We shall study now the exact null controllability of equation (3.1) with L^∞–controllers u.

Theorem 3.3. *Let $F = fe^{2s\alpha}\varphi^3$, $y_0 \in L^1(\mathcal{O})$ and $a \in L^\infty(Q)$, $b \in C^1(\overline{Q};\mathbb{R}^d)$. Then there are $\lambda_0 > 0$, $\mu = \mu(\lambda)$ and $s_0 = s_0(\lambda)$ such that for $\lambda \ge \lambda_0$ and $s \ge s_0(\lambda)$ there is $u^* \in L^\infty(Q)$ such that $y^{u^*}(T) = 0$ and for all $\delta > 0$,*

$$\begin{aligned} I(u^*) &= \int_Q e^{-2s\alpha}\varphi^{-3}|u^*|^2 dx\,dt + \|e^{(-s(1-\eta(\lambda))+\delta(1+\eta(\lambda)))\alpha}u^*\|^2_{L^\infty(Q)}\\ &\le C\left(e^{2\mu s}|y_0|_2^2 + s^6\int_Q e^{2s\alpha}\varphi^3 f^2 dx\,dt\right)\end{aligned} \tag{3.17}$$

if $y_0 \in L^2(\mathcal{O})$ and

$$I(u^*) \le Ce^{2\mu s}\left(|y_0|_1^2 + \int_Q e^{2s\alpha}\varphi^3 f^2 dx\,dt\right) \tag{3.18}$$

if $y_0 \in L^1(\mathcal{O})$ and $\alpha_1 = 0$. Here $\eta(\lambda) = e^{-\lambda\|\psi\|_{C(\overline{\mathcal{O}})}}$ and $C = C_\lambda^\delta$.

Proof. We shall assume first that $y_0 \in L^2(\mathcal{O})$ and consider the optimal control problem

$$\text{Minimize } \int_Q e^{-2s\alpha}\varphi^{-3}u^2 dx\,dt + \varepsilon^{-1}\int_{\mathcal{O}} y^2(x,T)dx \ \text{ subject to (3.1).} \tag{3.19}$$

Let $(y_\varepsilon, u_\varepsilon)$ be optimal in (3.19). Then, once again by the maximum principle, we have

$$u_\varepsilon = mp_\varepsilon e^{2s\alpha}\varphi^3 \text{ a.e. in } Q \tag{3.20}$$

where

$$\begin{aligned}
&(p_\varepsilon)_t + \Delta p_\varepsilon + \operatorname{div}(bp_\varepsilon) - ap_\varepsilon = 0 \quad \text{in } Q,\\
&\alpha_1 \frac{\partial p_\varepsilon}{\partial \nu} + (\alpha_2 + b\cdot\nu)p_\varepsilon = 0 \quad \text{on } \Sigma,\\
&p_\varepsilon(T) = -\frac{1}{\varepsilon} y_\varepsilon(T) \quad \text{in } \mathcal{O}.
\end{aligned} \tag{3.21}$$

By (3.9) and (3.20), (3.21) it follows that (see, e.g., (3.13))

$$\int_{Q_\omega} p_\varepsilon^2 e^{2s\alpha}\varphi^3 dx\,dt + \int_Q f e^{2s\alpha}\varphi^3 p_\varepsilon dx\,dt + \varepsilon^{-1}\int_{\mathcal{O}} y_\varepsilon^2(x,T)dx = -\int_{\mathcal{O}} y_0(x)p_\varepsilon(x,0)dx.$$

This yields

$$\frac{1}{2}\int_{Q_\omega} p_\varepsilon^2 e^{2s\alpha}\varphi^3 dx\,dt + \varepsilon^{-1}\int_{\mathcal{O}} y_\varepsilon^2(x,T)dx \le \frac{1}{2}\int_Q f^2 e^{2s\alpha}\varphi^3 dx\,dt + |y_0|_2|p_\varepsilon(0)|_2$$

and so, by Corollary 3.1, we obtain

$$\begin{aligned}
&\int_{Q_\omega} p_\varepsilon^2 e^{2s\alpha}\varphi^3 dx\,dt + \varepsilon^{-1}\int_{\mathcal{O}} y_\varepsilon^2(x,T)dx\\
&\qquad \le C\left(e^{2\mu s}|y_0|_2^2 + \int_Q e^{2s\alpha}\varphi^3 f^2 dx\,dt\right)
\end{aligned} \tag{3.22}$$

for $\lambda \ge \lambda_0,\ s \ge s_0(\lambda)$. Equivalently,

$$\begin{aligned}
&\int_Q e^{-2s\alpha}\varphi^{-3} u_\varepsilon^2 dx\,dt + \varepsilon^{-1}\int_{\mathcal{O}} y_\varepsilon^2(x,T)dx\\
&\qquad \le C\left(e^{2\mu s}|y_0|_2^2 + \int_Q e^{2s\alpha}\varphi^3 f^2 dx\,dt\right).
\end{aligned} \tag{3.23}$$

We shall use in the following a *bootstrap* argument to improve estimate (3.23).

Let $\delta \in (s_0, s)$ be arbitrary but fixed. We set

$$\alpha_0(t) = \frac{1 - e^{2\lambda\|\psi\|_{C(\overline{\mathcal{O}})}}}{t(T-t)}, \quad \varphi_0(t) = \frac{1}{t(T-t)}.$$

Clearly, we have

$$\alpha_0 \le \alpha \le \frac{e^{\lambda\|\psi\|_{C(\overline{\mathcal{O}})}}}{1 + e^{\lambda\|\psi\|_{C(\overline{\mathcal{O}})}}}\alpha_0, \quad \varphi_0 \le \varphi \le e^{\lambda\|\psi\|_{C(\overline{\mathcal{O}})}}\varphi_0. \tag{3.24}$$

Let $\delta > 0$ be arbitrarily small but fixed and let $\{\delta_j\}$ be an increasing sequence such that $0 < \delta_j < \delta$ for all j. For each j, we set

$$v_j(x,t) = e^{(s+\delta_j)\alpha_0(t)}\varphi_0^3(t)p_\varepsilon(x,t).$$

We have

$$\begin{aligned} &(v_j)_t + \Delta v_j + \operatorname{div}(bv_j) - av_j = g_j && \text{in } Q,\\ &\alpha_1 \frac{\partial v_j}{\partial \nu} + (\alpha_2 + b\cdot\nu)v_j = 0 && \text{on } \Sigma,\\ &v_j(x,0) = v_j(x,T) = 0 && \text{in } \mathcal{O}, \end{aligned} \tag{3.25}$$

where $g_j = p_\varepsilon(e^{(s+\delta_j)\alpha_0}\varphi_0^3)_t$. By (3.9) and by (3.24), we see that

$$\begin{aligned} &\int_Q ((g_1)_t^2 + g_1^2 + |\nabla g_1|^2)dx\,dt\\ &\quad \le C\int_Q e^{2(s+\delta_1)\alpha}\varphi^7((s+\delta_1)^4(p_\varepsilon^2 + |\nabla p_\varepsilon|^2) + (s+\delta_1)^2(p_\varepsilon)_t^2)dx\,dt\\ &\quad \le C_1 s^3 \int_Q e^{2s\alpha}(s^3\varphi^3 p_\varepsilon^2 + s\varphi|\nabla p_\varepsilon|^2 + (s\varphi)^{-1}(p_\varepsilon)_t^2)dx\,dt\\ &\quad \le C_1 s^6 \int_{Q_\omega} e^{2s\alpha}\varphi^3 p_\varepsilon^2 dx\,dt. \end{aligned} \tag{3.26}$$

Then (3.22) yields

$$\|g_1\|_{H^1(Q)}^2 \le C\left(e^{2\mu s}|y_0|_2^2 + s^6\int_Q e^{2s\alpha}\varphi^3 f^2 dx\,dt\right). \tag{3.27}$$

Recalling that, by the Sobolev embedding theorem, $H^1(Q) \subset L^{p_1}(Q)$ for $p_1 = \frac{2(d+1)}{d-1}$, we obtain by (3.27) that

$$|g_1|_{L^{p_1}(Q)} \le C\left(e^{\mu s}|y_0|_2 + s^3\left(\int_Q e^{2s\alpha}\varphi^3 f^2 dx\,dt\right)^{1/2}\right). \tag{3.28}$$

(Here C is a generic positive constant independent of ε, s, p_i and f.)

Then, by the parabolic regularity (see [83], p. 341), we have

$$\|v_1\|_{W^{2,1}_{p_1}(Q)} \le C\left(e^{\mu s}|y_0|_2 + s^3\left(\int_Q e^{2s\alpha}\varphi^3 f^2 dx\,dt\right)^{1/2}\right), \tag{3.29}$$

$$W_p^{2,1}(Q) \subset L^q(Q) \text{ for } \frac{1}{p} - \frac{1}{q} \le \frac{2}{d+2}. \tag{3.30}$$

Then we obtain by (3.29) that

$$\|v_1\|_{L^{p_2}(Q)} \le C\left(e^{\mu s}|y_0|_2 + s^3\left(\int_Q e^{2s\alpha}\varphi^3 f^2 dx\, dt\right)^{1/2}\right),$$

where $p_2 = p_1 + \frac{dp_1^2}{d+2-dp_1}$ (if $d+2-2p_1 \le 0$, then $W^{2,1}_{p_1}(Q) \subset L^\infty(Q)$). This implies that

$$\|g_2\|_{L^{p_2}(Q)} \le C\left(e^{\mu s}|y_0|_2 + s^3\left(\int_Q e^{2s\alpha}\varphi^3 f^2 dx\, dt\right)^{1/2}\right)$$

and, therefore, $v_2 \in W^{2,1}_{p_2}(Q) \subset L^{p_3}(Q)$ satisfies the estimate

$$\|v_2\|_{W^{2,1}_{p_2}(Q)} \le C\left(e^{\mu s}|y_0|_2 + s^3\left(\int_Q e^{2s\alpha}\varphi^3 f^2 dx\, dt\right)^{1/2}\right).$$

In general, it follows that

$$\|v_j\|_{W^{2,1}_{p_j}(Q)} \le C\left(e^{\mu s}|y_0|_2 + s^3\left(\int_Q e^{2s\alpha}\varphi^3 f^2 dx\, dt\right)^{1/2}\right), \tag{3.31}$$

where $p_j = p_{j-1} + \frac{dp_{j-1}^2}{d+2-dp_{j-1}}$.

Thus, there is N such that $d+2-2p_N \le 0$ and for such an p_N we have, therefore, $W^{2,1}_{p_N}(Q) \subset L^\infty(Q)$. Moreover, by (3.31), we have

$$\|v_N\|_{L^\infty(Q)} \le C\left(e^{\mu s}|y_0|_2 + s^3\left(\int_Q e^{2s\alpha}\varphi^3 f^2 dx\, dt\right)^{1/2}\right)$$

and, therefore,

$$\|e^{(s+\delta)\alpha_0}\varphi_0 p_\varepsilon\|_{L^\infty(Q)} \le C\left(e^{\mu s}|y_0|_2 + s^3\left(\int_Q e^{2s\alpha}\varphi^3 f^2 dx\, dt\right)^{1/2}\right).$$

By (3.24), we obtain that

$$\|e^{(s+\delta)(1-\eta(\lambda))\alpha}\varphi^3 p_\varepsilon\|_{L^\infty(Q)} \le C_\lambda e^{\mu s}\left(|y_0|_2 + s^3\left(\int_Q e^{2s\alpha}\varphi^3 f^2 dx\, dt\right)^{1/2}\right)$$

and, by (3.17), it follows that

$$\|e^{(-s(1-\eta(\lambda))+\delta(1+\eta(\lambda)))\alpha}u_\varepsilon\|_{L^\infty(Q)} \le C_\lambda\left(e^{\mu s}|y_0|_2 + s^3\left(\int_Q e^{2s\alpha}\varphi^3 f^2 dx\,dt\right)^{1/2}\right) \tag{3.32}$$

for $s \ge s_0,\ \lambda \ge \lambda_0$ and $s_0 \le \delta < s$.

By estimates (3.23) and (3.32), it follows that on a subsequence, for simplicity again denoted $\{\varepsilon\} \to 0$, we have

$$\begin{aligned}
u_\varepsilon &\longrightarrow u^* \quad \text{weak star in } L^\infty(Q),\\
y_\varepsilon &\longrightarrow y^* \quad \text{weakly in } L^2(0,T;H^1(\mathcal{O})),\ \text{weak star in } L^\infty(0,T;L^2(\mathcal{O})),\\
\sqrt{t}y_\varepsilon &\longrightarrow \sqrt{t}y^* \text{ weakly in } L^2(0,T;H^2(\mathcal{O})),\\
\sqrt{t}(y_\varepsilon)_t &\longrightarrow \sqrt{t}y^*_t \text{ weakly in } L^2(Q),\\
y_\varepsilon(t) &\longrightarrow y^*(t) \text{ uniformly in } L^2(\mathcal{O}) \text{ on every compact interval.}
\end{aligned}$$

Clearly, $y^* = u^*,\ y^*(T) = 0$ and u^* satisfies all the requirements of Theorem 3.2 (i.e., estimate (3.17)).

Let us assume now that $\alpha_1 = 0$. Let $S(t)$ be the C_0 semigroup generated on $L^p(\mathcal{O})$ by the Laplace operator with the Dirichlet homogeneous boundary value conditions (see Section 1.5). Then the solution y to (3.1) can be represented by the variation of the constant formula

$$y(t) = S(t)y_0 + \int_0^t S(t-s)(F(x,s) - b(x,s)\cdot\nabla y(x,s) - a(x,s)y(x,s))ds.$$

We recall that

$$|S(t)z|_p \le Ct^{-\frac{d}{2}(q^{-1}-p^{-1})}|z|_q,\ \forall t > 0, \tag{3.33}$$

for all $1 \le q \le p \le \infty$. Moreover, we recall that (see, e.g., [18])

$$t|\nabla S(t)(z)|_2^2 \le C|z|_2^2,\quad \forall z \in L^2(\mathcal{O}),\ t > 0.$$

For $p = 2$ and $q = 1$, we get by (3.33)

$$|y(t)|_2 \le C\left(t^{-\frac{d}{4}}|y_0|_1 + \int_0^t (|t-s|^{-\frac{1}{2}}|y(s)|_2 + |y(s)|_2 + |F(s)|_2)ds\right)$$

and, therefore, for $\eta > 0$ and sufficiently small, we have

$$|y(t)|_2 \le C\left(t^{-\frac{d}{2}}|y_0|_1 + \int_0^t |F(s)|_2 ds\right),\ \forall t \in (0,\eta).$$

By density, this implies that equation (3.1) has, for each $y_0 \in L^1(\mathscr{O})$, a unique solution $y \in C([0,T]; L^1(\mathscr{O})) \cap C(]0,T]; H_0^1(\mathscr{O}))$ which satisfies the estimate

$$|y(\eta)|_2 \leq C\left(|y_0|_1 + \int_0^\eta |fe^{s\alpha}\varphi^{3/2}|_2 dt\right).$$

Let $\eta \in (0,T)$ be arbitrary but fixed. According to the first part of the proof, there are $(\widetilde{y}, \widetilde{u})$ which satisfy equation (3.1) on $\mathscr{O}\times(\eta,T)$, $\widetilde{y}(\eta) = y(\eta)$, $\widetilde{y}(T) = 0$ and

$$\int_\eta^T \int_{\mathscr{O}} e^{-2s\alpha}\varphi^{-3}(\widetilde{u})^2 dx\, dt + \|e^{(-s(1-\eta(\lambda))+\delta(1+\eta(\lambda)))\alpha}\widetilde{u}\|^2_{L^\infty(Q)}$$
$$\leq C\left(e^{2\mu s}|y(\eta)|_2^2 + \int_Q e^{2s\alpha}\varphi^3 f^2 dx\, dt\right) \leq Ce^{2\mu s}\left(|y_0|_1^2 + \int_Q e^{2s\alpha}\varphi^3 f^2 dx\, dt\right).$$

The function

$$u^*(t) = \begin{cases} 0 & \text{if } 0 < t \leq \eta \\ u^*(t) & \text{if } \eta < t \leq T \end{cases}$$

clearly satisfies the estimate (3.17) and steers y_0 into origin in the time T. This completes the proof of Theorem 3.3.

From the proof of Theorems 3.1 and 3.2, one might suspect that there is an equivalence between Carleman's inequality (2.2) for the dual equation and the exact null controllability of the linear equation (3.1) with $F = e^{2s\alpha}\varphi^3 f$.

We shall see below that this is, indeed, the case in a certain precise sense.

Theorem 3.4. *The Carleman inequality* (2.2) *holds for* (3.8) *if and only if for each* $f \in L^2(Q)$ *there are* $u \in L^2(Q)$ *and* $y \in C([0,T]; L^2(\mathscr{O})) \cap L^2(0,T; H^1(\mathscr{O}))$ *satisfying* (3.1) *where* $y_0 = 0$, $F = e^{2s\alpha}\varphi^3 f$, *and such that*

$$\int_Q e^{-2s\alpha}\varphi^{-3}u^2 dx\, dt + s^3 \int_Q e^{-2s\alpha}y^2 dx\, dt \leq C\int_Q f^2\varphi^3 e^{2s\alpha} dx\, dt, \tag{3.34}$$

for $s \geq s_0(\lambda)$, $\lambda \geq \lambda_0$.

Proof. The *only if* part was proved in Theorem 3.2, so we shall confine to prove the *if* part. To this end, consider u and y satisfying (3.1) and (3.34). Multiplying (3.1) by p and (3.8) by y, we get

$$\int_Q mu\, p\, dxdt + \int_Q gy\, dx\, dt + \int_Q f\varphi^3 e^{2s\alpha} p\, dxdt + \int_{\mathscr{O}} y_0(x)p(x,0)dx = 0, \tag{3.35}$$

for all $f \in L^2(Q)$ and $y_0 \in L^2(\mathscr{O})$.

For $y_0 = 0$ and $f = p$, it follows by (3.35) that

$$\int_Q e^{2s\alpha}\varphi^3 p^2 dx\,dt \le \left(\int_{Q_\omega} e^{2s\alpha}\varphi^3 p^2 dx\,dt\right)^{1/2}\left(\int_Q u^2 e^{-2s\alpha}\varphi^{-3}dx\,dt\right)^{1/2} + \left(\int_Q e^{-2s\alpha}y^2 dx\,dt\right)^{1/2}\left(\int_Q e^{2s\alpha}g^2 dx\,dt\right)^{1/2}. \tag{3.36}$$

Then, by (3.34), it follows that

$$\int_Q e^{2s\alpha}\varphi^3 p^2 dx\,dt \le C\left(\int_{Q_\omega} e^{2s\alpha}\varphi^3 p^2 dx\,dt + s^{-3}\int_Q e^{2s\alpha}g^2 dx\,dt\right). \tag{3.37}$$

Next, we multiply equation (3.8) by $\varphi e^{2s\alpha}p$ and integrate on Q. After some calculation involving Green's formula, we get the estimate

$$\int_Q e^{2s\alpha}\varphi|\nabla p|^2 dx\,dt \le Cs^2\int_Q e^{2s\alpha}\varphi^3 p^2 dx\,dt + Cs^{-1}\int_Q e^{2s\alpha}g^2 dx\,dt. \tag{3.38}$$

Finally, we multiply (3.8) by $e^{2s\alpha}\varphi^{-1}(\Delta p + p_t)$ and integrate on Q.

Again using Green's formula, we obtain that

$$\begin{aligned}\int_Q & e^{2s\alpha}\varphi^{-1}(|\Delta p|^2 + |p_t|^2)dx\,dt\\ &\le Cs\int_Q e^{2s\alpha}|p_t||\nabla p|dx\,dt + C\int_Q e^{2s\alpha}(p^2 + |\nabla p|^2 + g^2)dx\,dt\\ &\le C\left(s\int_Q e^{2s\alpha}\varphi|\nabla p|^2 dx\,dt + \frac{1}{2}\int_Q e^{2s\alpha}\varphi^{-1}|p_t|^2 dx\,dt\right)\\ &+Cs^4\int_Q e^{2s\alpha}\varphi^3 p^2 dx\,dt + C\int_Q e^{2s\alpha}g^2 dx\,dt.\end{aligned} \tag{3.39}$$

By (3.37), (3.38), and (3.39), we get inequality (3.34), as claimed.

By formula (3.35), we see that more regular is the control u sharper is the Carleman inequality for the dual equation (3.8). In particular, by Theorem 3.4, we get the following Carleman inequality in $L^1(\mathcal{O})$.

Corollary 3.3. *Let $g \equiv 0$. Then, under the assumptions of Theorem 3.2, for $\lambda \ge \lambda_0$, $s \ge s_0(\lambda)$, the following inequality holds*

$$\left(\int_Q e^{2s\alpha}\varphi^3 p^2 dx\,dt\right)^{1/2} \le C_\delta^\lambda s^3\int_{Q_\omega} e^{(s(1-\eta(\lambda))-\delta(1+\eta(\lambda)))\alpha}|p|dx\,dt \tag{3.40}$$

and, if $\alpha_1 = 0$,

$$|p(0)|_\infty \le C_\delta^\lambda e^{\mu s}\int_{Q_\omega} e^{(s(1-\eta(\lambda))-\delta(1+\eta(\lambda)))\alpha}|p|dx\,dt \tag{3.41}$$

for $0 < \delta < s_0(\lambda)$ and all weak solutions $p \in L^1(Q)$ to (3.8).

Proof. In (3.35), we take $f = p$ and $y_0 = 0$. Then, by (3.17), we see that

$$\int_Q e^{2s\alpha}\varphi^3 p^2 dx\, dt \leq C_\delta^\lambda s^3 \int_{Q_\omega} e^{(s(1-\eta(\lambda))-\delta(1+\eta(\lambda))\alpha}|p|dx\, dt.$$

Similarly, for $f = 0$, estimates (3.35) and (3.18) imply that

$$|p(0)|_\infty \leq C_\delta^\lambda e^{\mu s} \int_{Q_\omega} e^{(s(1-\eta(\lambda))-\delta(1+\eta(\lambda))\alpha}|p|dx\, dt,$$

as claimed. In particular, we deduce by Corollary 3.3 that the homogeneous system (3.8) is $L^1 - L^\infty$ observable. We have a similar result for the nonhomogeneous equation (3.8).

Corollary 3.4. *Let* $g \in L^1(Q)$. *Then*

$$|p(0)|_1 \leq C_s \left(\int_{Q_\omega} e^{(s(1-\eta(\lambda))-\delta(1+\eta(\lambda)))\alpha}|p|dxdt + \int_Q |g|dxdt\right) \tag{3.42}$$

for $0 < \delta < s_0$ *and all solutions* p *to* (3.8).

Proof. We write $p = p_1 + p_2$ where p_1 is the solution to the homogeneous equation (3.8) and p_2 is the solution to (3.8) with the final value $p_2(T)$=0. Then we apply inequality (3.40) to p_1 and use the obvious inequality

$$|p_2(t)|_1 \leq C\int_t^T |g(\theta)|d\theta$$

to obtain the following estimate

$$\int_Q e^{2s\alpha}\varphi^3|p|dx\, dt \leq C_s\left(\int_{Q_\omega} e^{s(1-\eta(\lambda))-\delta(1+\eta(\lambda)))\alpha}|p|dx\, dt + \int_Q |g|dx\, dt\right),$$

which clearly implies (3.42), as claimed.

Remark 3.1. By inequality (3.40), it follows via the *bootstrap* argument developed earlier in the proof of Theorem 3.4 the following sharper Carleman inequality

$$\|e^{s\alpha_0}p\|_{L^\infty(Q)} \leq C\int_{Q\omega} |p|e^{s\delta\alpha_0}dx\, dt$$

for some μ suitable chosen. This implies an inequality of the form (3.41) in L^∞ norm for the left-hand side.

Remark 3.2. One might ask if the above controllability results remain true if int ω is empty. The following example shows that in general the answer is negative. For instance let us take ω to be a smooth boundary of a simple connected subdomain $\mathcal{O}_\omega$ of $\mathcal{O}$. Consider the equation

$$y_t + Ay = \mu \text{ in } Q = \mathcal{O} \times (0, T),$$

where $\mu \in H^{-1}(\mathcal{O})$ is defined by

$$\mu(\varphi) = \int_\omega u\varphi dx, \forall \varphi \in H_0^1(\mathcal{O})$$

and

$$(Ay, \varphi) = \int_{\mathcal{O}} \nabla y \cdot \nabla \varphi dx, \forall \varphi \in H_0^1(\mathcal{O}).$$

Here u is a given L^2 function on ω. This is just system (3.1), where $b = 0$, $a = 0$, $F = 0$ and with a distributed control u with the support in ω. We may, equivalently, write it as

$$\begin{aligned} &y_t - \Delta y = 0 \text{ in } Q, \\ &\frac{\partial^+ y}{\partial \nu} - \frac{\partial^- y}{\partial \nu} = u \text{ on } \omega \times (0, T), \ \ y = 0 \text{ on } \partial\mathcal{O} \times (0, T). \end{aligned}$$

As noticed earlier, the exact null controllability of the above system is equivalent with the observability inequality

$$|p(0)|_2^2 \le C \int_{\omega \times (0,T)} p^2 dx dt$$

for all the solutions to the equation

$$p_t + \Delta p = 0 \text{ in } Q; \ p = 0 \text{ on } \Sigma,$$

which, obviously, is false.

3.2 Controllability of Semilinear Parabolic Equations

We shall study in this section the exact null controllability of the equation

$$\begin{aligned} &y_t - \Delta y + b \cdot \nabla y + f(x, t, y) = mu + F(x, t) \ \text{ in } Q, \\ &y(x, 0) = y_0(x) \ \text{ in } \mathcal{O}, \\ &\alpha_1 \frac{\partial y}{\partial \nu} + \alpha_2 y = 0 \ \text{ on } \Sigma, \end{aligned} \tag{3.43}$$

where $\alpha_1, \alpha_2 \geq 0$ are nonnegative constants such that $\alpha_1 + \alpha_2 > 0$, $m = \mathbf{1}_\omega$ is the characteristic function of some open subset $\omega \subset \mathcal{O}$ and $f : \mathcal{O} \times (0, T) \times \mathbb{R} \to \mathbb{R}$ is continuous in y, measurable in (x, t) and satisfies the following conditions

$$|f(x, t, r)| \leq L|r|(\eta(|r|) + 1), \text{ a.e. } (x, t) \in Q, \ r \in \mathbb{R}, \tag{3.44}$$

$$f(x, t, r)r \geq -\gamma_0 r^2, \ \forall (x, t, r) \in Q \times \mathbb{R}, \tag{3.45}$$

$$|F(x, t)| \leq |F_0(x, t)| e^{s\alpha} \varphi^{\frac{3}{2}}, \ \forall (x, t) \in Q. \tag{3.46}$$

Here $L > 0$, $\gamma_0 \geq 0$, $F_0 \in L^2(Q)$, η is a nonnegative, continuous, and increasing function and α, $\lambda \geq \lambda_0$, $s \geq s_0$ are fixed as in Theorem 2.1 and α, φ are defined by equations (2.6), where $k = 1$.

It should be said that the sign condition (3.45) precludes the *blow up* of solutions and so it implies the existence of a global solution for the Cauchy problem (3.43) by a standard continuation argument. Note also that, if for some γ the function $y \to f(x, t, y) + \gamma y$ is monotonically increasing, then by Theorem 1.4 equation (3.43) has a unique global solution.

We shall see here that the exact null controllability is possible for nonlinear function $r \to f(\cdot, r)$ with *mild* growth to $+\infty$.

To begin with, we shall consider first the case where f is sublinear as a function of r.

Theorem 3.5. *Assume that $b \in C^1(\overline{Q}; \mathbb{R}^d)$ and $|f(t, x, r)| \leq L|r|$, a.e. $(x, t) \in Q, \forall r \in \mathbb{R}$. Then for each $y_0 \in L^2(\mathcal{O})$ there are $u \in L^2(Q)$ and $y \in C([0, T]; L^2(\mathcal{O})) \cap L^2(0, T; H_0^1(\mathcal{O}))$, $\sqrt{t} y_t \in L^2(Q)$, $\sqrt{t} y \in L^2(0, T; H^2(\mathcal{O}))$ satisfying* (3.43) *and such that $y(T) = 0$ in $\mathcal{O}$.*

Proof. The argument is standard for this type of controllability result and will be frequently used in the sequel. Namely, consider the set

$$K = \{z \in L^2(Q); \ \|z\|_{L^2(Q)} \leq \rho\}$$

and the linear system

$$\begin{cases} y_t - \Delta y + b \cdot \nabla y + g(x, t, z)y = mu + F & \text{in } Q, \\ y(x, 0) = y_0(x) & \text{in } \mathcal{O}, \\ \alpha_1 \dfrac{\partial y}{\partial \nu} + \alpha_2 y = 0 & \text{on } \partial\mathcal{O}, \end{cases} \tag{3.47}$$

where

$$g(x, t, r) = \begin{cases} \dfrac{f(x, t, r)}{r} & \text{if } |r| > 0, \\ \lim\limits_{\theta \to 0} \dfrac{f(x, t, \theta)}{\theta} & \text{if } r = 0. \end{cases} \tag{3.48}$$

(Without loss of generality we may assume that the previous limit exists; otherwise we approximate g by a family of smooth functions g_ε and let $\varepsilon \to 0$ in the corresponding controllability result.)

By Theorem 3.1, for each $z \in K$, there is at least one controller $u \in L^2(Q)$ such that $y^u(T) = 0$ and

$$\int_Q u^2 dx\, dt \le \gamma(|y_0|_2^2 + \|F_0\|^2_{L^2(Q)}), \tag{3.49}$$

where γ is independent of z (because $|g(x,t,z)| \le L$). Here y^u_z is the solution to (3.47).

Define the mapping $\Phi : K \to 2^{L^2(Q)}$ by

$$\Phi(z) = \{y^u;\ y^u_z(T) = 0;\ u \text{ satisfies (3.49)}\}.$$

By (3.47) and (3.48), we see that y^u_z satisfies the estimates

$$\begin{aligned} |y^u_z(t)|_2^2 + \int_0^T \int_{\mathcal{O}} |\nabla y^u_z(x,t)|^2 dx\, dt &\le C_1 \left(\int_Q (u^2 + F^2) dx\, dt + |y_0|_2^2 \right) \\ &\le C_2 \left(\int_Q F^2 dx\, dt + |y_0|_2^2 \right),\ t \in [0,T] \end{aligned} \tag{3.50}$$

and, therefore, by (3.47) we have

$$\|(y^u_z)_t\|_{L^2(0,T;(H^1(\mathcal{O}))')} \le C_3(|y_0|_2 + 1),\ \forall z \in K. \tag{3.51}$$

This implies that, for ρ sufficiently large Φ maps K into itself and for each $z \in K$, $\Phi(z)$ is a convex and compact subset of $L^2(Q)$. It is also easily seen that this multivalued map is upper semicontinuous on $L^2(Q)$, i.e., if $z_k \to z$ strongly in $L^2(Q)$ and $y_k \in \Phi(z_k)$ then $y \in \Phi(z)$. Indeed, $y_k = y^{u_k}_{z_k}$ where (see (3.49)) $\|u_k\|_{L^2(Q)} \le C$. By estimates (3.50) and (3.51), it follows via the Aubin–Lions compactness theorem (see, e.g., [89], Theorem 5.1) that $\{y_k\}$ is compact in $L^2(Q)$ and, therefore, selecting a subsequence

$$\begin{aligned} u_k &\longrightarrow u \text{ weakly in } L^2(Q) \\ y_k &\longrightarrow y \text{ strongly in } L^2(Q) \text{ and weakly in } L^2(0,T;H^1(\mathcal{O})) \end{aligned}$$

Moreover, since $g(z_k)y_k \to g(z)y$, a.e. on Q and $\{g(z_k)y_k\}$ is bounded in $L^2(Q)$, we infer that

$$g(z_k)y_k \longrightarrow g(z)y \ \text{ weakly in } L^2(Q).$$

Hence

$$(y_k)_t - \Delta y_k + b \cdot \nabla y_k \longrightarrow y_t - \Delta y + b \cdot \nabla y \text{ weakly in } L^2(Q)$$

and, therefore, $y = y^u_z$ as claimed.

By the infinite dimensional Kakutani theorem (see, e.g., [18]) and [65], p. 310, we may conclude that there is $y \in K$ such that $y \in \Phi(y)$. In other words, there is $u \in L^2(Q)$ such that $y^u(T) = 0$, as claimed. This completes the proof.

In spite of the restrictive condition (3.45), Theorem 3.5 is, however, applicable to a large class of physical processes. For instance, this is the case with the thermostat control model or with the kinetics of enzymatic reactions (the Michaelis–Menten model). In the latter case, $f(y) = a_1 y(a_2 y + a_3)^{-1}$, $\forall y \in \mathbb{R}$.

Analyzing the proof of Theorem 3.5, one sees that the sublinearity condition was essential to proving that $\Phi(K) \subset K$. However, it turns out that the conclusions of Theorem 3.5 still remain true for nonlinearities f with *mild* growth to $+\infty$. Namely, one has

Theorem 3.6. *Assume that the functions f, F satisfy conditions* (3.44), (3.45), *and* (3.46) *where $F_0 \in L^\infty(Q)$,*

$$|\eta(r)| \leq \mu(r)((\log r + 1)^{3/2} + 1), \quad \forall r \in \mathbb{R}^+, \tag{3.52}$$

and $\lim\limits_{r\to\infty} \mu(r) = 0$. Then, for each $y_0 \in L^2(\mathcal{O})$, there are $u \in L^\infty(Q)$ and $y \in C([0, T];\ L^2(\mathcal{O})) \cup L^2(0, T; H^1(\mathcal{O}))$ which satisfy (3.43) *and such that*

$$y(T) \equiv 0, \ \|u\|_{L^\infty(Q)} \leq C(|y_0|_2 + \|F_0\|_{L^\infty(Q)}).$$

Proof. We set

$$K_\infty = \{z \in L^\infty(Q);\ \|z\|_{L^\infty(Q)} \leq \rho\} \tag{3.53}$$

and, for each $z \in K_\infty$, denote by y_z^u the solution to equation (3.47). Lemma 3.1 below is the main ingredient of the proof.

Lemma 3.1. *Under the assumptions of Theorem* 3.6, *for each $z \in K_\infty$ there are $u \in L^\infty(Q)$ and $y = y_z^u$ such that $y_z^u(T) = 0$ and*

$$\|u\|_{L^\infty(Q)} \leq C(e^{\mu_0 \eta^{\frac{2}{3}}(\rho)}|y_0|_2 + \|F_0\|_\infty), \tag{3.54}$$

where C and μ_0 are independent of ρ and z and $\|F_0\|_\infty$ is the norm of F_0 in $L^\infty(Q)$.

Proof. Arguing as in the proof of Theorem 3.2, consider the optimal control problem

$$\text{Minimize} \int_Q e^{-2s\alpha}\varphi^{-3}u^2 dx\, dt + \frac{1}{\varepsilon}\int_{\mathcal{O}} y^2(x, T)dx \text{ subject to (3.47).} \tag{3.55}$$

We have

$$u_\varepsilon = mp_\varepsilon e^{2s\alpha}\varphi^3, \text{ a.e. in } Q,$$

where p_ε is the solution to linear system

$$\begin{cases} (p_\varepsilon)_t + \Delta p_\varepsilon - g(x,t,z)p_\varepsilon + \mathrm{div}\,(b \cdot p_\varepsilon) = 0 & \text{in } Q, \\ \alpha_1 \dfrac{\partial p_\varepsilon}{\partial \nu} + (\alpha_2 + b \cdot \nu)p_\varepsilon = 0 & \text{on } \Sigma, \\ p_\varepsilon(x,T) = -\dfrac{1}{\varepsilon} y_\varepsilon(x,T) & \text{in } \mathcal{O}. \end{cases} \tag{3.56}$$

Multiplying (3.56) by y_ε, (3.47) by p_ε, and integrating on Q, we obtain

$$\begin{aligned} \int_{Q_\omega} e^{2s\alpha}\varphi^3 p_\varepsilon^2 + \varepsilon^{-1}\int_{\mathcal{O}} y_\varepsilon^2(x,T)dx &= -\int_{\mathcal{O}} y_0(x)p_\varepsilon(x,0)dx - \int_Q F p_\varepsilon dx\,dt \\ &\le |y_0|_2|p_\varepsilon(0)|_2 + \delta\int_Q e^{2s\alpha}\varphi^3 p_\varepsilon^2 dx\,dt + C_\delta\|F_0\|_\infty^2 \end{aligned} \tag{3.57}$$

for each $\delta > 0$.

Next, applying the Carleman inequality (2.7) into equation (3.57) (i.e., g is replaced by $g(x,t,z)p_\varepsilon$) and using condition (3.44), we see that, for $\lambda = \lambda_0$ sufficiently large and $s \ge s_0$, we have

$$s^3\int_Q e^{2s\alpha}\varphi^3 p_\varepsilon^2 dx\,dt \le C\left(s^3\int_{Q_\omega} e^{2s\alpha}\varphi^3 p_\varepsilon^2 dx\,dt + \eta^2(\rho)\int_Q e^{2s\alpha} p_\varepsilon^2 dx\,dt\right).$$

This means that, for $s \ge C_0\eta^{\frac{2}{3}}(\rho) + s_0$, we have

$$\int_Q e^{2s\alpha}\varphi^3 p_\varepsilon^2 dx\,dt \le C\int_{Q_\omega} e^{2s\alpha}\varphi^3 p_\varepsilon^2 dx\,dt, \tag{3.58}$$

where C is independent of z and ε.

On the other hand, multiplying equation (3.56) by p_ε and recalling that, by condition (3.45),

$$g(x,t,z) \ge -\gamma_0, \quad \forall\,(x,t) \in Q,\ z \in K,$$

we obtain after some calculation involving Green's formula that

$$\frac{d}{dt}|p_\varepsilon(t)|_2^2 \ge -C_1|p_\varepsilon(t)|_2^2, \quad \text{a.e. } t \in (0,T),$$

where C_1 is independent of z and ρ. This yields

$$|p_\varepsilon(0)|^2{}_2 \le C_3 e^{\frac{\mu s}{t(T-t)}}\int_{\mathcal{O}} e^{2s\alpha}\varphi^3|p_\varepsilon(x,t)|^2dx, \quad \forall\, t \in (t_1,t_2),$$

where C_3 is independent of ρ, z and s. From now on, we shall argue as in the proof of Corollary 2.1. Integrating the latter on (t_1, t_2) and using estimate (3.58), we obtain, for some C and $\mu_0 > 0$ independent of ρ,

$$|p_\varepsilon(0)|_2 \le Ce^{\mu_0\eta^{2/3}(\rho)}\left(\int_{Q_\omega} e^{2s\alpha}\varphi^3 p_\varepsilon^2 dx\, dt\right)^{1/2}$$

for $s = C\eta^{\frac{2}{3}}(\rho) + s_0$ and so (3.57) yields

$$\begin{aligned}&\int_{Q_\omega} e^{2s\alpha}\varphi^3 p_\varepsilon^2 dx\, dt + \varepsilon^{-1}\int_{\mathcal{O}} y_\varepsilon^2(x,T)dx\\ &\quad \le C|y_0|_2^2 e^{2\mu_0\eta^{2/3}(\rho)} + \delta\int_Q e^{2s\alpha}\varphi^3 p_\varepsilon^2 dx\, dt + C_\delta\|F_0\|_\infty^2.\end{aligned} \tag{3.59}$$

By the Carleman inequality (3.9) applied to equation (3.56), it follows also that

$$\begin{aligned}&\int_Q e^{2s\alpha}(s^3\varphi^2 p_\varepsilon^2 + s\varphi|\nabla p_\varepsilon|^2 + (s\varphi)^{-1}(|\Delta p_\varepsilon|^2 + (p_\varepsilon)_t^2))\, dxdt\\ &\quad \le C\left(s^3\int_{Q_\omega} e^{2s\alpha}\varphi^3 p_\varepsilon^2 dx\, dt + \eta^2(\rho)\int_Q e^{2s\alpha}\varphi^3 p_\varepsilon^2 dx\, dt\right).\end{aligned}$$

Then, by (3.59), we get for δ suitable chosen that

$$\begin{aligned}&\int_Q e^{2s\alpha}(\varphi^3 p_\varepsilon^2 + \varphi|\nabla p_\varepsilon|^2 + \varphi^{-1}(p_\varepsilon)_t^2 + |\Delta p_\varepsilon|^2)dx\, dt + \varepsilon^{-1}\int_{\mathcal{O}} y_\varepsilon^2(x,T)dx\\ &\quad \le C\int_{Q_\omega} e^{2s\alpha}\varphi^3 p_\varepsilon^2 dx\, dt \le C(e^{2\mu_0\eta^{\frac{2}{3}}(\rho)}|y_0|_2^2 + \|F_0\|_\infty^2)\end{aligned} \tag{3.60}$$

for $s = C\eta^{\frac{2}{3}}(\rho) + s_0$ and $\lambda = \lambda_0$ large enough.

We shall continue the proof with a *bootstrap* argument already used in the proof of Theorem 3.3. Namely, we fix $0 < \delta < s = C\eta^{2/3}(\rho) + s_0$ and consider an increasing sequence $0 < \delta_j < \delta,\ j = 1, \dots$. We set

$$v_j = e^{(s+\delta_j)\alpha_0}\varphi_0^3 p_\varepsilon,\ g_j = p_j(e^{(s+\delta_j)\alpha_0}\varphi_0^3)_t,\ j = 1, \dots,$$

where α_0 and φ_0 are defined as in the proof of Theorem 3.3. We recall that, by (3.26), we have the estimate

$$\|g_1\|_{H^1(Q)}^2 \le Cs^6\int_{Q_\omega} e^{2s\alpha}\varphi^3 p_\varepsilon^2 dx\, dt.$$

Then, by (3.60), we obtain

$$\|g_1\|_{H^1(Q)}^2 \le Cs^6(e^{2\mu_0\eta^{2/3}(\rho)}|y_0|_2^2 + \|F_0\|_\infty)$$

for $s = C\eta^{2/3}(\rho) + s_0$ and $\lambda = \lambda_0$ fixed but large enough. This implies that (see (3.28))

$$\|g_1\|_{L^{p_1}(Q)} \leq C(e^{\mu_0 \eta^{\frac{2}{3}}}|y_0|_2 + \|F_0\|_\infty)$$

for $p_1 = \frac{2(d+1)}{d-1}$ and, therefore, by the parabolic regularity and by (3.29), we obtain the estimate

$$\|v_1\|_{W^{2,1}_{p_1}(Q)} \leq C(e^{\mu_0 \eta^{2/3}(\rho)}|y_0|_2 + \|F_0\|_\infty)$$

for $p_1 = \frac{2(d+1)}{d-1}$.

In general, we have (see (3.31))

$$\|v_j\|_{W^{2,1}_{p_j}(Q)} \leq C(e^{\mu_0 \eta^{2/3}(\rho)}|y_0|_2 + \|F_0\|_\infty),$$

where $p_j = p_{j-1} + \frac{dp^2_{j-1}}{d+2-dp_{j-1}}$.

For $j = N$ sufficiently large but finite, we have, therefore,

$$\|v_N\|_{L^\infty(Q)} \leq C(e^{\mu_0 \eta^{2/3}(\rho)}|y_0|_2 + \|F_0\|_\infty),$$

where C is independent of ε. Hence

$$\|e^{(s+\delta)\alpha_0}\varphi_0^3 p_\varepsilon\|_{L^\infty(Q)} \leq C(e^{\mu_0 \eta^{2/3}(\rho)}|y_0|_2 + \|F_0\|_\infty)$$

and, therefore, (see ((3.32))

$$\|u_\varepsilon\|_{L^\infty(Q)} \leq \|e^{(-s(1-\eta(\lambda))+\delta(1+\eta(\lambda)))\alpha}u_\varepsilon\|_\infty \leq C(e^{\mu_0 \eta^{2/3}(\rho)}|y_0|_2 + \|F_0\|_\infty), \tag{3.61}$$

for $s = C\eta^{2/3}(\rho) + s_0$ and $0 < \delta < s$. Here C is independent of ε and, redefining μ_0 if necessary, we may assume also that it is independent of ρ, too.

Now, multiplying equation (3.47), where $u = u_\varepsilon$ and $y = y_\varepsilon$, by y_ε and integrating on $\mathcal{O}\times(0,t)$, it follows by (3.45) that (see (3.50))

$$|y_\varepsilon(t)|_2^2 + \int_0^T \int_{\mathcal{O}} |\nabla y_\varepsilon|^2 dx\, dt \leq C\left(\|F_0\|^2_\infty + |y_0|^2_2\right) + C\int_Q u^2_\varepsilon dx\, dt. \tag{3.62}$$

Similarly, multiplying by $t(y_\varepsilon)_t$ and integrating on $\mathcal{O}\times(0,t)$, we get after some calculation that

$$\int_Q t(y^2_\varepsilon + |\Delta y_\varepsilon|^2)dx\, dt + t\|y_\varepsilon(t)\|^2_{H^1(\mathcal{O})} \leq C\left(\|F_0\|^2_\infty + |y_0|^2_2 + \int_Q u^2_\varepsilon dx\, dt\right). \tag{3.63}$$

In particular, it follows by estimate (3.62) that

$$\|(y_\varepsilon)_t\|^2_{L^2(0,T;(H^1(\mathscr{O}))')} \le C\left(\|F_0\|^2_\infty + |y_0|_2^2 + \int_Q |u_\varepsilon|^2 dx\,dt\right). \tag{3.64}$$

(The constant $C = C(\rho)$ arising in (3.63), (3.64) is independent of ε.) Then we conclude by (3.61)–(3.64) that $\{u_\varepsilon\}$ is weak-star compact in $L^\infty(Q)$ and $\{y_\varepsilon\}$ is compact in $L^2(Q)$ and weakly compact in $L^2(0,T;H^1(\mathscr{O})) \cap L^\infty(0,T;L^2(\mathscr{O}))$. Thus, selecting a subsequence if necessary, we have

$$\begin{aligned} u_\varepsilon &\longrightarrow u \text{ weak star in } L^\infty(Q) \\ y_\varepsilon &\longrightarrow y \text{ weakly in } L^2(0,T;H^1(\mathscr{O})) \text{ and weak star in } L^\infty(0,T;L^2(\mathscr{O})). \end{aligned}$$

Moreover, by estimate (3.63), it follows that

$$\begin{aligned} y_\varepsilon &\longrightarrow y \quad \text{weak star in } L^\infty(\eta,T;H^1(\mathscr{O})) \cap L^2(0,T;H^2(\mathscr{O})), \\ (y_\varepsilon)_t &\longrightarrow y_t \quad \text{weakly in } L^2(\eta,T;L^2(\mathscr{O})), \\ y_\varepsilon(t) &\longrightarrow y(t) \text{ strongly in } L^2(\mathscr{O}), \text{ uniformly on } [\eta,T] \end{aligned}$$

for each $0 < \eta < T$. Then, letting ε tend to zero into equation (3.43), where $u = u_\varepsilon$, we infer that $y = y_z^u$, and by (3.61) it follows that

$$\|u\|_{L^\infty(Q)} \le C(e^{\mu_0\eta^{2/3}(\rho)}|y_0|_2 + \|F_0\|_\infty),$$

where C is independent of ρ and z. This is precisely the desired estimate (3.54). We note also for later use that, by (3.62) and (3.64), it follows that y_z^u satisfies the estimates

$$\begin{aligned} |y_z^u(t)|_2^2 + \int_0^T \|y_z^u(t)\|^2_{H^1(\mathscr{O})}dt + \int_0^T \|(y_z^u)_t(t)\|^2_{(H^1(\mathscr{O}))'}dt \\ \le C(e^{\mu_0\eta^{2/3}(\rho)}|y_0|_2^2 + \|F_0\|^2_\infty), \end{aligned} \tag{3.65}$$

where C is independent of z and ρ. This completes the proof of Lemma 3.1.

Proof of Theorem 3.6 (Continued). For each $z \in K_\infty$ defined above denote by $\Phi(z) \subset L^2(Q)$ the set of all $y_z^u \in L^2(Q)$ such that

$$y_z^u(T) = 0, \ \|u\|_{L^\infty(Q)} \le C(e^{\mu_0\eta^{\frac{2}{3}}(\rho)}|y_0|_2 + \|F_0\|_\infty),$$

where C and μ_0 are as in Lemma 3.1. Hence $\Phi(z) \ne \emptyset$, $\forall z \in K_\infty$. It is also easily seen by estimates (3.65) that $\Phi(z)$ is a closed, convex, and compact subset of $L^2(Q)$. Moreover, arguing as in the proof of Theorem 3.5, it follows that Φ is upper semicontinuous on $L^2(Q)\times L^2(Q)$. To conclude the proof, it remains to be shown that $\Phi(K_\infty)\subset K_\infty$ and to apply the Kakutani fixed point theorem as in the previous theorem. We shall assume first that $y_0 \in L^\infty(Q)$. We set

$$M_0 = |y_0|_\infty, \quad M_1 = \|mu\|_{L^\infty(Q)} + \|F\|_{L^\infty(Q)}.$$

Let $w = y_z^u - M_0 - Mt$, where M will be made precise below. By (3.48) and (3.44), we have

$$|g(z)| \le L(\eta(\rho) + 1), \text{ a.e. in } Q.$$

We chose T sufficiently small such that

$$L(\eta(\rho) + 1)T \le 1. \tag{3.66}$$

Now, we choose $M = 2(M_1 + L(\eta(\rho) + 1)M_0)$. We have

$$\begin{aligned} &w_t - \Delta w + b\cdot\nabla w + g(z)w = mu + F - M - g(z)(M_0 + Mt) \le 0,\\ &\alpha_1 \frac{\partial w}{\partial \nu} + \alpha_2 w = -\alpha_2(M_0 + Mt) \le 0 \text{ on } \Sigma,\\ &w(x,0) = y_0 - M_0 \le 0. \end{aligned} \tag{3.67}$$

If we multiply (3.67) by w^+ and integrate on Q, we see via Gronwall's lemma that $w^+ = 0$ and, therefore, $w \le 0$ in Q. Hence $y_z^u \le M_0 + MT$, a.e. on Q. Similarly, it follows that $y_z^u \ge -M_0 - MT$. Recalling estimate (3.54), we obtain, therefore, that

$$\begin{aligned} \|y_z^u\|_{L^\infty(Q)} &\le 2(M_1 + L(\eta(\rho) + 1)M_0)T + M_0\\ &\le 2C(e^{\mu_0 \eta^{\frac{2}{3}}(\rho)}|y_0|_\infty + L|y_0|_\infty(\eta(\rho) + 1) + \|F_0\|_\infty), \quad \forall z \in K_\infty. \end{aligned}$$

Then, by (3.52), it follows that for ρ large enough and for T chosen as in (3.66) we have $y_z^u \in K_\infty$ for all $z \in K_\infty$. This completes the proof in the case $y_0 \in L^\infty(\mathcal{O})$ and T sufficiently small. Obviously, this implies the controllability for any $T > 0$ along with the conclusions of Theorem 3.6.

Now, let $y_0 \in L^2(\mathcal{O})$ and let y be the solution to the uncontrolled equation

$$\begin{aligned} &y_t - \Delta y + b\cdot\nabla y + f(x,t,y) = F && \text{in } Q,\\ &y(x,0) = y_0(x) && \text{in } \mathcal{O},\\ &\alpha_1 \frac{\partial y}{\partial \nu} + \alpha_2 y = 0 && \text{on } \Sigma. \end{aligned} \tag{3.68}$$

As noticed earlier, problem (3.68) has at least one solution

$$y \in C([0,T]; L^2(\mathcal{O})) \cap C(]0,T[; H^1(\mathcal{O})) \cap L^2(\eta, T; H^2(\mathcal{O}))$$

for each $0 < \eta < T$. By the Sobolev embedding theorem, $H^1(\mathcal{O}_\omega) \subset L^{\frac{2d}{d-2}}(\mathcal{O}_\omega)$ and, by condition (3.52), it follows that

$$b \cdot \nabla y + f(x,t,y) \in L^{p_1}(\mathcal{O} \times (\eta, T)), \ p_1 < \frac{2d}{d-2},$$

and so, by the parabolic regularity, we see that $y \in W_{p_1}^{2,1}(\mathcal{O} \times (\eta, T))$. Hence

$$b \cdot \nabla y \in W_{p_1}^{1,1}(\mathcal{O} \times (\eta, T)) \subset L^{p_2}(\mathcal{O} \times (\eta, T)), \ p_2 = \frac{dp_1}{d - p_1},$$

and

$$y \in L_1^q(Q), \ q_2 = p_1 + \frac{2p_1^2}{d + 2 - 2p_1}.$$

This implies that $f(y) \in L^{\bar{p}_2}(Q)$ for $\bar{p}_2 < q_2$.

If we continue the process, we get after a finite number of steps that $y \in C(\overline{Q}_\eta)$ where $Q_\eta = \mathcal{O} \times (\eta, T)$ and, therefore, $y(\eta) \in L^\infty(\mathcal{O})$. Then

$$u^*(t) = \begin{cases} 0, \ t \in (0, \eta), \\ u, \ t \in (\eta, T), \end{cases}$$

where u steers $y(\eta)$ into origin in the time interval (η, T), is the desired controller. This completes the proof. ∎

Remark 3.3. It turns out the above controllability result remains true in the absence of the sign condition (3.45), that is, for blowing-up semilinear systems of the form (3.43). (We refer the reader to the work [72].)

We note also that the controllability result established in Theorem 3.6 is *almost* optimal since, as shown by S. Anita and D. Tataru in [9], if the function f satisfies the condition

$$f(s) \leq C(s \log^p s + 1),$$

where $p > 2$, then for each $y_0 \in L^2(\mathcal{O})$ such that $y_0 > 0$ on an open subset of $\mathcal{O}$ there is $T_1 > 0$ such that the system is not null controllable in T for each $T > T_1$. Previously, weaker versions of this result were established by A. Fursikov and O.Yu. Imanuvilov [76]. (See also [71].) This result suggests that one cannot expect null controllability for f which grows to $+\infty$ faster than $\log^2(|s| + 1)$ though it does not preclude the exact null controllability for some special classes of nonlinearities f. Anyway, it seems that, so far, the case $\frac{3}{2} < p < 2$ remained open.

Remark 3.4. By Theorem 3.6, under conditions (3.44)–(3.46) and (3.52), it follows that system (3.43) is exactly null boundary controllable, i.e., for each $y_0 \in H^1(\mathcal{O})$

there are $v \in L^2(\Sigma)$ and $y \in C([0, T; L^2(\mathcal{O})) \cap L^2(0, T; H^1(\mathcal{O}))$ which satisfy system (3.43) with $u = 0$ and

$$y(T) = 0, \quad y = v \text{ on } \Sigma.$$

The proof is exactly the same as that given in Corollary 3.2, i.e., one *inflates* a little bit the domain $\mathcal{O}$ and one applies Theorem 3.6 on the new domain. The details are left to the reader.

Note also that, if $f(x, t, y) \equiv f(x, y)$ and if the function

$$r \longrightarrow f(x, r + y_e) - f(x, y_e)$$

satisfies conditions (3.43)–(3.45) and (3.52), then one might derive by Theorem 3.6 the exact controllability of the steady-state solutions y_e to system (3.43). In fact, by the substitution $z = y - y_e$, one reduces the exact controllability of y_e to the null controllability of a system of form (3.43) and one might apply Theorem 3.6 to obtain the desired result.

3.3 Approximate Controllability

System (3.43) is said to be approximately controllable if for all $y_0, y_1 \in L^2(\mathcal{O})$ there are $u \in L^2(Q)$ and $y \in L^2(Q)$ satisfying (3.43) and such that

$$|y(T) - y_1|_2 \leq \varepsilon. \tag{3.69}$$

It should be recalled that, for linear systems, the approximate controllability is implied by exact controllability and it is an immediate consequence of a unique continuation property for the dual equation, but this does not happen for the nonlinear systems considered here. As a matter of fact, this weaker concept of controllability has the same degree of difficulty as the exact controllability, though it can be deduced by similar arguments and, essentially, under the same conditions. We shall illustrate this for the case described in Theorem 3.6 only, though it extends *mutatis–mutandis* to all the situations considered above. (See also [70], [121], [122] for other results of this type.)

Theorem 3.7. *Assume that $F \equiv 0$ and that, for each $y_1 \in L^\infty(\mathcal{O})$, the function $(x, t, r) \to f_1(x, t, r) = f(x, t, r + y_1) - f(x, t, y_1)$ satisfies conditions* (3.44), (3.45). *Then system* (3.43) *is approximately controllable.*

Proof. By density, it suffices to prove (3.69) for each $y_1 \in W^{2,\infty}(\mathcal{O}) \cap H_0^1(\mathcal{O})$. We set $w = y - y_1$ and note that w satisfies a system of form (3.43), i.e.,

$$\begin{aligned} &w_t - \Delta w + b \cdot \nabla w + f_1(w) = mu + F_1 \text{ in } Q, \\ &w(x, 0) = y_0(x) - y_1(x) \text{ in } \mathcal{O};\ w = 0 \text{ on } \Sigma, \end{aligned} \tag{3.70}$$

where

$$f_1(x,t,w) = f(x,t,w+y_1) - f(x,t,y_1),\ F_1 = \Delta y_1 + b\cdot\nabla y_1 - f(x,t,y_1).$$

We set

$$F_\varepsilon(x,t,w) = e^{2s\alpha}(\varepsilon + e^{2s\alpha})^{-1}F_1(x,t),$$

where $\varepsilon \geq 0$ and s, α are chosen as in Theorem 3.6. Hence, there are u_ε, w_ε such that

$$\begin{aligned}(w_\varepsilon)_t - \Delta w_\varepsilon + b\cdot\nabla w_\varepsilon + f_1(x,t,w_\varepsilon) &= mu_\varepsilon + F_\varepsilon \text{ in } Q\\ w_\varepsilon(x,0) = y_0 - y_1,\ w_\varepsilon(x,T) = 0 \text{ in } \mathcal{O},\ w_\varepsilon &= 0 \text{ on } \Sigma.\end{aligned} \tag{3.71}$$

Let w^ε be a solution to equation (3.70), where $u = u_\varepsilon$. We set $\theta_\varepsilon = w^\varepsilon - w_\varepsilon$ and notice that

$$\begin{aligned}(\theta_\varepsilon)_t - \Delta\theta_\varepsilon + b\cdot\nabla\theta_\varepsilon + f_1(x,t,\theta_\varepsilon + w_\varepsilon) - f_1(x,t,w_\varepsilon) &= (\varepsilon + e^{2s\alpha})^{-1}F_1 \ \text{ in } Q,\\ \theta_\varepsilon(x,0) = 0 \ \text{ in } \mathcal{O},\quad \theta_\varepsilon &= 0 \ \text{ on } \Sigma.\end{aligned}$$

Since the right-hand side is a.e. convergent to 0 on Q and is bounded by $|F_1|_\infty$, we infer by virtue of conditions (3.44), (3.45) that

$$|\theta_\varepsilon(T)|_2 \longrightarrow 0 \text{ as } \varepsilon \to 0$$

and, therefore,

$$w^\varepsilon(T) \longrightarrow 0 \text{ strongly in } L^2(\mathcal{O})$$

as $\varepsilon \to 0$. This implies (3.69) for a suitable chosen controller u, thereby completing the proof.

3.4 Local Controllability of Semilinear Parabolic Equations

For general systems of form (3.43), the best one can expect is the *local null controllability* i.e., the exact null controllability for the initial data in a neighborhood of the origin.

Theorem 3.8. *Let* $F \equiv 0$, $b \in C^1(\overline{Q}; \mathbb{R}^)$, *and let* $f : Q\times R \to R$ *satisfy conditions* (3.44), (3.45). *Then there is* $\rho_0 > 0$ *such that for all* $y_0 \in L^\infty(\mathcal{O})$, $|y_0|_\infty \leq \rho_0$ *there are* $y \in C([0,T]; L^2(\mathcal{O})) \cap L^2(0,T; H^1(\mathcal{O})) \cap L^\infty(Q)$ *and* $u \in L^\infty(Q)$ *which satisfy* (3.43) *and such that* $y(T) \equiv 0$.

Proof. Consider the set

$$K = \{z \in L^2(Q);\ \|z\|_{L^\infty(Q)} \le \rho\}$$

and define the mapping $\Phi : K \to 2^{L^2(Q)}$

$$\Phi(z) = \left\{ y_z^u \in L^2(Q);\ y_z^u(T) = 0,\ \|u\|_{L^\infty(Q)} \le Ce^{\mu_0 \eta^{\frac{2}{3}}(\rho)} |y_0|_2 \right\},$$

where y_z^u is the solution to (3.47) and C, μ_0 are as in Lemma 3.1. We have (recall that $F \equiv 0$)

$$\|y_z^u\|_{L^\infty(Q)} \le C(\rho)(|y_0|_\infty + \|u\|_{L^\infty(Q)}) \le C_1(\rho)|y_0|_\infty,\ \forall z \in K,$$

and so $\Phi(K) \subset K$ for $|y_0|_\infty \le \rho_0$ sufficiently small. We recall also the estimate

$$|y_z^u(t)|_2^2 + \int_0^t \|y_z^u(s)\|^2_{H^1(\mathcal{O})} ds \le C\left(|y_0|_2^2 + \eta(\rho) \int_0^t |y_z^u(s)|^2 ds + \|u\|^2_{L^2(Q)} \right),$$

which implies that

$$|y_z^u(t)|_2^2 + |y_z^u|^2_{L^2(0,T;H^1(\mathcal{O}))} + \|(y_z^u)_t\|^2{}_{L^2(0,T;(H^1(\mathcal{O}))'} \le C(\rho)|y_0|_2^2.$$

Hence, $\Phi(K)$ is compact in $L^2(Q)$ and, arguing as in the previous cases, we see also that Φ is upper semicontinuous on $L^2(Q)$. Thus, there is $y \in K$ such that $y \in \Phi(y)$ and this completes the proof.

In particular, we find by Theorem 3.8 the following local boundary controllability result.

Corollary 3.5. *Let f satisfy conditions* (3.44) *and* (3.45). *Then there is $\rho_0 > 0$ such that, for all $y_0 \in L^\infty(\mathcal{O})$, $|y_0|_\infty \le \rho_0$, there are $y \in C([0,T]; L^2(\mathcal{O})) \cap L^2(0,T; H^1(\mathcal{O})) \cap L^\infty(Q)$ and $u \in L^2(\Sigma)$ such that*

$$\begin{aligned} &y_t - \Delta y + b \cdot \nabla y + f(x,t,y) = 0 \ \text{ in } Q,\\ &y = u \ \text{ on } \Sigma,\\ &y(x,0) = y_0(x),\ y(x,T) = 0 \ \text{ in } \mathcal{O}. \end{aligned}$$

The proof is the same as that used in the previous cases (see Corollary 3.1), i.e., one extends the domain $\mathcal{O}$ near $\partial\mathcal{O}$ and one applies Theorem 3.8 on $Q_1 = \mathcal{O}_1 \times (0,T)$ where $\mathcal{O}_1$ is the extension of $\mathcal{O}$.

If $f(x,t,y) \equiv f(x,y)$, $b \equiv b(x)$ and $y_e \in H^1(\mathcal{O}) \cap L^\infty(\mathcal{O})$ is a steady-state solution to system (3.43), that is,

$$\begin{aligned} &-\Delta y_e + b \cdot \nabla y_e + f(x,y_e) = 0 \ \text{ in } \mathcal{O}\\ &\alpha_1 \frac{\partial y_e}{\partial \nu} + \alpha_2 y_e = 0 \qquad\qquad \text{on } \Sigma, \end{aligned}$$

then, by applying Theorem 3.8 to the equation obtained by the substitution $y \to y - y_e$, we are led to the following controllability result.

Corollary 3.6. *Let f satisfy the assumptions of Theorem* 3.8. *Then there is $\rho_0 > 0$ such that, for all $y_0 \in L^\infty(\mathcal{O})$ such that $|y_0 - y_e|_\infty \le \rho_0$, there are (y, u) which satisfy* (3.43) *and $y(T) = y_e$.*

In other words, the steady-state solution is locally exactly controllable.

One of the main consequences of this corollary is that each equilibrium solution $y_0 \in H^1(\mathcal{O}) \cap L^\infty(\mathcal{O})$ to equation (3.43) (which, in general, is unstable) is *stabilizable* by an open loop internal distributed controller with the support in an arbitrary open subset ω of $\mathcal{O}$ (or by boundary controllers).

More generally, we may replace in Corollary 3.6 the steady-state solution y_e by the final value $y^*(T)$ of a solution y^* to the equation

$$\begin{aligned} & y_t^* - \Delta y^* + b \cdot \nabla y^* + f(x, t, y^*) = 0 \ \text{in Q}\,, \\ & \alpha_1 \frac{\partial y}{\partial \nu} + \alpha_2 y = 0 \qquad\qquad\qquad\qquad \text{on } \Sigma. \end{aligned}$$

Then, it follows as above that, if y^* is regular enough, then for $|y_0 - y^*(0)|_\infty$ sufficiently small there are (y, u), which satisfy equation (3.43) with the right-hand side $mu + f_0$ and $y(T) = y^*(T)$. In particular, this means that the nonlinear diffusion equation (3.43) excited by a distributed control with the support in Q_ω is locally reversible. It should be also said that, as in the case of Theorem 3.6, condition $y_0 \in L^\infty(\mathcal{O})$ can be relaxed to $y_0 \in L^2(\mathcal{O})$ due to smoothing effect of solutions to parabolic equations on initial data.

The previous treatment suggests a general strategy to obtain the local (global) controllability of nonlinear distributed systems of the form

$$\begin{aligned} & y_t + Ly + f(y) = mu \ \ \text{in } Q, \\ & y(0) = y_0 \qquad\qquad\quad \text{in } \mathcal{O}, \end{aligned}$$

where L is a linear partial differential operator with homogeneous boundary value conditions. Namely, one check first (via Carleman inequality) the (global) exact null controllability of the linearized system

$$\begin{aligned} & y_t + Ly + g(z)y = mu \ \ \text{in } Q, \\ & y(0) = y_0 \qquad\qquad\qquad \text{in } \mathcal{O}, \end{aligned}$$

where $f(y) = g(y)y$ and one applies the Kakutani fixed point theorem to the map $z \to y_z^u$ on a suitable chosen set $K \subset L^2(Q)$.

A quite different approach, systematically used in the works of Fursikov and Imanuvilov (see, e.g., [76]), is to invoke the implicit function theorem in a convenient function space but we did not pursue this way here.

Now, we shall study with the above methods the local controllability of the phase field system

$$
\begin{aligned}
&u_t(x,t) + \ell\varphi_t(x,t) - k\Delta u(x,t) = m(x)w(x,t) + f_1(x),\\
&\varphi_t(x,t) - \alpha\Delta\varphi(x,t) - \beta(\varphi(x,t) - \varphi^3(x,t)) + \gamma u(x,t)\\
&\qquad = m(x)v(x,t) + f_2(x),\ (x,t) \in Q = \mathscr{O}\times(0,T),\\
&u(x,0) = u_0(x),\ \varphi(x,0) = \varphi_0(x),\\
&u(x,t) = \bar{u}(x),\ \varphi(x,t) = \overline{\varphi}(x),\ \forall\,(x,t) \in \Sigma = \partial\mathscr{O}\times(0,T),
\end{aligned}
\tag{3.72}
$$

where w, v are input controllers and $\alpha, \beta, \gamma, \ell, k$ are positive constants, m is the characteristic function of an open set $\omega\subset\mathscr{O}$, $f_1, f_2 \in L^2(\mathscr{O})$, and $u_0, \varphi_0 \in H_0^1(\mathscr{O})$, $\bar{u}, \overline{\varphi}$ are given functions.

Let $(u^*, \varphi^*) \in H^2(\mathscr{O})\times H^2(\mathscr{O})$ be a steady-state solution to (3.72), that is, the solution of the Landau–Ginzburg equations

$$
\begin{aligned}
&\ell\varphi^* - k\Delta u^* = f_1 && \text{in } \mathscr{O}\\
&-\alpha\Delta\varphi^* - \beta(\varphi^* - (\varphi^*)^3) + \gamma u^* = f_2 && \text{in } \mathscr{O}\\
&u^* = \bar{u},\ \varphi^* = \overline{\varphi} && \text{on } \partial\mathscr{O}.
\end{aligned}
\tag{3.73}
$$

It is well known that, for each w and v in $L^2(Q)$ such that $u_0 - u^*,\ \varphi_0 - \varphi^* \in H_0^1(\mathscr{O})$, problem (3.72) has a unique solution $(u, \varphi) \in (H^{2,1}(Q))^2$. (See, e.g., [26], p 235.) We recall that $H^{2,1}(Q) = \{y \in L^2(Q);\ H_0^1(\mathscr{O}) \cap H^2(\mathscr{O}),\ y_t \in L^2(Q)\}$.

System (3.72) models the phase transition of several physical processes including the melting and solidification. The two phase Stefan problem as well as other classical mathematical models of phase transition are limiting cases of system (3.72). (See [57].) Here u is the temperature while φ is the phase field function defining the liquid or the solid phase.

As a matter of fact, in phase transition models $\varphi = \varphi(t,x)$ is an order parameter which takes values is a specific interval $[\varphi_0, \varphi_1]$, describing the phase evolution. For instance, if $[(t,x); \varphi(t,x) = \varphi_0]$ and $[(t,x);\ \varphi(t,x) = \varphi_1]$ in the first and, respectively, second phase, then the domain $\{(t,x);\ \varphi(t,x) = \varphi^*\}$, where $\varphi^* \in (\varphi_0, \varphi_1)$ may define the interface. It should be said, however, that the value of the temperature u does not determine the phase and so, if wants to steer the phase system in a certain steady state, one must act not only on the temperature u, but also on the phase state φ. For $f_2 \equiv 0$, the second equation in (3.72) is derived from the kinetic equation $\frac{\partial\varphi}{\partial t} + \nabla_\varphi E(u,\varphi) = 0$, where E is the Landau energy functional

$$
E(u,\varphi) = \int_{\mathscr{O}} \left(\frac{1}{2}\left(\alpha|\nabla\varphi|^2 + \beta\left(\frac{1}{2}\varphi^4 - \varphi^2\right)\right) + (\gamma u - mv)\varphi\right) dx.
$$

So, the controller v, likes w in the first equation, is acting on temperature in order to restore (u, φ) to the equilibrium state (u^*, φ^*). Throughout in the sequel, $\alpha_0(d)$ is any real number which satisfies the following conditions:

$$
\alpha_0(d) > \frac{d+2}{2} \ \text{ if } d = 2, 3;\ \ \alpha_0(d) = 2 \ \text{ if } d = 1. \tag{3.74}
$$

We set

$$\|u\|_{\alpha_0(d)} = \|u\|_{W_0^{2\left(1-\frac{1}{\alpha_0(d)}\right),\alpha_0(d)}(\mathcal{O})}.$$

Theorem 3.9 below is the main result.

Theorem 3.9. *Let* $(u^*, \varphi^*) \in H^2(\mathcal{O})\times H^2(\mathcal{O})$ *be any steady-state solution to* (3.72). *Then there is* $\delta > 0$ *such that, for all* $(u_0, \varphi_0) \in H^1(\mathcal{O})\times H^1(\mathcal{O})$ *satisfying the conditions*

$$\|u^* - u_0\|_{\alpha_0(d)} + \|\varphi^* - \varphi_0\|_{H_0^1(\mathcal{O})} \le \delta, \tag{3.75}$$

there is $(w, v) \in L^2(Q)\times L^2(Q)$ *such that*

$$u^{w,v}(T) = u^*, \quad \varphi^{w,v}(T) = \varphi^*.$$

Here $(u^{w,v}, \varphi^{w,v}) \in H^{2,1}(Q)\times H^{2,1}(Q)$ is the solution to (3.72).

A similar result remains true for the boundary control system

$$\begin{array}{ll} u_t + \ell\varphi_t - k\Delta u = f_1 & \text{in } Q,\\ \varphi_t - \alpha\Delta\varphi - \beta(\varphi - \varphi^3) + \gamma u = f_2 & \text{in } Q,\\ u(0) = u_0,\ \varphi(0) = \varphi_0 & \text{in } \mathcal{O},\\ u = \bar{u},\ \varphi = \overline{\varphi} & \text{on } \Sigma. \end{array} \tag{3.76}$$

Namely, we have

Corollary 3.7. *Let* $(u^*, \varphi^*) \in H^2(\mathcal{O})\times H^2(\mathcal{O})$ *be any steady-state solution to* (3.72). *Then there is* $\delta > 0$ *such that, for all* $u_0, \varphi_0 \in H^1(\mathcal{O})$ *satisfying* (3.75), *there are* $\bar{u}, \overline{\varphi} \in L^2(\Sigma)\times L^2(\Sigma)$ *such that*

$$u_{\bar{u},\overline{\varphi}}(T) = u^*, \quad \varphi_{\bar{u},\overline{\varphi}}(T) = \varphi^*.$$

Here $(u_{\bar{u},\bar{\varphi}}, \varphi_{\bar{u},\overline{\varphi}}) \in H^{2,1}(Q)\times H^{2,1}(Q)$ is the solution to (3.76).

As noted earlier, the function φ determines the phase of physical system (liquid or solid, for instance) and the above result amounts to saying that the system can be steered in equilibrium state by prescribing a certain values of temperature and phase functions on the boundary.

Proof of Theorem 3.9. Let $y = \varphi - \varphi^*,\ z = u - u^*$. Then by redefining the controllers v, w, and the state (y, z), after some elementary calculation, Theorem 3.9 reduces to the local null controllability of the system

$$\begin{array}{ll} y_t - \Delta y + ay + by^2 + cy^3 + dz = mw & \text{in } Q,\\ z_t - \Delta z + \ell\Delta y + a_1 y + b_1 y^2 + c_1 y^3 + d_1 z = mv & \text{in } Q,\\ y(0) = y_0,\ z(0) = z_0 & \text{in } \mathcal{O},\\ y = 0,\ z = 0 & \text{on } \Sigma, \end{array} \tag{3.77}$$

where $a, b, c, d, a_1, b_2, c_1, d_1 \in L^\infty(\mathcal{O})$.

According to the general scheme presented above, we shall consider first the linearized system

$$\begin{aligned}
&y_t - \Delta y + \xi y + dz = mw && \text{in } Q,\\
&z_t - \Delta z + \ell\Delta y + \eta y + d_1 z = mv && \text{in } Q,\\
&y(0) = y_0,\ z(0) = z_0 && \text{in } \mathcal{O},\\
&y = 0,\ z = 0 && \text{on } \Sigma,
\end{aligned} \tag{3.78}$$

where $\xi, \eta \in L^\infty(Q)$ are given functions. We have,

Lemma 3.2. *There are $(w, v) \in L^{\alpha_0(d)}(Q) \times L^{\alpha_0(d)}(Q)$ such that*

$$\|w\|_{L^{\alpha_0(d)}(Q)} + \|v\|_{L^{\alpha_0(d)}(Q)} \le \mu(|\xi|_\infty, |\eta|_\infty)(|y_0|_2 + |z_0|_2)$$

and

$$y(T) = 0,\ z(T) = 0.$$

Here $\mu : \mathbb{R} \to \mathbb{R}$ is a positive function which is bounded on bounded subsets.

To prove this lemma we shall consider the corresponding dual backward system

$$\begin{aligned}
&p_t + (\Delta - \xi)p - (\ell\Delta + \eta)q = 0 && \text{in } Q,\\
&q_t + (\Delta - d_1)q - dp = 0 && \text{in } Q,\\
&p = q = 0 && \text{on } \Sigma,
\end{aligned} \tag{3.79}$$

and prove a Carleman estimate for (3.79).

Lemma 3.3. *There are $s, \lambda > 0$ such that*

$$\int_Q e^{2s\alpha}\varphi(p^2 + \varphi^2 q^2)dx\,dt + \int_Q e^{2s\alpha}(\varphi^{-1}|\nabla p|^2 + \varphi|\nabla q|^2)dx\,dt \\
+ \int_Q e^{2s\alpha}\varphi^{-1}|\Delta q|^2 dx\,dt \le C(|\xi|_\infty, |\eta|_\infty)\int_{Q_\omega} \varphi e^{2s\alpha}(p^2 + \varphi^2 q^2)dx\,dt.$$

Proof. We shall apply the Carleman estimate (2.2) in the second equation of (3.79). We obtain

$$\begin{aligned}
&\int_Q e^{2s\alpha}((s\varphi)^{-1}|\Delta q|^2 + s\varphi|\nabla q|^2 + s^3\varphi^3 q^2)dx\,dt\\
&\quad\le C(\lambda, |\xi|_\infty)\left(\int_Q e^{2s\alpha}p^2 dx\,dt + \int_{Q_\omega} s^3\varphi^3 e^{2s\alpha}q^2 dx\,dt\right)\\
&\qquad\qquad\text{for } \lambda \ge \lambda_0 \text{ and } s \ge s_0(\lambda).
\end{aligned} \tag{3.80}$$

On the other hand, we have

$$(t(T-t)p)_t + (\Delta - \xi)(t(T-t)p) = t(T-t)(\ell\Delta + \eta)q + (t(T-t))_t p \text{ in } Q,$$
$$t(T-t)p = 0 \text{ on } \Sigma.$$

Then, again applying the Carleman inequality in the latter equation, we get

$$s^3 \int_Q \varphi p^2 e^{2s\alpha} dx\, dt + s \int_Q \varphi^{-1} |\nabla p|^2 e^{2s\alpha} \le C s^3 \int_{Q_\omega} \varphi p^2 e^{2s\alpha} dx\, dt$$
$$+ C \int_Q \varphi^{-2} ((\ell\Delta + \eta)q)^2 e^{2s\alpha} dx\, dt + C \int_Q e^{2s\alpha} p^2 dx\, dt.$$

Thus, for $\lambda = \lambda_0$ and $s \ge s_0$ large enough, we have

$$s^3 \int_Q \varphi p^2 e^{2s\alpha} dx\, dt + s \int_Q \varphi^{-1} |\nabla p|^2 e^{2s\alpha} dx\, dt$$
$$\le C s^3 \int_{Q_\omega} \varphi p^2 e^{2s\alpha} dx\, dt + C \int_Q \varphi^{-2} ((\ell\Delta + d)q)^2 e^{2s\alpha} dx\, dt$$

and so, by (3.80), we obtain

$$s^3 \int_Q \varphi p^2 e^{2s\alpha} dx\, dt + s \int_Q \varphi^{-1} |\nabla p|^2 e^{2s\alpha} dx\, dt \le C s^3 \int_{Q_\omega} \varphi p^2 e^{2s\alpha} dx\, dt$$
$$+ C \left(s \int_Q e^{2s\alpha} p^2 dx\, dt + s^4 \int_{Q_\omega} \varphi^3 e^{2s\alpha} q^2 \right) dx\, dt,$$

where $C = C(|\xi|_\infty, |\eta|_\infty)$. Therefore, for s large enough, we have

$$\int_Q \varphi p^2 e^{2s\alpha} dx\, dt + \int_Q \varphi^{-1} |\nabla p|^2 e^{2s\alpha} dx\, dt$$
$$\le C(|\xi|_\infty, |\eta|_\infty) \int_{Q_\omega} \varphi e^{2s\alpha} (p^2 + \varphi^2 q^2) dx\, dt$$

and

$$\int_Q e^{2s\alpha} (\varphi |\nabla q|^2 + \varphi^{-1} |\Delta q|^2 + \varphi^3 q^2) dx\, dt$$
$$\le C(|\xi|_\infty, |\eta|_\infty) \int_{Q_\omega} \varphi e^{2s\alpha} (p^2 + \varphi^2 q^2) dx\, dt,$$

as claimed. (Here and everywhere in the following, $C(r_1, r_2)$ is a positive function bounded on bounded subsets.)

This completes the proof of Lemma 3.3.

By Lemma 3.3, we get the following observability inequality for the solution (p, q) to system (3.79)

$$|p(0)|_2^2 + |q(0)|_2^2 \le C(|\xi|_\infty, |\eta|_\infty) \int_{Q\omega} \varphi e^{2s\alpha}(p^2 + \varphi^2 q^2) dx\, dt. \tag{3.81}$$

In fact, multiplying the first equation of (3.79) by ρp, where $\rho > 0$ is suitable chosen, the second by q and integrating on $\mathcal{O}$, we see after some calculation involving Green's formula that

$$(|p(t)|_2^2 + |q(t)|_2^2)_t \ge -C(|\xi|_\infty, |\eta|_\infty)(|p(t)|_2^2 + |q(t)|_2^2.$$

This yields

$$\begin{aligned} |p(0)|_2^2 + |q(0)|_2^2 &\le C(|\xi|_\infty, |\eta|_\infty)(|p(t)|_2^2 + |q(t)|_2^2) \\ &\le C(|\xi|_\infty, |\eta|_\infty) \sup\,(e^{-2s\alpha}\varphi^{-1}) \int_Q e^{2s\alpha}\varphi(p^2 + \varphi^2 q^2) dx,\ \forall t \in [0, T]. \end{aligned}$$

Then, integrating on some subinterval (t_1, t_2) and using the Carleman inequality given in Lemma 3.3, we obtain (3.81), as claimed.

Proof of Lemma 3.2. As in the previous situations (see (1.84)), we shall consider the optimal control problem

$$\text{Min} \left\{ \int_Q e^{-2s\alpha}(\varphi^{-1} w^2 + \varphi^{-3} v^2) dx\, dt + \frac{1}{\varepsilon} \int_{\mathcal{O}} (y^2(x, T) + z^2(x, T)) dx \right\}$$

subject to (3.78).

Let $(y_\varepsilon, z_\varepsilon)$, $(w_\varepsilon, v_\varepsilon)$ be an optimal solution. Then, by the maximum principle, we have

$$w_\varepsilon = m e^{2s\alpha} \varphi p_\varepsilon,\quad v_\varepsilon = m e^{2s\alpha} \varphi^2 q_\varepsilon\,,\quad \text{a.e. in } Q,$$

where $(p_\varepsilon, q_\varepsilon)$ is the solution to (3.79) with the final conditions

$$p_\varepsilon(T) = -\frac{1}{\varepsilon} y_\varepsilon(T),\quad q_\varepsilon(T) = -\frac{1}{\varepsilon} z_\varepsilon(T).$$

Thus, taking into account (3.81), we obtain that

$$\begin{aligned} &\int_{Q_\omega} e^{2s\alpha}(\varphi p_\varepsilon^2 + \varphi^3 q_\varepsilon^2) dx\, dt + \frac{1}{\varepsilon} \int_{\mathcal{O}} (y_\varepsilon^2(x, T) + z_\varepsilon^2(x, T)) dx \\ &\quad = -\int_{\mathcal{O}} (y_0(x) p_\varepsilon(x, 0) + z_0(x) q_\varepsilon(x, 0)) dx \\ &\quad \le C(|\xi|_\infty, |\eta|_\infty)(|y_0|_2^2 + |z_0|_2^2) + \tfrac{1}{2} \int_{Q_\omega} \varphi e^{2s\alpha}(p_\varepsilon^2 + \varphi^2 q_\varepsilon^2) dx\, dt. \end{aligned}$$

Hence

$$\int_Q e^{-2s\alpha}(\varphi^{-1}w_\varepsilon^2+\varphi^{-3}v_\varepsilon^2)dx\,dt+\frac{1}{\varepsilon}\int_{\mathcal{O}}(y_\varepsilon^2(x,T)+z_\varepsilon^2(x,T))dx \\ \leq C(|\xi|_\infty,|\eta|_\infty)(|y_0|_2^2+|z_0|_2^2),\ \forall\,\varepsilon>0. \tag{3.82}$$

Thus, letting ε tend to 0, we infer that $\exists(w,v)\in L^2(Q)\times L^2(Q)$ such that

$$\varphi^{-1/2}e^{-s\alpha}w\in L^2(Q),\ \ \varphi^{-3/2}e^{-s\alpha}v\in L^2(Q)$$

and $y(T)=0,\ z(T)=0$. By (3.82), it follows that

$$\int_{Q_\omega} e^{2s\alpha}\varphi(p_\varepsilon^2 dx\,dt+\varphi^2q_\varepsilon^2)dx\,dt\leq C(|\xi|_\infty,|\eta|_\infty)(|y_0|_2^2+|z_0|_2^2) \tag{3.83}$$

and so, by Lemma 3.3,

$$\int_Q e^{2s\alpha}\varphi(p_\varepsilon^2+\varphi^2q_\varepsilon^2)dx\,dt+\int_Q e^{2s\alpha}\varphi^{-1}|\Delta q_\varepsilon|^2dx\,dt \\ +\int_Q e^{2s\alpha}\varphi^{-1}|\nabla p_\varepsilon|^2dx\,dt\leq C(|\xi|_\infty,|\eta|_\infty)(|y_0|_2^2+|z_0|_2^2). \tag{3.84}$$

We set $\theta_\varepsilon=e^{\frac{3s}{2}\alpha}\varphi^{-1}p_\varepsilon$ and notice that, by (3.79), we have

$$(\theta_\varepsilon)_t+(\Delta-\xi)\theta_\varepsilon=\varphi^{-1}e^{\frac{3s}{2}\alpha}(\ell\Delta+\eta)q_\varepsilon+p_\varepsilon(e^{\frac{3s}{2}\alpha}\varphi^{-1}) \\ +p_\varepsilon\Delta(e^{\frac{3s}{2}\alpha}\varphi^{-1})+2\nabla p_\varepsilon\cdot\nabla(e^{\frac{3s}{2}\alpha}\varphi^{-1}).$$

Then, by parabolic regularity and by estimate (3.84), we get

$$\|\theta_\varepsilon\|_{H^{2,1}(Q)}\leq C(|\xi|_\infty,|\eta|_\infty)(|y_0|_2+|z_0|_2).$$

Since $H^{2,1}(Q)\subset L^{\alpha_0(d)}(Q)$ (we recall that, by the Sobolev embedding theorem, $H^{2,1}(Q)\subset L^{\frac{2(d+2)}{d-2}}(Q)$), we infer that

$$\|p_\varepsilon\varphi e^{2s\alpha}\|_{L^{\alpha_0(d)}(Q)}\leq C(|\xi|_\infty,|\eta|_\infty)(|y_0|_2+|z_0|_2).$$

Hence

$$\|w_\varepsilon\|_{L^{\alpha_0(d)}(Q)}\leq C(|\xi|_\infty,|\eta|_\infty)(|y_0|_2+|z_0|_2). \tag{3.85}$$

Similarly, it follows that

$$\|v_\varepsilon\|_{L^{\alpha_0(d)}(Q)}\leq C(|\xi|_\infty,|\eta|_\infty)(|y_0|_2+|z_0|_2). \tag{3.86}$$

Letting ε tend to zero into (3.85), (3.86), we conclude that (w, v) satisfy the conditions of Lemma 3.2.

Proof of Theorem 3.9 (Continued). We set

$$K = \{\zeta \in L^\infty(Q);\ \|\zeta\|_{L^\infty(Q)} \leq R\} \tag{3.87}$$

and let $\Phi : K \to L^2(Q)$ be defined by

$$\Phi(\zeta)=\{y\in L^2(Q);\ y(T)=0,\ z(T)=0;\ \|w\|_{L^{\alpha_0(d)}(Q)}+\|v\|_{L^{\alpha_0(d)})Q)}\leq M\}, \tag{3.88}$$

where $(y, z) \in H^{2,1}(Q)\times H^{2,1}(Q)$ is the solution to the linearized system

$$\begin{aligned}
&y_t - \Delta y + (a + b\zeta + c\zeta^2)y + dz = mv && \text{in } Q,\\
&z_t - \Delta z + \ell\Delta y + (a_1 + b_1\zeta + c_1\zeta^2)y + d_1 z = mv && \text{in } Q,\\
&y(0) = y_0,\ z(0) = z_0 && \text{in } \mathcal{O},\\
&y = z = 0 && \text{on } \Sigma.
\end{aligned} \tag{3.89}$$

We shall assume that

$$y_0 \in W_0^{2\left(1-\frac{1}{\alpha_0(d)}\right),\alpha_0(d)}(\mathcal{O}),\ z_0 \in H_0^1(\mathcal{O}).$$

We note first that, by Lemma 3.2, $\Phi(\zeta) \neq \emptyset$ for $\zeta \in K$ if

$$M \leq \mu(R, R)(|y_0|_2 + |z_0|_2). \tag{3.90}$$

Let us show now that for $\|y_0\|_{\alpha_0(d)}+\|z_0\|_{H_0^1(\mathcal{O})}$ sufficiently small $\Phi(K)\subset K$. Indeed, by (3.89), we have the standard parabolic estimates

$$\|y\|_{H^{2,1}(Q)} + \|z\|_{H^{2,1}(Q)} \leq C(M + \|y_0\|_{H_0^1(\mathcal{O})} + \|z_0\|_{H_0^1(\mathcal{O})}). \tag{3.91}$$

Next, by the first equation in (3.89) and the L^p- theory of parabolic boundary value problems, we have

$$\begin{aligned}
\|y\|_{W^{2,1}_{\alpha_0(d)}(Q)} &\leq C(R)(\|z\|_{H^{2,1}(Q)} + \|y_0\|_{\alpha_0(d)})\\
&\leq C(R)(M + \|y_0\|_{\alpha_0(d)} + \|z_0\|_{H_0^1(\mathcal{O})}).
\end{aligned} \tag{3.92}$$

(Here $C = C(r)$ is a positive and continuous and monotonically increasing function.) The last estimate follows by the following argument. By the first equation, we have

$$\|y\|^2_{H^{2,1}(Q)} \leq C(R)(M + \|z\|^2_{L^2(Q)} + \|y_0\|^2_{H_0^1(\mathcal{O})})$$

and, by the second equation,

$$\|z\|^2_{H^{2,1}(Q)} \leq C(R)(M + \|y\|^2_{H^{2,1}(Q)} + \|z_0\|^2_{H^1_0(\mathscr{O})}).$$

Next, we multiply the first equation by y and the second by δz, where $\delta > 0$ is sufficiently small. Then, after some calculation involving Gronwall's lemma, we get the desired estimate.

Since $\alpha_0(d) > \frac{d+2}{2}$, $\alpha_0(d) \neq d+2$ for $d = 1, 2, 3$, we infer by (3.92) that

$$\|y\|_{L^\infty(Q)} \leq C(r)(M + \|y_0\|_{\alpha_0(d)} + \|z_0\|_{H^1_0(\mathscr{O})}). \tag{3.93}$$

Hence, by (3.90) and (3.93), we obtain that

$$\|y\|_{L^\infty(Q)} \leq C(R)(\mu(R, R) + 1)(\|y_0\|_{\alpha_0(d)} + \|z_0\|_{H^1_0(\mathscr{O})}) \leq R$$

if $\|y_0\|_{\alpha_0(d)} + \|z_0\|_{H^1_0(\mathscr{O})}$ is sufficiently small.

Moreover, it follows by (3.91) via standard compactness results in $L^2(Q)$ that $\Phi(K)$ is a compact subset of $L^2(Q)$. It is also easily seen that $\Phi(\zeta)$ is a closed convex subset for each $\zeta \in K$ and that the map $\zeta \to \Phi(\zeta)$ is upper semicontinuous from $L^2(\mathscr{O})$ to itself. Then, once again, by the Kakutani fixed point theorem, it follows that Φ has a fixed point y, i.e., there is $(w, v) \in L^2(Q) \times L^2(Q)$ such that $y(T) = 0$, $z(T) = 0$. This completes the proof of Theorem 3.9. ∎

Proof of Corollary 3.7. By the substitution $y = \varphi - \varphi^*$, $z = u - u^*$ the problem reduces to that of boundary controllability of system (3.77). Then the conclusions of Corollary 3.7 follow from Theorem 3.9, as in the previous situations. Namely, let $\widetilde{\mathscr{O}}$ be an open smooth subset of $\mathbb{R}^d$ such that $\mathscr{O} \subset \widetilde{\mathscr{O}}$. We extend, by standard procedure, the functions $a, b, c, d, a_1, b_1, c_1, d_1$ on $\widetilde{Q} = \widetilde{\mathscr{O}} \times [0, T)$ and y_0, z_0 on $\widetilde{\mathscr{O}}$ to

$$(\widetilde{y}_0, \widetilde{z}_0) \in W_0^{2\left(1 - \frac{1}{\alpha_0(d)}\right)}(\widetilde{\mathscr{O}}) \times H^1_0(\widetilde{\mathscr{O}}).$$

Then, by Theorem 3.9 and its proof, we know that for $\|\widetilde{y}_0\|_{\alpha_0(d)} + \|\widetilde{z}_0\|_{H^1_0(\widetilde{\mathscr{O}})}$ sufficiently small there are $(\widetilde{u}, \widetilde{w}) \in L^2(\widetilde{Q}) \times L^2(\widetilde{Q})$ such that

$$\widetilde{y}(T) = 0, \quad \widetilde{z}(T) = 0,$$

where $(\widetilde{y}, \widetilde{z})$ is the solution to (3.77).

Clearly, $y = \widetilde{y}|_Q$, $z = \widetilde{z}|_Q$ is the solution to system (3.77), where $v = w = 0$ and with the boundary conditions

$$y|_\Sigma = \widetilde{y}|_\Sigma, \quad z|_\Sigma = \widetilde{y}|_\Sigma$$

and $y(T) = z(T) = 0$. This completes the proof. ∎

Remark 3.5. Theorem 3.9 is, in particular, important because the Stefan two phase problem (see, e.g., [18, 26]), $y_t - \Delta\beta(y) = mu$ in Q, which describes the dynamic of melting processes, is so far beyond the controllability approach developed in the previous section. (See, however, [50] for a controllability result in $1 - D$.)

Remark 3.6. Theorem 3.9, which was established in [23], was meantime extended to the controllability of phase-field system (3.72) using an internal control only (see [3, 10, 77]).

3.5 Controllability of the Kolmogorov Equation

On unbounded domains $\mathcal{O}$ a controllability result similar to Theorem 3.1 is not true in general. (See examples in [93, 94].) We shall present, however, in the following such a result for the linear parabolic equation with a drift term,

$$\begin{aligned} &\frac{\partial y}{\partial t} - \frac{1}{2}\,\Delta y + F(x)\cdot\nabla y = \mathbf{1}_{\mathcal{O}_0} u \ \text{ in } \ (0,T)\times\mathcal{O}, \\ &\frac{\partial y}{\partial \nu} = 0 \ \text{ on } \ (0,T)\times\partial\mathcal{O}, \\ &y(0) = y_0(x),\ x\in\mathcal{O}, \end{aligned} \tag{3.94}$$

where $\mathcal{O}$ is an open and convex set in $\mathbb{R}^d$ (eventually unbounded), $d \geq 1$, $\mathcal{O}_0$ is an open subset of $\mathcal{O}$, and $F : \mathbb{R}^d \to \mathbb{R}^d$ is a C^1-continuous, coercive, and bounded mapping of gradient type.

The linear parabolic equation (3.94) is known in literature as the *Kolmogorov equation* and arises in the theory of stochastic processes on the domain $\mathcal{O}$. In the special case where $\operatorname{div} F = 0$, (3.94) reduces to the Fokker-Planck equation on $\mathcal{O}$.

Given a closed convex set $K \subset \mathbb{R}^d$, the *recession cone* of K is defined by

$$\operatorname{recc}(K) = \{y \in \mathbb{R}^d;\ x + \lambda y \in K,\ \forall x \in K,\ \forall \lambda \geq 0\}$$

or, equivalently,

$$\operatorname{recc}(K) = \bigcap_{\lambda>0} \lambda(K - y),\ \forall y \in K.$$

If K is bounded, then $\operatorname{recc}(K) = \{0\}$, but otherwise $\operatorname{recc}(K)$ is an unbounded set (cone).

Denote by p_K the *Minkowski functional* (*gauge*) associated with the closed convex set K, that is,

$$p_K(x) = \inf\left\{\lambda \geq 0;\ \frac{1}{\lambda}\,x \in K\right\},\ \forall x \in \mathbb{R}^d. \tag{3.95}$$

We recall that p_K is subadditive, positively homogeneous and, if $\overset{\circ}{K} \neq \emptyset$, then

$$\overset{\circ}{K} = \{x \in \mathbb{R}^d;\ p_K(x) < 1\},\ \partial K = \{x \in \mathbb{R}^d;\ p_K(x) = 1\}. \tag{3.96}$$

(Here, $\overset{\circ}{K}$ is the interior of K and ∂K is its boundary.)

If $0 \in \overset{\circ}{K}$, then

$$\text{recc}(K) = \{x \in K;\ p_K(x) = 0\}. \tag{3.97}$$

We shall assume in the following that

(i) $\mathscr{O}$ is an open convex set of $\mathbb{R}^d$ with C^2-boundary $\partial\mathscr{O}$, $0 \in \mathscr{O}$.
(ii) $F = \nabla g$, where $g \in C^2(\mathbb{R}^d)$ is convex and

$$\sup\{|F(x)|_d + \|DF(x)\|;\ x \in \mathbb{R}^d\} < \infty, \tag{3.98}$$

$$g(x) \geq \alpha_1 |x|_d + \alpha_2,\ \forall x \in \mathbb{R}^d, \tag{3.99}$$

where $\alpha_1 > 0$ and $\alpha_2 \in \mathbb{R}$. Here, DF stands for the differential of $F : \mathbb{R}^d \to \mathbb{R}^d$ and $\|\cdot\|$ is the norm in $L(\mathbb{R}^d, \mathbb{R}^d)$.
(iii) $\mathscr{O}_0$ is an open subset of $\mathscr{O}$ which contains an open subset $\mathscr{O}_1$ such that $\overline{\mathscr{O}}_1 \subset \mathscr{O}_0$ and

$$\inf\{|\nabla p_{\overline{\mathscr{O}}}(x)|_d;\ x \in \overline{\mathscr{O}} \setminus \mathscr{O}_1\} = \gamma > 0. \tag{3.100}$$

We note that $p_{\overline{\mathscr{O}}} \in C^2(\overline{\mathscr{O}} \setminus \text{recc}(\overline{\mathscr{O}}))$ and so (3.100) makes sense.

Taking into account (3.97), we see by (iii) that $\text{recc}(\overline{\mathscr{O}}) \subset \mathscr{O}_1 \subset \mathscr{O}_0$.

A convenient functional space to treat equation (3.94) is $L^2(\mathscr{O}; \mu)$, where μ is a Gaussian measure with the density $\rho : \mathbb{R}^d \to [0, \infty)$ defined by

$$\rho(x) = \begin{cases} \exp(-2g(x)) \left(\int_{\mathscr{O}} \exp(-2g(x))dx\right)^{-1}, & x \in \overline{\mathscr{O}}, \\ 0, \quad x \in \mathbb{R}^d \setminus \overline{\mathscr{O}}. \end{cases} \tag{3.101}$$

A simple calculation shows that, if $g \in C^2(\mathbb{R}^d)$, then ρ is a solution to the Neumann problem

$$\begin{aligned} &\frac{1}{2}\,\Delta\rho + \text{div}\,(F\rho) = 0 \ \text{ in } \ \mathscr{O}, \\ &\frac{\partial\rho}{\partial\nu} + (F \cdot \nu)\rho = 0 \quad \text{ on } \partial\mathscr{O}. \end{aligned} \tag{3.102}$$

(Otherwise, (3.102) holds in the weak distributional sense.)

We consider the probability measure μ defined by

$$d\mu = \rho\, dx \tag{3.103}$$

and consider in the space $L^2(\mathscr{O}; d\mu) = L^2(\mathscr{O}; \mu)$ the operator

$$\begin{aligned} Ny &= -\frac{1}{2}\,\Delta y + F \cdot \nabla y,\ y \in D(N),\\ D(N) &= \left\{ y \in W^{2,2}(\mathscr{O}; \mu);\ \frac{\partial y}{\partial \nu} = 0 \text{ on } \partial\mathscr{O} \right\}. \end{aligned} \tag{3.104}$$

In fact, N is the Kolmogorov operator associated with the stochastic reflection equation

$$\begin{aligned} &dX + F(X)dt + N_K(X)dt \ni dW_t,\\ &X(0) = x, \end{aligned} \tag{3.105}$$

where N_K is the normal cone to K and W_t is a Wiener d-dimensional process. As shown in [38], the operator N is m-accretive in $L^2(\mathscr{O}; \mu)$ and so it generates a C_0-semigroup of contractions e^{-tN} in $L^2(\mathscr{O}; \mu)$, given by

$$(e^{-tN}y_0)(x) = \mathbb{E}y_0(X(x,t)),\ x \in \mathscr{O},\ \forall\, y_0 \in L^2(\mathscr{O}; d\mu),$$

where $X(x,t)$ is the solution to (3.105).

In particular, this implies that problem (3.94), or equivalently

$$\frac{dy}{dt} + Ny = \mathbf{1}_{\mathscr{O}_0}u,\ t \in [0,T],\ y(0) = y_0, \tag{3.106}$$

has, for each $y_0 \in L^2(\mathscr{O}; \mu)$ and all $u \in L^2(0,T; L^2(\mathscr{O}; \mu))$, a unique mild solution $y^u \in C([0,T]; L^2(\mathscr{O}; d\mu))$, that is,

$$y^u(t) = e^{-tN}y_0 + \int_0^t e^{-(t-s)N}(\mathbf{1}_{\mathscr{O}_0}u)(s)ds,\ t \in [0,T]. \tag{3.107}$$

Now, we can formulate the main controllability result.

Theorem 3.10. *Under assumptions* (i)–(iii), *for each* $0 < T < \infty$ *and all* $y_0 \in L^2(\mathscr{O}; \mu)$ *there is at least one controller* $u \in L^2(0,T; L^2(\mathscr{O}; \mu))$ *such that* $y^u(T) \equiv 0$.

Theorem 3.10 will be proved as Theorem 3.1 via the Carleman inequality for the backward dual equation associated with (3.106) similar to Theorem 2.1. The main difference is that here the Lebesgue measure will be replaced by the probability measure μ. It should be said that in equation (3.94) (respectively (3.104)) the coefficient $\frac{1}{2}$ in front of Δ was taken for the sake of symmetry only. Of course, one can replace it by any constant $a > 0$. As a matter of fact, the operator $\frac{1}{2}\,\Delta$ can be replaced by any second order elliptic operator with constant coefficients.

In the following, we discuss the form of $\mathscr{O}_0$ arising in (iii) in some special cases. Assume that

$$\mathscr{O} = \{(x', x_d) \in \mathbb{R}^d;\ x_d > \phi(x') - b\}, \tag{3.108}$$

where $b > 0$ and $\phi \in C^2(\mathbb{R}^{d-1})$ is a convex function satisfying $\phi(0) = 0$ and

$$\phi(u) \geq a|u|_{d-1}^m,\ \forall u \in \mathbb{R}^d, \tag{3.109}$$

where $a > 0$ and $1 \leq m < \infty$. (We have always such a local representation of $\mathscr{O}$.) It is readily seen that

$$\begin{aligned} \mathrm{recc}(\overline{\mathscr{O}}) &= \{(0, x_d);\ x_d \geq 0\} && \text{for } m > 1, \\ \mathrm{recc}(\overline{\mathscr{O}}) &= \{(x', x_d);\ x_d \geq a|x'|_{d-1}\} && \text{for } m = 1. \end{aligned}$$

(Here, $x' = x_1, \ldots, x_{d-1})$.) We have

Proposition 3.1. *Let $\eta > 0$ be the solution to the equation*

$$x_d = \eta\phi\left(\frac{x'}{\eta}\right) - b\eta. \tag{3.110}$$

Then, $\eta = p_{\overline{\mathscr{O}}}$ and any set $\mathscr{O}_0 = \overline{\mathscr{O}} \setminus G_\alpha^\varepsilon$,

$$G_\alpha^\varepsilon := \{(x', x_d) \in \overline{\mathscr{O}};\ |x'|_{d-1} \leq \alpha p_{\overline{\mathscr{O}}}(x) - \varepsilon\}, \alpha > 0,\ 0 < \varepsilon < b,$$

satisfies (iii).

Proof. By (3.95), we see that, for each $x \in \overline{\mathscr{O}}$, $p_{\overline{\mathscr{O}}}(x) = \eta(x)$ is the unique positive solution to (3.110). We have

$$\frac{\partial \eta}{\partial x_2} = \frac{\eta}{\eta\phi\left(\frac{x'}{\eta}\right) - \nabla\phi\left(\frac{x'}{\eta}\right) \cdot x' - b}$$

and, since ϕ and $\nabla\phi$ are bounded on bounded sets, we infer that, for each $\alpha > 0$,

$$\left|\frac{\partial \eta}{\partial x_2}\right| \geq \gamma(\alpha) > 0, \ \text{ for } |x'|_{d-1} \leq \alpha\eta,$$

which implies that

$$\inf\{|\nabla p_{\overline{\mathscr{O}}}(x)|;\ x \in \overline{\mathscr{O}} \setminus \mathscr{O}_1\} > 0,$$

where $\mathscr{O}_1 = \{(x', x_d);\ |x'|_{d-1} > \alpha p_{\overline{\mathscr{O}}}(x)\} \subset \mathscr{O}_0$.

Example 3.1. Let $\phi(u) = a|u|_{d-1}^m$, where $a > 0$ and $m \geq 2$. Then (3.110) reduces to

$$x_d = a\eta^{1-m}|x'|_{d-1}^m - b\eta. \tag{3.111}$$

Equivalently,

$$by^m + \frac{x_d}{|x'|_{d-1}^{m-1}}\, y^{m-1} - a = 0, \tag{3.112}$$

where $y = \frac{x_d}{|x'|_{d-1}}$. A simple analysis of equation (3.112) reveals that, for each $\alpha > 0$,

$$y \geq \alpha \ \text{ if } \ 0 < x_d \leq \zeta(\alpha)|x'|_d^{m-1}.$$

Then, by Proposition (3.1), it follows that, for each $\gamma > 0$, $0 < \varepsilon < b$, and

$$G_\gamma^\varepsilon = \{x'x_d) \in \mathcal{O};\ x_d \leq \gamma|x'|_d^{m-1} - \varepsilon\}, \tag{3.113}$$

the set $\mathcal{O}_0 \subset \overline{\mathcal{O}} \setminus G_\gamma^\varepsilon$ satisfies (iii).

Then, Theorem 3.10 implies that

Corollary 3.8. *Let $\mathcal{O} = \{(x', x_d);\ x_d > a|x'|_d^m - b\}$ for $a, b > 0$, $2 \leq m < \infty$. Then, we may take $\mathcal{O}_0$, any set of the form*

$$\{(x', x_d);\ x_d > \gamma|x'|_d^{m-1} - \varepsilon\}, \text{ where } \gamma > 0 \text{ and } 0 < \varepsilon < b. \tag{3.114}$$

In particular, it follows by Theorem 3.10 that (3.94) is exactly null controllable with controllers $v = \mathbf{1}_{\mathcal{O}_0} u$ in any set $\mathcal{O}_0$ of the form (3.114). At finite distance, this set can be taken as close as we want of the recession cone $\{(0, x_d);\ x_d \geq 0\}$.

Remark 3.7. The conclusion of Corollary 3.8 remains true if $\mathcal{O}$ is of the form (3.108) away from origin, that is,

$$\phi(u) = a|u|_{d-1}^m \text{ for } |u|_{d-1} \geq \lambda > a,\ 1 < m < \infty,\ a > 0,\ b > 0. \tag{3.115}$$

Indeed, the calculation in Example 3.1 shows that (3.113) holds because only the values $|x'|_{d-1} + x_d$ large enough are relevant. This extends to the case $m = 1$, where $p_{\overline{\mathcal{O}}}(x) = b^{-1}(a|x'|_{d-1} - x_d)^-$ for $|x'|_{d-1} \geq \lambda > 0$, and so

$$\mathcal{O}_0 = \{(x', x_d);\ x_d - a|x'|_{d-1} \geq -\varepsilon\}, \tag{3.116}$$

where $\varepsilon > 0$ is arbitrarily small.

Proof of Theorem 3.10. Denote by N^* the dual operator of N in the space $L^2(\mathscr{O};\mu)$, that is,

$$\langle N^* p, y\rangle_{L^2(\mathscr{O};\mu)} = \langle p, Ny\rangle_{L^2(\mathscr{O};\mu)},$$

for all $y \in D(N)$ and $p \in D(N^*)$. A simple calculation involving (3.102) and (3.104) shows that

$$\begin{aligned} N^* p &= -\frac{1}{2}\,\Delta p - F\cdot\nabla p - \nabla(\log\rho)\cdot\nabla p,\\ D(N^*) &= \left\{p \in W^{2,2}(\mathscr{O});\ \frac{\partial p}{\partial \nu} = 0 \text{ on } \partial\mathscr{O}\right\}. \end{aligned} \tag{3.117}$$

Moreover, taking into account that $F = \nabla g = -\frac{1}{2}\,\nabla(\log\rho)$, we see by (3.117) that $N^* = N$.

As it is well known, for the exact controllability of (3.106) we need the observability inequality

$$\|p(0)\|^2_{L^2(\mathscr{O};\mu)} \le C\int_0^T dt\int_{\mathscr{O}_0}|p(x,t)|^2 d\mu = C\int_0^T\int_{\mathscr{O}_0}\rho(x)|p(x,t)|^2 dx\,dt, \tag{3.118}$$

for any solution p to the backward equation

$$\frac{dp}{dt} - N^* p = 0,\ t \in (0,T),$$

or, equivalently,

$$\begin{aligned} &\frac{\partial p}{\partial t} + \frac{1}{2}\,\Delta p - F\cdot\nabla p = 0 \text{ in } (0,T)\times\mathscr{O},\\ &\frac{\partial p}{\partial \nu} = 0 \qquad\qquad \text{on } (0,T)\times\partial\mathscr{O}. \end{aligned} \tag{3.119}$$

To get (3.118), we prove first a Carleman-type inequality for solutions p to equation (3.119). To this end, proceeding as in Section 2.1, we consider an open set $\mathscr{O}_1, \overline{\mathscr{O}_1} \subset \mathscr{O}_0$ as in assumption (iii), and set

$$\alpha(x,t) = \frac{e^{-\lambda\psi(x)} - e^{2\lambda\|\psi\|_{C(\overline{\mathscr{O}})}}}{t(T-t)},\quad \varphi(x,t) = \frac{e^{-\lambda\psi(x)}}{t(T-t)},\quad x \in \overline{\mathscr{O}},\ t \in (0,T),$$

where ψ is the function given by Lemma 3.4 below.

Lemma 3.4. *There is* $\psi \in C^2(\overline{\mathscr{O}})$ *such that*

$$\psi(x) > 0,\ \forall x \in \mathscr{O},\ \psi(x) = 0,\ \forall x \in \partial\mathscr{O}, \tag{3.120}$$

$$|\nabla\psi(x)|_d \geq \gamma > 0,\ \forall x \in \overline{\mathscr{O}} \setminus \mathscr{O}_1, \tag{3.121}$$

$$\sup\{|\nabla\psi(x)|_d + |D^2_{x_i x_j}\psi(x)|;\ i, j = 1, \dots, d\} < \infty. \tag{3.122}$$

Proof. Let $\mathscr{O}_2$ be an open subset of $\mathscr{O}_1$ such that $\overline{\mathscr{O}}_2 \subset \mathscr{O}_1$ and $\text{dist}(\partial\mathscr{O}_2, \partial\mathscr{O}_1) > 0$ is sufficiently small. Then, consider a function $\mathscr{X} \in C_b^\infty(\mathbb{R}^d)$ such that $0 \leq \mathscr{X} \leq 1$, $\mathscr{X} = 0$ on $\overline{\mathscr{O}} \setminus \mathscr{O}_1$ and $\mathscr{X} = 1$ in $\overline{\mathscr{O}}_2$. (This function can be constructed in a standard way via mollifiers technique.) Then, we set $\psi = 1-(1-\mathscr{X})p_{\overline{\mathscr{O}}}$. Taking into account assumption (iii), we see that ψ satisfies (3.120), (3.121) (because $\nabla\psi = -\nabla p_{\overline{\mathscr{O}}}$ on $\overline{\mathscr{O}} \setminus \mathscr{O}_1$). Moreover, by Lemma 3.5 below, (3.122) follows, too.

The following Carleman inequality is exactly of the same form as that given in Theorem 2.1, so we give it without proof (see [31] for the proof).

Proposition 3.2. *There are $\lambda_0 > 0$ and a function $s_0 : \mathbb{R}^+ \to \mathbb{R}^+$ such that, for $\lambda \geq \lambda_0$ and $s \geq s_0$,*

$$\int_0^T \int_{\mathscr{O}} e^{2s\alpha}(s^3\varphi^3 p^2 + s\varphi|\nabla p|^2 d\mu\, dt \leq C_\lambda s^3 \int_0^T \int_{\mathscr{O}_0} e^{2s\alpha}\varphi^3 p^2 d\mu\, dt, \tag{3.123}$$

for all the solutions p to (3.119).

By (3.123), we obtain estimate (3.118). Namely,

Corollary 3.9. *The observability inequality* (3.118) *holds for all the solutions p to* (3.119).

Proof. We note first that

$$\langle p, Np\rangle_{L^2(\mathscr{O};\mu)} = \frac{1}{2}\int_{\mathscr{O}} |\nabla p|^2 d\mu,$$

which yields

$$\frac{d}{dt}\int_{\mathscr{O}} p^2(x,t)d\mu - \int_{\mathscr{O}} |\nabla p(x,t)|^2 d\mu = 0,\ \forall t \geq 0.$$

We have, for all $0 \leq \tau \leq t < \infty$,

$$\int_{\mathscr{O}} p^2(x,\tau)d\mu + \int_\tau^t \int_{\mathscr{O}} |\nabla p(x,\theta)|^2 d\mu\, d\theta = \int_{\mathscr{O}} p^2(x,t)d\mu, \tag{3.124}$$

and, therefore,

$$\int_{\mathscr{O}} p^2(x,0)d\mu \leq \int_{\mathscr{O}} p^2(x,t)d\mu \leq \gamma(t)\int_{\mathscr{O}} e^{2s\alpha}\varphi^3 p^2 d\mu, \tag{3.125}$$

where $\gamma(t) = \sup\{e^{-2s\alpha}\varphi^{-3}(x,t);\ x \in \mathscr{O}\} \leq C\exp\left(\frac{\mu s}{t(T-t)}\right)$, $\mu = 2e^{2\lambda}\|\psi\|_{C(\overline{\mathscr{O}})}$.

Integrating on $(t_1, t_2) \subset (0, T)$, we obtain by (3.123) the desired inequality (3.118) with a suitable constant C. We note also that (3.124) implies that $p \in W^{1,2}(0, T; L^2(\mathscr{O}; \mu))$. Without loss of generality, we may assume in the following that $p \in L^2(0, T; W^{2,2}(\mathscr{O}; \mu))$. (Taking into account the structure of the domain $D(N)$ this happens if $p(T) \in D(N)$ which, without any loss of generality, we may assume.) Also, for the sake of simplicity, we shall prove Proposition 3.2 for the equation

$$\begin{aligned} &\frac{\partial p}{\partial t} + \Delta p - F \cdot \nabla p = 0 \text{ in } (0, T) \times \mathscr{O}, \\ &\frac{\partial p}{\partial \nu} = 0 \qquad\qquad\qquad\quad \text{on } (0, T) \times \partial\mathscr{O}, \end{aligned} \tag{3.126}$$

because (3.119) is obtained from (3.126) by rescaling the time t.

Lemma 3.5. *Let $\mathscr{O}$ be an open and convex set with C^2-boundary $\partial\mathscr{O}$ and $0 \in \mathscr{O}$. Let $p_{\overline{\mathscr{O}}}$ be the gauge of $\overline{\mathscr{O}}$. Then*

$$p_{\overline{\mathscr{O}}} \in C^2(\overline{\mathscr{O}} \setminus \mathrm{recc}(\overline{\mathscr{O}})), \tag{3.127}$$

$$\sup\{|\nabla p_{\overline{\mathscr{O}}}(x)|_d;\ x \in \overline{\mathscr{O}} \setminus \mathrm{recc}(\overline{\mathscr{O}})\} < \infty, \tag{3.128}$$

$$\sup\{|D^2_{x_i x_j} p_{\overline{\mathscr{O}}}(x)|;\ x \in \overline{\mathscr{O}} \setminus \mathrm{recc}(\overline{\mathscr{O}})\} < \infty,\ i, j = 0, 1, 2 \dots, d. \tag{3.129}$$

Proof. We set $\eta = p^{-1}_{\overline{\mathscr{O}}}$ on $\overline{\mathscr{O}} \setminus \mathrm{recc}(\overline{\mathscr{O}})$. Then, $\eta(x)x \in \partial\mathscr{O}$ and, since $\partial\mathscr{O}$ is of class C^2, it is locally represented as $x_d = \phi(x')$, where ϕ is convex and of class C^2. We have, therefore,

$$x_d \eta = \phi(\eta x'),\ \forall x \in \overline{\mathscr{O}} \setminus \mathrm{recc}(\overline{\mathscr{O}}),$$

and, since $\nabla\phi$ is monotone and of class C^1 in $\mathbb{R}^{d-1}$, $\alpha I + \nabla\phi$, is invertible for all $\alpha > 0$. So, we conclude that η is of class C^2 on $\overline{\mathscr{O}} \setminus \mathrm{recc}(\overline{\mathscr{O}})$.

Since $p_{\overline{\mathscr{O}}}$ is positively homogeneous, subadditive, and $0 \in \mathscr{O}$, we have, for all $y \in \mathscr{O}$ with $|y|_d \le \varepsilon$ sufficiently small,

$$\nabla p_{\overline{\mathscr{O}}}(x) \cdot y \le p_{\mathscr{O}}(y),\ \forall x \in \overline{\mathscr{O}} \setminus \mathrm{recc}(\overline{\mathscr{O}}),$$

and this, clearly, implies (3.128). Now, if we differentiate two times with respect to x the equation $p_{\overline{\mathscr{O}}}(\lambda x) = \lambda p_{\overline{\mathscr{O}}}(x)$, we obtain

$$\lambda D^2_{ij} p_{\overline{\mathscr{O}}}(\lambda x) = D^2_{ij} p_{\overline{\mathscr{O}}}(x),\ \forall x \in \overline{\mathscr{O}} \setminus \mathrm{recc}(\overline{\mathscr{O}}).$$

This yields

$$D^2_{ij} p_{\overline{\mathscr{O}}}(y) = \frac{1}{\lambda}\, D^2_{ij} p_{\overline{\mathscr{O}}}\left(\frac{y}{\lambda}\right),\ y \in \overline{\mathscr{O}} \setminus \mathrm{recc}(\overline{\mathscr{O}}),$$

and, for $\lambda = |y|_d/\varepsilon$, we get

$$D^2_{ij} p_{\overline{\mathcal{O}}}(y) = \frac{\varepsilon}{|y|_d}\, D^2 p_{\overline{\mathcal{O}}}\left(\varepsilon\, \frac{y}{|y|_d}\right),\ \forall\, y \in \overline{\mathcal{O}} \setminus \mathrm{recc}(\overline{\mathcal{O}}),$$

which, clearly, implies (3.129), as desired.

3.6 Exact Controllability of Stochastic Parabolic Equations

The linear parabolic *stochastic equation* is a model for linear diffusion dynamic perturbed by a Gaussian random driving force. In most situations, this force can be viewed as a linear multiplicative *Gaussian noise* and so the dynamic can be represented as the stochastic parabolic equation

$$\begin{aligned} &d_t X(\xi,t) - \Delta_\xi X(\xi,t)dt + a(\xi,t)X(\xi,t)dt + (b(\xi,t)\cdot \nabla_\xi X(\xi,t))dt \\ &\qquad = \mathbf{1}_{\mathcal{O}_0} u(\xi,t)dt + \sum_{k=1}^N \mu_k(\xi,t)X(\xi,t)d\beta_k(t) \text{ in } (0,T)\times\mathcal{O}, \\ &X(\xi,0) = x(\xi), \quad \xi \in \mathcal{O}, \\ &X(\xi,t) = 0 \quad \text{on } (0,T)\times\partial\mathcal{O}, \end{aligned} \tag{3.130}$$

where $\{\beta_k\}_{k=1}^N$ is a linear independent system of *Brownian motion* in a probability space $\{\mathcal{O}, \mathcal{F}, \mathbb{P}\}$, $\mathcal{O}$ is a bounded and open subset of $\mathbb{R}^d$, $d \geq 1$, with smooth boundary $\partial\mathcal{O}$ and $\mathcal{O}_0$ is any open subset of $\mathcal{O}$. By $\mathbf{1}_{\mathcal{O}_0}$ we have denoted, as usually, the characteristic function of $\mathcal{O}_0$. As regard the functions a, b and μ_k, we assume that the following hypothesis holds:

(i) $a \in L^\infty(\mathcal{O}\times(0,T))$, $b \in C^1(\overline{\mathcal{O}}\times[0,T]; \mathbb{R}^d)$, $\mu_k \in C^2(\overline{\mathcal{O}}\times[0,T])$, $k = 1, 2, \ldots, N$.

Here and everywhere in the following, the controller $u : \mathcal{O}\times(0,T)\times\Omega \to \mathbb{R}$, $u = u(\xi,t)$, is taken in the class of all adapted $L^2(\mathcal{O})$-valued processes with respect to the natural filtration $\{\mathcal{F}_t\}_{t\geq 0}$ induced by the Brownian motions $\{\beta_k\}_{k=1}^N$. It should be said that also in this case, the exact null controllability is related and, in a certain sense, which will be made precise later on, equivalent with the observability of the dual backward stochastic equation

$$\begin{aligned} &d_t p(\xi,t) + \Delta_\xi(\xi,t)dt - a(\xi,t)p(\xi,t)dt + \mathrm{div}(b(\xi,t)p(\xi,t))dt \\ &\quad + \Big(\sum_{k=1}^N \mu_k(\xi,t)q_k(\xi,t)\Big)dt = \sum_{k=1}^N q_k(\xi,t)d\beta_k(t),\ (\xi,t) \in \mathcal{O}\times(0,T), \\ &p(\xi,t) = 0,\ (\xi,t) \in \partial\mathcal{O}\times(0,T). \end{aligned} \tag{3.131}$$

Given the stochastic basis $\{\Omega, \mathscr{F}, \mathbb{P}, \{\mathscr{F}_t\}_{t\geq 0}\}$ and H a Hilbert space with the norm $|\cdot|_H$ and $1 \leq p < \infty$, we denote by $M^2_{\mathbb{P}}(0, T; H)$ the space of all H-valued progressively measurable processes $X : \Omega \times (0, T) \to H$, such that

$$\|X\|^2_{M^2_{\mathbb{P}}(0,T;H)} = \mathbb{E}\int_0^T |X(t)|^2_H dt < \infty,$$

where $\mathbb{E}$ is the expectation.

Denote by $C_{\mathbb{P}}([0, T]; H)$ the space of all $\mathscr{F}_t$-adapted processes $X \in M^2_{\mathbb{P}}(0, T; H)$ which have a modification in $C([0, T]; L^2(\Omega))$. We denote by $L^2_{ad}(\Omega; C([0, T]; L^2(\mathscr{O})))$ the space of all $\mathscr{F}_t$-adapted processes $X : \mathscr{O}\times[0, T] \to L^2(\mathscr{O})$, $X \in C([0, T]; L^2(\mathscr{O}))$, $\mathbb{P}$-a.s., and

$$\mathbb{E}\|X\|^2_{C([0,T];L^2(\mathscr{O}))} < \infty.$$

We denote by $L^2_T(\Omega, \mathscr{O})$ the space of all random variables $\eta : \Omega \to L^2(\mathscr{O})$ such that Z is $\mathscr{F}_T$ measurable and

$$\mathbb{E}\int_{\mathscr{O}} |Z(\xi)|^2 d\xi < \infty.$$

By solution to (3.130), we mean a process $X \in L^2_{\rm ad}(\Omega; C([0, T]; L^2(\mathscr{O})))$ which satisfies $\mathbb{P}$-a.s. the equation

$$\begin{aligned} X(\xi, t) = x &+ \int_0^t \Delta_\xi X(\xi, s)ds - \int_0^t (a(s)X(\xi, s) + b(s)\cdot\nabla_\xi X(\xi, s))ds \\ &+ \int_0^t \mathbf{1}_{\mathscr{O}_0} u(\xi, s)ds + \sum_{k=1}^N \int_0^t \mu_k(\xi, s)X(\xi, s)d\beta_k(s),\ \forall t \in (0, T),\ \xi \in \mathscr{O}. \end{aligned} \tag{3.132}$$

Here the integral arising in the right-hand side of (3.132) is taken in sense of Itô with values in $H^{-1}(\mathscr{O})$. (We refer to [64] for existence and uniqueness of such a solution.) The solution X to (3.130) is denoted by X^u.

In order to study the exact null controllability of (3.130), we shall reduce it by a rescaling procedure presented below to a random parabolic equation and apply to this equation the controllability results established in Section 3.1. Namely, we set

$$\begin{aligned} W(\xi, t) &= \sum_{k=1}^N \mu_k(\xi, t)\beta_k(t),\quad (\xi, t) \in (0, T)\times\mathscr{O}, \\ \mu(\xi, t) &= \sum_{k=1}^N \left(\frac{\partial\mu_k}{\partial t}(t, \xi)\beta_k(t) + \frac{1}{2}\mu_k^2(\xi, t)\right),\quad (\xi, t) \in (0, T)\times\mathscr{O}. \end{aligned}$$

By the substitution

$$X(\xi, t) = e^{W(\xi,t)} y(\xi, t), \tag{3.133}$$

equation (3.130) reduces to the random parabolic equation

$$\begin{aligned}
&\frac{\partial y}{\partial t} - e^{-W}\Delta(e^W y) + (a+\mu)y + e^{-W} b\cdot\nabla(e^W y) = \mathbf{1}_{\mathcal{O}_0} v \text{ in } \mathcal{O}\times(0,T),\\
&y(\xi,0) = x(\xi) \quad \text{in } \mathcal{O},\\
&y = 0 \quad \text{on } \partial\mathcal{O}\times(0,T),
\end{aligned} \tag{3.134}$$

where $v = e^{-W}u$. Indeed, if y is a regular solution to (3.134) (for instance, absolutely continuous in t), which is progressively measurable in (t,ω) in the probability space $\{\Omega, \mathbb{P}, \mathcal{F}, \mathcal{F}_t\}$ and $y \in M^2_{\mathbb{P}}(0,T; H^1_0(\mathcal{O}) \cap H^2(\mathcal{O}))$, then, by *Itô's formula* in $(0,T)\times\Omega\times\mathcal{O}$, we have

$$dX = y d(e^W) + e^W \frac{\partial y}{\partial t}\, dt \qquad \text{in } (0,T)\times\mathcal{O}$$

and

$$d(e^W) = e^W dW + \frac{1}{2}\, e^W \sum_{k=1}^N \mu_k^2 dt = e^W \sum_{k=1}^N \mu_k d\beta_k + \frac{1}{2}\, e^W \sum_{k=1}^N \mu_k^2 dt + e^W \sum_{k=1}^N \frac{\partial \mu_k}{\partial t} \beta_k dt.$$

Then, we obtain for y the random equation (3.134), as claimed. On the other hand, any $\mathcal{F}_t$-adapted solution $t \to y(t)$ to equation (3.134) leads via transformation (3.133) to a solution X to (3.130) in the sense of the above definition. We write (3.134) as

$$\begin{aligned}
&\frac{\partial y}{\partial t} - \Delta y + Fy + G\cdot\nabla y = \mathbf{1}_{\mathcal{O}_0} v \text{ in } (0,T)\times\mathcal{O}\\
&y(0) = x \text{ in } \mathcal{O}, \quad y = 0 \text{ in } (0,T)\times\partial\mathcal{O},
\end{aligned} \tag{3.135}$$

where

$$F = a + \mu + b\cdot\nabla W - |\nabla W|^2 - \Delta W, \qquad G = -2\nabla W + b. \tag{3.136}$$

We may, equivalently, rewrite (3.135) as

$$y(t) = U(t,0)x + \int_0^t U(t,s)\mathbf{1}_{\mathcal{O}_0} v(s) ds,\ t\in[0,T], \tag{3.137}$$

where $U(t,s) : [0,T] \to L(L^2(\mathcal{O}), L^2(\mathcal{O}))$ is the evolution generated by the operator $z \to -\Delta z + Fz + G\cdot\nabla z$ with the domain $H^1_0(\mathcal{O})\cap H^2(\mathcal{O})$ (see Section 1.5).

It is easily seen that the operator function $t \to U(t,s)$ is $\mathcal{F}_t$-adapted on $[0,T]$. We note also that, if $\mu_k(t,\xi) \equiv \mu_k(t)$, then (3.135) is a deterministic linear parabolic system which, by virtue of Theorem 3.1, is exactly null controllable.

Now, we define the controllability target set S_T for system (3.130) as

$$S_T = \left\{ \eta = \int_0^T U(T,t)v(t)dt,\ v \in M_{\mathbb{P}}^2(0,T;L^2(\mathcal{O})), \right. \\ \left. \mathbb{E}[v(t)|\mathcal{F}_t] = 0,\ \forall t \in [0,T] \right\}. \tag{3.138}$$

(For each σ-algebra $\mathcal{G} \subset \mathcal{F}_0$, $\mathbb{E}(Y|\mathcal{G})$ denotes the conditional expectation of the random variable Y with respect to $\mathcal{G}$.) We note that S_T is a linear closed subspace of $L_T^2(\Omega;\mathcal{O})$.

Theorem 3.11 below is the main controllability result.

Theorem 3.11. *Let $T > 0$ and let $\mathcal{O}_0$ be an arbitrary open subset of $\mathcal{O}$. Then, under hypothesis* (i)*, for each $x \in L^2(\mathcal{O})$, there is $u^* \in M_{\mathbb{P}}^2(0,T;L^2(\mathcal{O}))$ such that*

$$\mathbb{E}\int_{\mathcal{Q}_0} |u^*|^2 dt\, d\xi \le C|x|_2^2, \tag{3.139}$$

$$X^{u^*}(T,\xi) \in S_T. \tag{3.140}$$

Here C is independent of x.

Theorem 3.11 amounts to saying that system (3.130) is exactly S_T controllable on an interval $[0,T]$ by an internal $M_{\mathbb{P}}^2$-controller u^* with support in $\mathcal{O}_0$.

Denote by $S_T^\perp$ the orthogonal of S_T in $L_T^2(\Omega;\mathcal{O})$. We consider, now, the boundary control stochastic system

$$\begin{aligned} & d_tX - \Delta X\,dt + aX\,dt + (b\cdot\nabla X)dt = \sum_{k=1}^N X\mu_k d\beta_k \text{ in } (0,T)\times\mathcal{O}, \\ & X = u \text{ on } (0,T)\times\partial\mathcal{O},\ \mathbb{P}\text{-a.s.} \\ & X(\xi,0) = x(\xi),\quad \forall \xi \in \mathcal{O}. \end{aligned} \tag{3.141}$$

We have

Theorem 3.12. *Under assumption* (i)*, for each $T > 0$ and $x \in L^2(\mathcal{O})$, system* (3.141) *is exactly S_T-controllable, that is, there are $u \in M_{\mathbb{P}}^2(0,T;L^2(\partial\mathcal{O}))$ and $X \in C_{\mathbb{P}}([0,T];L^2(\mathcal{O}))$, which satisfy* (3.141) *and*

$$X(T) \in S_T,\quad \mathbb{E}\int_\Sigma |u|^2 dt\, d\xi \le C|x|_2^2. \tag{3.142}$$

Theorem 3.12 follows by Theorem 3.11 by the following simple argument, already used in the controllability of deterministic systems (see Corollary 3.2).

Namely, one applies Theorem 3.11 on the domain $\widetilde{\mathcal{O}} \supset \mathcal{O}$ and for $\mathcal{O}_0 = \widetilde{\mathcal{O}} \setminus \overline{\mathcal{O}}$ to equation (3.130), where a, b, μ_k are replaced by $\widetilde{a} : \widetilde{\mathcal{O}}\times(0,T) \to \mathbb{R}$, $\widetilde{b} : \widetilde{\mathcal{O}}\times(0,T) \to \mathbb{R}^d$, $\widetilde{\mu}_k : (0,T) \to \mathbb{R}$ which are extensions of a, b, μ_k to

$\widetilde{\mathscr{O}}\times(0,T)$. The initial value $\widetilde{x}\in L^2(\mathscr{O},\mathscr{F}_0,\mathbb{P})$ is an extension of x to $\widetilde{\mathscr{O}}$. Then, by Theorem 3.11 there is $\widetilde{u}\in M^2_{\mathbb{P}}(0,T;L^2(\widetilde{\mathscr{O}}))$ and $\widetilde{X}\in C_{\mathbb{P}}([0,T],L^2(\mathscr{O}))\cap M^2_{\mathbb{P}}(0,T;H^1(\widetilde{\mathscr{O}}))$, which satisfy equation (3.130) (respectively, (3.131)) on $(0,T)\times\widetilde{\mathscr{O}}$ and such that $\widetilde{X}(T)\in S_T$.

If we take $u=\widetilde{X}|_{\partial\mathscr{O}}$, we see by the trace theorem that $u\in M^2_{\mathbb{P}}(0,T;L^2(\partial\mathscr{O}))$ and $X=\widetilde{X}|_{\widetilde{\mathscr{O}}}\in C_{\mathbb{P}}([0,T];L^2(\mathscr{O}))$ satisfy the conditions of Theorem 3.12. (Estimate (3.142) follows by (3.139) via trace theorem.) In this case, too, the solution $X=X^u$ satisfies (3.141).

Remark 3.8. A similar result remains true for the boundary control u on some segment $\Gamma\subset\partial\mathscr{O}$ of the boundary with nonempty interior.

By virtue of the duality relation between exact controllability and observability, by Theorem 3.11 we derive the following internal observability result for the backward stochastic equation (3.131).

Theorem 3.13. *Under assumption* (i)*, there is C independent of p such that*

$$|p(0)|_2\le C\mathbb{E}\left(\int_{\mathscr{O}_0}p^2\,dt\,d\xi\right)^{\frac12},\qquad Q_0=(0,T)\times\mathscr{O}_0,\tag{3.143}$$

for all the solutions $\{p\in C_{\mathbb{P}}([0,T];L^2(\mathscr{O}))\cap M^2_{\mathbb{P}}(0,T;H^1_0(\mathscr{O})),\ q_k\in M^2_{\mathbb{P}}(0,T;L^2(\mathscr{O})),\ k=1,\dots,N\}$ to the dual backward equation (3.131) *such that $p(T)\in S_T^{\perp}$.*

(We note that, since the filtration $\{\mathscr{F}_t\}_{t\ge0}$ is natural, that is, it is induced by the system $\{\beta_j\}_{j=1}^N$, the process $p(0)$ is deterministic.)

We have also the following boundary observability inequality.

Theorem 3.14. *There is C independent of p such that*

$$|p(0)|_2\le C\mathbb{E}\left(\int_{\Sigma}\left|\frac{\partial p}{\partial\nu}\right|^2dt\,d\xi\right)^{\frac12},\tag{3.144}$$

for all solutions $p\in M^2_{\mathbb{P}}(0,T;H^2(\mathscr{O}))$ to the dual backward stochastic equation (3.131) *with $p(T)\in S_T^{\perp}$.*

Proof of Theorem 3.11. By Theorem 3.2, for each $\omega\in\Omega$, equation (3.135) is exactly null controllable, that is, there are a controller $v\in C([0,T];L^2(\mathscr{O}))$ and $y^v\in C([0,T];\ L^2(\mathscr{O}))\cap L^2(0,T;H^1_0(\mathscr{O}))$ which satisfy (3.135) and such that $y(\xi,T)=0$, a.e. in $\mathscr{O}$. However, this does not complete the proof since *a priori* is not known that such a controller $v=v(\xi,t,\omega)$ is an $\mathscr{F}_t$-adapted process. (In fact, this happens if $W(t,\xi)\equiv\sum\limits_{k=1}^N\mu_k(t)\beta_k(t)$ because, in this case, equation (3.135) is deterministic.) We set

$$v^*(t)=\mathbb{E}[v(t)|\mathscr{F}_t],\ t\in[0,T],\tag{3.145}$$

and consider the solution y^* to equation (3.135) where v is replaced by v^*, that is,

$$\begin{aligned}
&\frac{\partial y^*}{\partial t} - \Delta y^* + F y^* + G \cdot \nabla y^* = \mathbf{1}_{\mathscr{O}_0} v^* \text{ in } (0,T)\times\mathscr{O},\\
&y^*(0) = x \text{ in } \mathscr{O},\\
&y^* = 0 \text{ on } (0,T)\times\partial\mathscr{O}.
\end{aligned} \tag{3.146}$$

We set $z = y^* - y^v$ and obtain

$$\begin{aligned}
&\frac{\partial z}{\partial t} - \Delta z + F z + G \cdot \nabla z = \mathbf{1}_{\mathscr{O}_0} (v^* - v) \text{ in } (0,T)\times\mathscr{O},\\
&z(0) = 0 \text{ in } \mathscr{O},\\
&z^* = 0 \text{ on } (0,T)\times\partial\mathscr{O}.
\end{aligned} \tag{3.147}$$

We have

$$z(T) = \int_0^T U(T,t)(\mathbf{1}_{\mathscr{O}_0}(v^*(t) - v(t)))dt = \int_0^T \zeta(t)dt, \ \mathbb{P}\text{-a.s.}$$

By (3.145), we see that

$$\mathbb{E}[\zeta(t)|\mathscr{F}_t] = 0, \ \forall t \in [0,T],$$

because $\mathbb{E}[v(t) - v^*(t)|\mathscr{F}_t] = 0$ and $t \to U(T,t)$ is $\mathscr{F}_t$-adapted. On the other hand, since $y^v(T) = 0$, this yields

$$y^*(T) = \int_0^T U(T,t)\zeta(t)dt.$$

Hence $y^*(T) \in S_T$, as claimed. Moreover, by (3.16), we have

$$\int_0^T \int_{\mathscr{O}} (u^*)^2 dt\, d\xi \le C_T |x|_2^2, \ \mathbb{P}\text{-a.s.},$$

where $C_T = C_T(\omega)$ is estimated by

$$C_T \le C_1 \sup_{t\in[0,T]} \sum_{k=1}^m |\beta_k(t)|^m,$$

for some natural m. This implies that $\mathbb{E}\int_0^T \int_{\mathscr{O}} |v^*(t)|^2 dt\, d\xi \le C_T^* |x|_2^2$ and, therefore, $\mathbb{E}\sup\{|y^*(t)|_2^2;\ t \in [0,T]\} \le C|x|_2^2$.

This completes the proof of Theorem 3.11. ■

Proof of Theorem 3.13. If p is the solution to the dual backward equation (3.131), with the final condition $p(T) = p_T$ in $\mathscr{O}$, then by Itô's formula we have

$$\mathbb{E}\int_{\mathscr{O}} X^u(\xi,T)p_T(\xi)d\xi - \int_{\mathscr{O}} x(\xi)p(\xi,0)d\xi = \mathbb{E}\left[\int_{Q_0} pu\,dt\,d\xi\right], \tag{3.148}$$

for all $u \in M^2_{\mathbb{P}}(0,T;L^2(\mathscr{O}))$. In (3.148), we take $u = u^*$ to obtain that, since the filtration $\{\mathscr{F}_t\}_{t\ge0}$ is the natural one, $p(0)$ is $\mathbb{P}$-a.s. deterministic),

$$\begin{aligned}\left|\int_{\mathscr{O}} x(\xi)p(\xi,0)d\xi\right| &\le \left(\mathbb{E}\int_{Q_0}(u^*)^2 dt\,d\xi\right)^{\frac12}\left(\mathbb{E}\int_{Q_0}(\mathbf{1}_{\mathscr{O}_0}p)^2 dt\,d\xi\right)^{\frac12}\\ &\le C|x|_2^2\left(\mathbb{E}\int_{Q_0} p^2\,dt\,d\xi\right)^{\frac12},\end{aligned}$$

because $p_T \in S_T^\perp$. This yields $|p(0)|_2 \le C\left(\mathbb{E}\int_{Q_0} p^2\,dt\,d\xi\right)^{\frac12}$, as claimed. ■

Proof of Theorem 3.14. Estimate (3.144) follows by (3.131) and (3.141) which imply via Itô's formula the equality $\int_{\mathscr{O}} x(\xi)p(\xi,0)d\xi = \mathbb{E}\int_\Sigma \frac{\partial p}{\partial \nu}\,u\,dt\,d\xi$, where u is a boundary controller provided by Theorem 3.13. ■

3.7 Approximate Controllability of Stochastic Parabolic Equation

We shall prove here an approximate controllability result for (3.130) obtained from exact controllability of the stochastic n-dimensional equation

$$dY + A(t)Y\,dt = B(t)u(t)dt + \sum_{k=1}^{N}\sigma_k(t,Y)d\beta_k, \tag{3.149}$$

where A and B are matrices.

Theorem 3.15. *Assume that hypothesis* (i) *holds. Then, for every* $\varepsilon > 0$, *there is an* $(\mathscr{F}_t)_{t\ge0}$*-adapted controller* $u_\varepsilon \in M^2_P(0,T;L^2(\mathscr{O}))$ *such that*

$$\mathbb{P}[\|X^{u_\varepsilon}(t)\|_{L^2(\mathscr{O})} \le \varepsilon,\ \forall t \ge T] \ge 1-\varepsilon. \tag{3.150}$$

Proof. For $n \in \mathbb{N}$, we set

$$\widetilde{X}^n = \sum_{i=1}^{n} X_i^n e_i,\ \widetilde{u}^n = \sum_{i=1}^{n} u_i^n e_i, \tag{3.151}$$

where $\{e_i\}_{i=1}^{\infty}$ is the orthonormal basis in $L^2(\mathscr{O})$ defined

$$-\Delta e_i = \lambda_i e_i \text{ in } \mathscr{O};\quad e_i = 0 \text{ on } \partial\mathscr{O}.$$

We approximate equation (3.130) by the finite dimensional control system

$$\begin{aligned} &dX^n + A_n X^n dt + D_n(t) X^n dt = \sum_{k=1}^{N} \sigma_k(t, X^n) d\beta_k + B_n u^n dt, \\ &X^n(0) = x_n = \{\langle x, e_j\rangle_2\}_{j=1}^n. \end{aligned} \tag{3.152}$$

Here

$$\begin{aligned} X^n &= \{X_i^n\}_{i=1}^n,\quad u^n = \{u_i^n\}_{i=1}^n, \\ A_n &= \operatorname{diag}\|\lambda_i\|_{i=1}^n, \\ B_n &= \left\| \int_{\mathscr{O}_0} e_i e_j d\xi \right\|_{i,j=1}^n, \\ D_n(t) &= \| \langle a(t) e_k + b(t)\cdot \nabla e_k, e_i\rangle_2 \|_{i,k=1}^n, \\ \sigma_k(t, X^n) &= \left\{ \sigma_k^i(t, X^n) = \sum_{j=1}^n X_j^n \langle \mu_k(t) e_j, e_i\rangle_2 \right\}_{i=1}^n \end{aligned}$$

and $\langle\cdot,\cdot\rangle_2$ is the scalar product of $L^2(\mathscr{O})$. We have

Lemma 3.6. *Let n arbitrary but fixed. For each $\varepsilon > 0$, there is an $(\mathscr{F}_t)_{t\ge 0}$-adapted controller $u_\varepsilon^n \in L^\infty((0,T)\times\Omega;\mathbb{R}^n)$ such that, if*

$$\tau_n = \inf\{t \ge 0;\ |X^n(t)|_n = 0\}, \tag{3.153}$$

then

$$\mathbb{P}[\tau_n \le T) \ge 1 - \varepsilon. \tag{3.154}$$

Proof. We consider in equation (3.152) the feedback controller

$$u^n = -\rho \operatorname{sign}(B_n X^n), \tag{3.155}$$

where $\operatorname{sign} v = \frac{v}{|v|_n}$ if $v \ne 0$, $\operatorname{sign} 0 = \{v \in \mathbb{R}^n;\ |v|_n \le 1\}$. By a standard device based on the approximation of the m-accretive mapping $y \to \operatorname{sign}(B_n y)$ by a Lipschitzian monotone mapping (Yosida approximation) (see, e.g., (1.7)), it follows that the corresponding closed loop system

$$\begin{aligned} &dX^n + A_n X^n dt + D_n(t) X^n dt = \sum_{k=1}^{N} \sigma_k(t, X^n) d\beta_k - \rho B_n \operatorname{sign}(B_n X^n) dt, \\ &X^n(0) = x_n, \end{aligned} \tag{3.156}$$

has a unique strong solution $X^n \in L^2(\Omega; C([0,T]; \mathbb{R}^n))$. (For a detailed proof, see [42].) As solution to (3.156), X^n as well as u^n are $(\mathscr{F}_t)_{t\geq 0}$-adapted $\mathbb{R}^n$-valued processes.

Let $\varphi_\varepsilon \in C^2(\mathbb{R}^+)$ be such that $\varphi_\varepsilon(r) = \frac{r}{\varepsilon}, \forall r \in [0,t]$, $\varphi'_\varepsilon(r) = 1+\varepsilon, \forall r \geq 2\varepsilon$, $|\varphi''_\varepsilon(r)| \leq \frac{c}{\varepsilon}, \forall r \in \mathbb{R}$. We set $\phi_\varepsilon(y) = \varphi_\varepsilon(|y|_n)$, $\forall y \in \mathbb{R}^n$, and note that

$$\begin{aligned} &\nabla\phi_\varepsilon(y) = \varphi'_\varepsilon(|y|_n)\text{sign}\, y,\ \nabla^2\phi_\varepsilon(y) = 0 \text{ for } |y|_n > 2\varepsilon,\\ &|\nabla^2\phi_\varepsilon(y)|_n \leq \frac{C}{\varepsilon},\ \ \forall y \in \mathbb{R}^n. \end{aligned} \tag{3.157}$$

We apply in (3.156) Itô's formula to the function $t \to \phi_\varepsilon(X^n(t))$. We obtain

$$\begin{aligned} &d\phi_\varepsilon(X^n(t)) + \langle A_nX^n(t) + D_n(t)X^n(t), \nabla\phi_\varepsilon(X^n(t))\rangle_n\, dt\\ &\qquad + \rho\,\langle B_n\,\text{sign}(B_nX^n(t)), \nabla\phi_\varepsilon(X^n(t))\rangle_n\, dt\\ &= \frac{1}{2}\sum_{i,j=1}^{n} \alpha^n_{ij}(t)(\nabla^2\phi_\varepsilon(X^n(t)))_{ij}dt + \sum_{j=1}^{N}\big\langle\sigma_j(t, X^n(t))d\beta_j(t), \nabla\phi_\varepsilon(X^n(t))\big\rangle_n, \end{aligned} \tag{3.158}$$

where $\alpha^n_{ij} = \sum\limits_{\ell=1}^{m} \sigma^i_\ell(t, X^n(t))\sigma^j_\ell(t, X^n(t))$.

Here $\langle\cdot,\cdot\rangle_n$ is the scalar product in $\mathbb{R}^n$, $|\cdot|_n$ is the Euclidean norm. We have

$$|\alpha^n_{ij}(t)| \leq L_n|X^n(t)|^2_n,\ \ i,j = 1,\ldots,n. \tag{3.159}$$

On the other hand, by the unique continuation property of the eigenfunctions e_j (see Theorem 2.2), we have $\det B_n \neq 0$, and so,

$$|B_nX|_n \geq \gamma_n|X|_n,\ \ \forall X \in \mathbb{R}^n. \tag{3.160}$$

Integrating (3.158) on (s,t), $0 < s < t < T$, and using (3.158)–(3.160), we get $\mathbb{P}$-a.s.,

$$\begin{aligned} &\phi_\varepsilon(X^n(t)) + \rho\int_s^t \big\langle\text{sign}(B_nX^n(\theta)), B_n\nabla\phi_\varepsilon(X^n(\theta))\big\rangle_n\, d\theta\\ &\quad \leq \phi_\varepsilon(X^n(s)) + C_n\int_s^t |X^n(\theta)|_n\|\nabla^2\phi_\varepsilon(X^n(\theta))\|_{L(\mathbb{R}^n,\mathbb{R}^n)}d\theta\\ &\quad + \int_s^t\sum_{j=1}^{N}\big\langle\sigma_j(\theta, X^n(\theta))d\beta_j(\theta), \nabla\phi_\varepsilon(X^n(\theta))\big\rangle_n. \end{aligned} \tag{3.161}$$

Taking into account (3.160) and that, for $\varepsilon \to 0$, $\nabla\phi_\varepsilon(y) \to \eta \in \text{sign}\, y, \forall y \in \mathbb{R}^n$, by letting $\varepsilon \to 0$ in (3.161), we get

$$
\begin{aligned}
&|X^n(t)|_n + \rho\gamma_n \int_s^t \mathbf{1}_{[X_n \neq 0]} d\theta \\
&\quad \leq |X^n(s)|_n + \int_s^t \sum_{j=1}^N \left\langle \sigma_j(\theta, X^n(\theta)) d\beta_j(\theta), \operatorname{sign} X^n(\theta) \right\rangle_n ,
\end{aligned}
$$

and so, by the stochastic Gronwall's lemma,

$$
\begin{aligned}
|X_n(t)|_n + \rho\gamma_n \int_s^t e^{C_n(t-\theta)} \mathbf{1}_{[X_n>0]} d\theta &\leq e^{C_n(t-s)} |X_n(s)|_n \\
+ \int_s^t e^{C_n(t-\theta)} \sum_{j=1}^N & \left\langle \sigma_j(\theta, X^n(\theta)) d\beta_j(\theta), X_n(\theta) \right\rangle_n ,
\end{aligned}
\tag{3.162}
$$

$0 \leq s < t < T$. In particular, it follows by (3.162) that $t \to e^{-C_n t}|X_n(t)|_n$ is a nonnegative supermartingale, that is,

$$
\mathbb{E}\left[|X_n(t)|_n e^{-C_n t} \mid \mathscr{F}_s \right] = |X_n(s)|_n e^{-C_n s}, \ \forall t \geq s.
$$

Hence, if τ_n is the stopping time (3.153), we have $|X_n(t)|_n e^{-C_n t} \leq |X_n(\tau_n)|_n e^{-C_n \tau_n}$ and so the extinction of $X_n(t)$ occurs at the time τ_n. Moreover, taking the expectation in (3.162) with $s = 0$ yields

$$
\mathbb{E}|X_n(t)|_n + \rho\gamma_n \int_0^t e^{-C_n\theta} \mathbb{P}[\tau_n > \theta] d\theta \leq |x_n|_n
$$

and, therefore,

$$
\frac{\rho\gamma_n}{C_n} (1 - e^{-C_n t}) \mathbb{P}[\tau_n > t] \leq |x_n|_n, \ \forall t \in [0, T].
$$

If we take $\rho = \rho_n = C_n(\gamma_n \varepsilon |x_n|_n)^{-1}(1 - e^{-C_n T})^{-1}$, we get (3.154), as claimed.

Proof of Theorem 3.15 (Continued). We fix $\varepsilon > 0$ and take n sufficiently large such that $\|x - x^n\|_{L^2(\mathscr{O})} \leq \varepsilon$, and consider the solution $X^{\widetilde{u}^n}$ to (3.130), where $\widetilde{u}^n = \sum_{i=1}^n u_i^n e_i$ and $\{u_i^n\}_{i=1}^n = u^n$ is given by (3.155). Subtracting the equation in $(\widetilde{X}_n, \widetilde{u}_n)$ given by (3.150), we get

$$
\begin{aligned}
& d(X^{\widetilde{u}_n} - \widetilde{X}^n) - \Delta(X^{\widetilde{u}_n} - \widetilde{X}^n) dt \\
& \quad + \left(a X^{\widetilde{u}_n} + b \cdot \nabla \widetilde{X}^{u_n} - \sum_{i=1}^n D_n(t) X_i^n e_i \right) dt = \sum_{k=1}^N \left(\mu_k X^{\widetilde{u}_n} - \sum_{i=1}^n \sigma_k(t, X^n) e_i \right) d\beta_k, \\
& (X^{\widetilde{u}_n} - \widetilde{X}^n)(0) = x - x_n.
\end{aligned}
$$

By applying Itô's formula, we get after some calculation involving Gronwall's lemma that

$$\mathbb{E}\left[\sup\left\{\|X^{\widetilde{u}_n}(t)-\widetilde{X}^n(t)\|^2_{L^2(\mathscr{O})};\ t\in[0,T]\right\}\right]\le C\|x-x^n\|_{L^2(\mathscr{O})}\le C\varepsilon,$$

where C is independent of n. Taking into account (3.154), we obtain

$$\mathbb{P}[\|X^{\widetilde{u}_n}(t)\|_{L^2(\mathscr{O})}\le(1+C)\varepsilon,\ \forall t\ge T]\ge 1-2\varepsilon.$$

Then, redefining ε, we see that the controller $\widetilde{u}_n=u_\varepsilon$ satisfies (3.150), as claimed.

3.8 Notes on Chapter 3

Theorem 3.1 was first established by A.V. Fursikov and O.Yu. Imanuvilov [76]. (See also O.Yu. Imanuvilov and Iamamoto [81].) However, for the heat equation the internal exact controllability was previously proved by G. Lebeau and L. Robbiano [85] via the Carleman inequality for elliptic equations. The boundary controllability of the heat equation with zero potential was first proved by D.L. Russell [108] (see also the survey [109]), via harmonic analysis. (We refer also to [96] and [69] for previous results in $1-D$.) Theorems 3.2, 3.3 and Corollaries 3.3, 3.4 closely follow the author's work [22]. In the special case of nonlinearities f satisfying the growth condition

$$\lim_{r\to\infty} f(|r|)/|r|(\log|r|+1)=0,$$

Theorem 3.2 was first proved by E. Fernandez-Cara [71]. For the more general nonlinearity of form (3.52), this theorem was proved by V. Barbu [20] in the case $1\le n\le 5$. Later, E. Fernandez-Cara and E. Zuazua [72] extended this result to the arbitrary n and to blowing up systems, that is, in the absence of sign condition (3.45). The proofs of Theorem 3.6 and Lemma 3.1 closely follow the treatment of [22]. The case of parabolic equations with discontinuous coefficients was studied in [67] and, more recently, by J. Le Rousseau and L. Robbiano [88] for more general systems. In [7], it is studied the exact controllability of the nonlinear heat (diffusion) processes with nonlinear flux of the form $q=-\nabla y+b(y)$, that is,

$$\begin{aligned}&y_t-\Delta y+\mathrm{div}(b(y))=mu\ \text{ in } Q,\\ &y=0 \text{ on } \Sigma;\qquad y(0)=y_0 \text{ on } \mathscr{O},\end{aligned}$$

where b satisfies the following growth conditions:

$$|b'(r)|\le\varphi(r)(1+\ \ln(|r|+1)),\ \ |b''(r)|\le\varphi_0(r)(1+\ \ln^\delta(|r|+1)(|r|+1)^{-1},$$

$\delta \in (0, \frac{1}{2})$ and $\varphi(r), \varphi_0(r) \to 0$ as $|r| \to \infty$. It is shown that if $d = 2, 3$, then the above system is exactly null controllable for all $y_0 \in H_0^1(\mathcal{O}) \cap H^2(\mathcal{O})$. For an extension of this result we refer to [68]. The exact controllability of the equation with impulse controllers was studied in [105]. We note also the works [110, 119] for sharper results concerning the exact and feedback controllability.

During the last decade or so, some important work was done toward the exact controllability of degenerate parabolic equations of the form $y_t - \text{div}(a(x)y_x) = mu$, and we refer to the works [58, 59] of P. Cannarsa et al. Such an equation is relevant in financial mathematics (the Black-Scholes equation).

As regards the approximate controllability, we cite the works [70, 72]. It should be said that Theorem 3.8 extends to systems with nonlinear flux boundary conditions of the form

$$y_t - \Delta y = mu \text{ in } Q, \qquad \frac{\partial}{\partial \nu} y + \beta(y) = 0 \text{ on } \partial\mathcal{O},$$

where β is a smooth monotonically nondecreasing function. In the case where β is Lipschitzian the controllability was studied by J.I. Diaz et al. [66]. The exact controllability of low diffusion porous media equations was studied in $1 - D$ in the works [50] and [63], but the general $3 - D$ case remains open. In [34], it is studied the controllability of the heat equation with memory, that is,

$$y_t(x, t) - \gamma_0 \Delta y(x, t) - \int_0^t a(t - s)\Delta y(x, s)ds = m(x)u(x, t) \text{ in } Q = \mathcal{O} \times (0, T),$$

where $\gamma_0 \geq 0$ and a is a completely monotone kernel. If $\gamma_0 > 0$, then this equation is of parabolic type and is approximately controllable ([34]). In this context, we mention also the works [102, 103] of L. Pandolfi. However, as recently shown by S. Guerrero and O. Imanuvilov [78], an equation of this type is not boundary controllable for all the initial data in $L^2(\mathcal{O})$. Moreover, for $\gamma_0 = 0$ the above equation is of hyperbolic type, and so the internal null controllability is not possible for all the intervals $(0, T)$.

Local controllability of semilinear parabolic equations was first studied in [76] via the implicit function theorem. However, Theorem 3.8 seems to be new in this context. Theorem 3.9 was taken from the author's work [23]. The local controllability results presented above were extended in several directions and we want to mention a few of them. For instance, in [8] it is studied the local internal controllability of the steady-state (equilibrium) solution to the *reaction–diffusion system*

$$y_t - k_1 \Delta y + ayz = f + mu \text{ in } Q, \qquad z_t - k_2 \Delta z + byz = g + mv \text{ in } Q.$$

Sharper results concerning one control controllability were also given in [73, 77]. It should be said, however, that the controllability of 2-D or 3-D parabolic systems by one control force is a difficult problem and it is only partially solved. For two and three heat equations coupled with cubic nonlinearities, this problem was solved by J.M. Coron, S. Guerrero, and L. Rossier [62] and in the work [61] by J.M. Coron and J.Ph. Guilleron by the so-called *return method* introduced by J.M. Coron in the controllability of hydrodynamic equations. (See also [60].) The results of Section 3.5 on exact controllability on unbounded domains are taken from author's work [31]. For related results, we mention the works [94, 95]. The exact controllability of the stochastic parabolic equation (3.130) remains so far an open problem, the results given in Sections 3.6 and 3.8 being only partial results in this direction. Theorem 3.11 was established in a slight different form in [43] along with a Carleman-type inequality for stochastic parabolic equations. For other results in this direction, we mention [90, 112] and [91]. (See, also, Zhang's survey [120].) As regards Theorem 3.14, it was established in [42] along with other results pertaining controllability of finite dimensional stochastic equations of the form (3.149), under Kalman's rank assumption.

Chapter 4
Internal Controllability of Parabolic Equations with Inputs in Coefficients

Very often, the input control arises in the coefficients of a parabolic equation and the exact controllability of initial data to origin or to a given stationary state is a delicate problem which cannot be treated by the linearization method developed in the previous chapter. However, in some situations, one can construct explicit feedback controllers which steer initial data to a given stationary state. In general, such a controller is nonlinear, eventually multivalued mapping, and its controllability effect is based on the property of solutions to certain nonlinear partial differential equations to have extinction in finite time. Here we shall study a few examples of this type.

4.1 The Exact Controllability via Self-Organized Criticality

Consider the control system

$$\begin{aligned}
&\frac{\partial y}{\partial t} - \Delta(uy) = 0 \text{ in } (0,T)\times\mathcal{O},\\
&y(x,0) = y_0(x),\ \ x\in\mathcal{O},\\
&y(x,t) = 0,\ \ \forall\,(t,x)\in(0,T)\times\partial\mathcal{O},
\end{aligned} \tag{4.1}$$

where $\mathcal{O}$ is a bounded and open set of $\mathbb{R}^d$, $d = 1, 2, 3$, with smooth boundary $\partial\mathcal{O}$, $u \in \mathcal{U}$, where the controller set $\mathcal{U}$ is defined by

$$\mathcal{U} = \{u \in L^\infty((0,T)\times\mathcal{O});\ u \geq 0,\ \text{a.e. in } (0,T)\times\mathcal{O}\}.$$

V. Barbu, *Controllability and Stabilization of Parabolic Equations*,
Progress in Nonlinear Differential Equations and Their Applications 90,
https://doi.org/10.1007/978-3-319-76666-9_4

As mentioned earlier, equation (4.1) models a large variety of nonisotropic diffusion processes (for instance, gas flow dynamics through a porous medium) and the control parameter u describes the physical properties of the medium.

The problem to be studied here is that of the structural controllability of the system via nonlinear feedback controller.

Problem 4.1. Let $y_1 \in L^\infty(\mathcal{O})$, $y_1 \geq 0$, a.e. in $\mathcal{O}$, be given. Find a feedback controller $u = \Phi(y, x)$ which steers y_0 in y_1 in time T.

In other words, one should find the map $y \to \Phi(y, x)$ such that the solution $y \in C([0, T]; L^2(\mathcal{O}))$ to the closed loop system

$$\begin{aligned} &\frac{\partial}{\partial t} y(x,t) - \Delta(\Phi(y(x,t),x)y(x,t)) = 0 \text{ in } (0,T) \times \mathcal{O}, \\ &y(x,0) = y_0(x) \text{ in } \mathcal{O}; \quad y = 0 \text{ on } (0,T) \times \partial\mathcal{O}, \end{aligned} \tag{4.2}$$

satisfies

$$y(T,x) = y_1(x), \text{ a.e. } x \in \mathcal{O}. \tag{4.3}$$

We shall see below that such a function $\Phi : \mathbb{R} \times \mathcal{O} \to \mathbb{R}$ can be taken of the form

$$\Phi(y,x) = \alpha H(y - y_1(x)), \ \forall\, y \in \mathbb{R},\ x \in \mathcal{O}, \tag{4.4}$$

where α is a positive constant and H is the Heaviside (multivalued) function

$$H(r) = \begin{cases} 1 & \text{if } r > 0, \\ [0,1] & \text{if } r = 0, \\ 0 & \text{if } r < 0. \end{cases} \tag{4.5}$$

More precisely, we shall prove that, for each $T > 0$, problem (4.2) is well posed in a weak sense to be made precise below and that, for α suitable chosen, the controllability condition (4.3) holds.

The feedback law $u = \Phi(y(x,t),x)$ has a special interest because it is a relay, bang-bang controller. On the other hand, the closed loop system (4.2) is physically motivated. In fact, the parabolic boundary value problem (4.2), (4.3) is a nonlinear diffusion equation with phase transition which models the *self-organized criticality* of diffusion processes. More precisely, for $d = 2$, (4.2) describes the dynamic of the sandpile model of self-organized criticality which we briefly present below (see, e.g., [14, 27, 56] for details). If $\rho = \rho(t, x_1, x_2)$ is the energy (or mass density) assigned to the site $x = (x_1, x_2)$ of a square lattice at time t, then, if $\rho(t, x_1, x_2)$ exceeds the critical value $\rho_c(x_1, x_2)$, the site becomes unstable and an avalanche develops according to the following evolution

$$\rho(t+1, x_1, x_2 \pm 1) = \rho(t, x_1, x_2 \pm 1) + \frac{1}{4}\,\rho(t, x_1, x_2),$$

$$\rho(t+1, x_1 \pm 1, x_2) = \rho(t, x_1 \pm 1, x_2) + \frac{1}{4}\,\rho(t, x_1, x_2).$$

If $\rho(t, j)$ is the energy of the j-th cell at time t, this dynamic can be described by the following discrete system of equations:

$$\begin{aligned}\rho(t+1,k) &= \rho(t,k) - \rho(t,k)H(\rho(t,k)-\rho_c(k))\\ &\quad +\frac{1}{4}\sum_{j\neq k}\rho(t,j)H(\rho(t,j)-\rho_c(j)).\end{aligned} \tag{4.6}$$

The continuous version of (4.6) is just equation (4.2) with Φ given by (4.4), $\alpha = 1$ and $y(t,x) = \rho(t,x)$, $\rho_c = y_1$. In cellular automata model presented above, the set $\{(t,x);\ \rho(t,x) = \rho_c(x)\}$ is the critical region while $\{(t,x);\ \rho(t,x) > \rho_c(x)\}$ is the supercritical region which is unstable and absorbed in finite time by the critical region (see [33]) and this is essentially the self-organized criticality mechanism. It should be emphasized that the diffusion dynamic induced by the avalanche process described above is not apparently generated by the Fick's first law, but seems to be the materialization of a more complex physical process.

By weak solution to problem (4.2) on $(0,T)\times\mathcal{O}$, we mean a function $y \in C([0,T];L^1(\mathcal{O}))$ which satisfies equation (4.2) in sense of distributions. An equivalent definition can be done in terms of *mild* solutions to nonlinear m-accretive Cauchy problem in the space $L^1(\mathcal{O})$.

Namely, setting $y - y_1 = z$, we reduce problem (4.2) to

$$\begin{aligned}&\frac{\partial z}{\partial t} - \alpha\Delta((z+y_1)H(z)) \ni 0 \text{ in } (0,T)\times\mathcal{O},\\ &z(x,0) = z_0(x) = y_0 - y_1 \text{ in } \mathcal{O},\\ &H(z)(z+y_1) = 0 \text{ on } (0,T)\times\partial\mathcal{O}.\end{aligned} \tag{4.7}$$

In the following, we shall assume that $z_0 = y_0 - y_1 \geq 0$, a.e. in $\mathcal{O}$. We may write problem (4.7) as an infinite-dimensional Cauchy problem in the space $L^1(\mathcal{O})$ (see Section 1.1)

$$\begin{aligned}&\frac{dz}{dt} + Az \ni 0 \text{ for } t \geq 0,\\ &z(0) = z_0,\end{aligned} \tag{4.8}$$

where the operator $A : D(A) \subset L^1(\mathcal{O}) \to L^1(\mathcal{O})$ is defined by

$$\begin{aligned}Az = \{&-\Delta\eta;\ \eta \in W_0^{1,1}(\mathcal{O}),\ \Delta\eta \in L^1(\mathcal{O}),\\ &\eta(x) \in \alpha(z(x)+y_1(x))H(z(x)),\ \text{a.e. } x \in \mathcal{O}\},\ \forall z \in D(A),\end{aligned} \tag{4.9}$$

where $D(A)$ consists of all $z \in L^1(\mathcal{O})$, $z \geq 0$, a.e. in $\mathcal{O}$, for which there is $\eta \in W_0^{1,1}(\mathcal{O})$ such that $\eta \in \alpha(z+y_1)H(z)$, a.e. in $\mathcal{O}$. We have

Lemma 4.1. *The operator A is accretive in $L^1(\mathcal{O})$ and*

$$\{z \in L^1(\mathcal{O});\ z \geq 0,\ \text{a.e. in } \mathcal{O}\} \subset \mathbb{R}(I+\lambda A),\ \forall\lambda > 0. \tag{4.10}$$

Proof. Taking into account that the multivalued operator $J : L^1(\mathscr{O}) \to L^\infty(\mathscr{O})$,

$$J(z)(x) = \{\zeta \in L^\infty(\mathscr{O});\ \zeta(x) \in \operatorname{sign} z(x) \text{ a.e. } x \in \mathscr{O}\},$$

where $\operatorname{sign} z = \frac{z}{|z|}$ for $z \neq 0$, $\operatorname{sign} 0 = [-1, 1]$, is the duality mapping of the space $L^1(\mathscr{O})$ (see, e.g., [26], p. 4), for the accretivity of A it suffices to show that

$$-\int_{\mathscr{O}} \Delta(H(z)(z + y_1) - H(\bar{z})(\bar{z} + y_1))\zeta dx \geq 0,\ \forall z_1, \bar{z}_1 \in D(A),$$

for some $\zeta \in L^\infty(\mathscr{O})$, $\zeta \in \operatorname{sign}(z - \bar{z})$, a.e. in $\mathscr{O}$. Taking into account that $\operatorname{sign}(z - \bar{z}) = \operatorname{sign}(H(z)(z+y_1) - H(\bar{z})(\bar{z}+y_1))$, a.e. in $\mathscr{O}$, the latter follows by the accretivity of the elliptic operator $A_0 y = -\alpha\Delta y$, with the domain $\{y \in W_0^{1,1}(\mathscr{O});\ \Delta y \in L^1(\mathscr{O})\} \subset W_0^{1,q}(\mathscr{O}) \subset L^2(\mathscr{O})$, $1 \leq q < \frac{d}{d-1}$. (See, e.g., [26], p. 110.) It remains to prove (4.10), for each $f \in L^1(\mathscr{O})$, $f \geq 0$, that is, the existence of a solution $z \in D(A)$ to the equation

$$z - \alpha\Delta((z + y_1)H(z)) \ni f \text{ in } \mathscr{D}'(\mathscr{O}). \tag{4.11}$$

Assume first that $f \in L^2(\mathscr{O})$ and approximate (4.11) by an appropriate convenient equation. Namely, we note that, for $\varepsilon > 0$, the operator $(A_0 + \varepsilon I)^{-1}$ is accretive and continuous in $L^1(\mathscr{O})$, while $Bz = \alpha(z + y_1)H(z)$ is m-accretive. Hence, $(A_0 + \varepsilon I)^{-1} + B$ is m-accretive (see [26], p. 104). This means that, for each $\varepsilon > 0$, there is $z_\varepsilon \in L^1(\mathscr{O})$ such that

$$\varepsilon z_\varepsilon + (A_0 + \varepsilon I)^{-1} z_\varepsilon + \alpha(z_\varepsilon + y_1)H(z_\varepsilon) \ni (A_0 + \varepsilon I)^{-1} f. \tag{4.12}$$

Equivalently,

$$(1 + \varepsilon^2)z_\varepsilon - \varepsilon\Delta z_\varepsilon - \alpha\Delta((z_\varepsilon + y_1)H(z_\varepsilon)) + \alpha\varepsilon(z_\varepsilon + y_1)H(z_\varepsilon) \ni f \text{ in } \mathscr{D}'(\mathscr{O}), \tag{4.13}$$

where $\varepsilon z_\varepsilon + \alpha(z_\varepsilon + y_1)H(z_\varepsilon) \subset D(A_0) \subset L^2(\mathscr{O})$. Hence $z_\varepsilon \in L^2(\mathscr{O})$ and, since $f \in L^2(\mathscr{O})$, by the elliptic regularity (see, e.g., [55], p. 297) it follows that $\varepsilon z_\varepsilon + \alpha(z_\varepsilon + y_1)H(z_\varepsilon) = g_\varepsilon \in H_0^1(\mathscr{O}) \cap H^2(\mathscr{O})$. Since $z_\varepsilon = (\varepsilon + \alpha)^{-1}(g_\varepsilon - \alpha y_1)^+$, this implies that $z_\varepsilon \in H_0^1(\mathscr{O})$. By (4.13), we see via the maximum principle that, for $f \geq 0$, we have $z_\varepsilon \geq 0$, a.e. in $\mathscr{O}$. Moreover, multiplying (4.13) by $\operatorname{sign} z_\varepsilon$ and integrating on $\mathscr{O}$, we get by the accretivity of A_0 that

$$\int_{\mathscr{O}} |z_\varepsilon| dx \leq \int_{\mathscr{O}} |f| dx,\ \forall \varepsilon > 0.$$

If we multiply (4.12) by $(z_\varepsilon + y_1)H(z_\varepsilon)$ and integrate on $\mathscr{O}$, we obtain via Green's formula

$$
\begin{aligned}
&(1+\varepsilon^2)\int_{\mathscr{O}} z_\varepsilon(z_\varepsilon+y_1)dx+\varepsilon\int_{\mathscr{O}}\nabla(z_\varepsilon)\cdot\nabla((z_\varepsilon+y_1)H(z_\varepsilon))dx\\
&+\alpha\int_{\mathscr{O}}|\nabla((z_\varepsilon+y_1)H(z_\varepsilon))|^2dx+\alpha\varepsilon\int_{\mathscr{O}}(z_\varepsilon+y_1)^2H(z_\varepsilon)dx\\
&=\int_{\mathscr{O}} f(z_\varepsilon+y_1)H(z_\varepsilon)dx.
\end{aligned}
$$

This yields

$$
\int_{\mathscr{O}}|z_\varepsilon|^2dx+\int_{\mathscr{O}}|\nabla((z_\varepsilon+y_1)H(z_\varepsilon))|^2dx\le C\int_{\mathscr{O}}(f^2+y_1^2)dx,\ \forall\,\varepsilon>0.
$$

Hence, on a subsequence $\{\varepsilon\}\to 0$ we have

$$
\begin{aligned}
(z_\varepsilon+y_1)H(z_\varepsilon)&\longrightarrow(z+y_1)H(z) &&\text{weakly in } H_0^1(\mathscr{O}) \text{ and strongly in } L^2(\mathscr{O}),\\
z_\varepsilon&\longrightarrow z &&\text{strongly in } L^2(\mathscr{O}),\\
\Delta(z_\varepsilon+y_1)H(z_\varepsilon)&\longrightarrow\Delta(z+y_1)H(z) &&\text{strongly in } H^{-1}(\mathscr{O}),
\end{aligned}
$$

where z is the solution to (4.11).

Now, if $f\in L^1(\mathscr{O})$, we choose a sequence $\{f_n\}\subset L^2(\mathscr{O})$ such that $f_n\to f$ strongly in $L^1(\mathscr{O})$. If $z_n\in L^2(\mathscr{O})$ are corresponding solutions to (4.11), we see once again by the accretivity of A_0 that

$$
\int_{\mathscr{O}}|z_n-z_m|dx\le\int_{\mathscr{O}}|f_n-f_m|dx,\ \forall\,n,m\in\mathbb{N}.
$$

If $z=\lim\limits_{n\to\infty}z_n$ in $L^1(\mathscr{O})$, it follows that z is the solution to (4.11), as claimed.

By Theorem 1.2, applied with $X=L^1(\mathscr{O})$ and $C=\{z\in L^1(\mathscr{O});\ z\ge 0, \text{ a.e. in } \mathscr{O}\}$, it follows that, for each $T\in(0,\infty)$ and $z_0\in\overline{D(A)}=\{z_0\in L^1(\mathscr{O}),\ z_0\ge 0$, a.e. in $\mathscr{O}\}$, equation (4.8) has a unique mild solution $z\in C([0,T];L^1(\mathscr{O}))$ given by Crandall & Liggett exponential formula (1.6), that is,

$$
z(t)=\lim_{n\to\infty}\left(I+\frac{t}{n}A\right)^{-n}z_0\quad\text{strongly in } L^1(\mathscr{O})
$$

uniformly on compact intervals. Equivalently,

$$
z(t)=\lim_{h\to 0}z^h(t)\text{ strongly in } L^1(\mathscr{O}),\tag{4.14}
$$

uniformly in t on every compact interval $[0,T]$, where z^h is the solution to the finite difference scheme

$$
z^h(t)=z_i^h,\ t\in[ih,(i+1)h),\ i=0,1,\ldots,N=\left[\frac{T}{h}\right]\tag{4.15}
$$

$$
z_{i+1}^h+hAz_{i+1}^h\ni z_i^h,\ i=0,1,\ldots,N,\ z_0^h=z_0.\tag{4.16}
$$

This existence result can be completed as follows.

Theorem 4.1. *Equation* (4.7) (*equivalently* (4.8)) *has a unique mild solution* $z \in C([0,T]; L^1(\mathscr{O}))$.

Moreover, if $z_0 \in L^2(\mathscr{O})$, *then* $z \in C([0,T]; L^1(\mathscr{O})) \cap W^{1,1}([0,T]; H^{-1}(\mathscr{O}))$, $z \geq 0$, *a.e.in* $(0,T) \times \mathscr{O}$ *and* z *is a strong solution to* (4.7), *that is,*

$$\begin{aligned} &\frac{dz}{dt}(t) - \Delta\eta(t) = 0, \;\; a.e.\ t \in (0,T), \\ &\eta(x,t) \in \alpha(z(x,t) + y_1(x))H(z(x,t)), \;\; a.e.\ (t,x) \in (0,T) \times \mathscr{O}, \end{aligned} \tag{4.17}$$

$$\eta \in L^2(0,T; H_0^1(\mathscr{O})). \tag{4.18}$$

If $z_0 \in L^\infty(\mathscr{O})$, *then* $z \in L^\infty((0,T) \times \mathscr{O})$.

Proof. If $z_0 \in L^2(\mathscr{O})$, then, by (4.15)–(4.16), we have

$$\begin{aligned} &z_{i+1}^h - h\Delta\eta_{i+1}^h = z_i^h \text{ in } \mathscr{O}, \;\; \eta_{i+1}^h \in H_0^1(\mathscr{O}), \\ &\eta_{i+1}^h \in \alpha(z_{i+1}^h + y_1)H(z_{i+1}^h), \text{ a.e. in } \mathscr{O}. \end{aligned} \tag{4.19}$$

We set

$$\widetilde{H}(x,r) = \alpha(r + y_1(x))H(r), \;\; \forall r \in \mathbb{R},$$

and $j : \mathscr{O} \times \mathbb{R} \to \mathbb{R}$ be defined by

$$j(x,r) = \begin{cases} \dfrac{\alpha}{2}r^2 + \alpha r y_1(x) & \text{if } r \geq 0, \\ 0 & \text{if } r < 0. \end{cases}$$

We have $\nabla_r j(x,r) = \widetilde{H}(x,r)$, $\forall x \in \mathscr{O}$, $r \in \mathbb{R}$, and so, by the convexity of j, we have

$$j(x, z_{i+1}^h) \leq j(x,z) + \widetilde{H}(x, z_{i+1}^h)(z_{i+1}^h - z), \;\; \forall z \in \mathbb{R}.$$

Then, (4.19) yields, via Green's formula, that

$$\begin{aligned} \int_{\mathscr{O}} j(x, z_{i+1}^h(x))dx + h\int_{\mathscr{O}} |\nabla\eta_{i+1}^h(x)|^2 dx &\leq \int_{\mathscr{O}} j(x, z_i^h(x))dx, \\ &\forall i = 0,1,\dots,N. \end{aligned} \tag{4.20}$$

Hence

$$\int_{\mathscr{O}} j(x, z_{i+1}^h)dx + h\sum_{k=0}^{i}\int_{\mathscr{O}} |\nabla\eta_k^h(x)|^2 dx \leq \int_{\mathscr{O}} j(x, z_0(x))dx, \;\; i = 0,1,\dots,N,$$

and, therefore,

$$\int_{\mathcal{O}} j(x, z^h(t,x))dx + \int_0^t \int_{\mathcal{O}} |\nabla\eta^h(x,s)|^2 ds\, dx \le \int_{\mathcal{O}} j(x, z_0(x))ds, \tag{4.21}$$

where $\eta^h(x,t) = \eta_i^h(x)$ for $t \in [ih, (i+1)h)$. Hence, for $h \to 0$,

$$\begin{aligned} z^h &\longrightarrow z \ \text{ strongly in } L^\infty(0,T;L^1(\mathcal{O})), \text{ weak star in } L^\infty(0,T;L^2(\mathcal{O})),\\ \eta^h &\longrightarrow \eta \ \text{ weakly in } L^2(0,T;H_0^1(\mathcal{O})). \end{aligned} \tag{4.22}$$

By (4.15), (4.16), and (4.19), we have

$$\int_0^T \left\langle z_h(t), \frac{d\varphi}{dt}(t)\right\rangle_2 dt + \int_0^T \int_{\mathcal{O}} \nabla\eta_h(t)\cdot\nabla\varphi(t)dx\,dt = 0, \ \forall\, h > 0,$$

for all $\varphi \in C^1([0,T];L^1(\mathcal{O})) \cap C([0,T];H_0^1(\mathcal{O}))$, $\varphi(0) = \varphi(T) = 0$. Then, by (4.22), we get, for $h \to 0$,

$$\int_0^T \left\langle z(t), \frac{d\psi}{dt}(t)\right\rangle_2 dt + \int_0^T \langle\eta(t), \Delta\varphi(t)\rangle_2\, dt = 0,$$

and this yields $z \in W^{1,2}([0,T];H^{-1}(\mathcal{O}))$, that is, $\frac{dz}{dt} \in L^2(0,T;H^{-1}(\mathcal{O}))$, and

$$\frac{dz}{dt} - \Delta\eta = 0 \text{ in } \mathcal{D}'(0,T;H^{-1}(\mathcal{O})). \tag{4.23}$$

Here $\langle\cdot,\cdot\rangle_2$ is the duality functional on $H_0^1(\mathcal{O}) \times H^{-1}(\mathcal{O})$ (respectively, the scalar product on $L^2(\mathcal{O})$), and $\langle\cdot,\cdot\rangle_{-1}$ is the scalar product of $H^{-1}(\mathcal{O})$, that is, $\langle u,v\rangle_{-1} = \big\langle(-\Delta)^{-1}u, v\big\rangle$, where $-\Delta$ is the Laplace operator with the domain $H_0^1(\mathcal{O}) \cap H^2(\mathcal{O})$, and $\mathcal{D}'(0,T;H^{-1}(\mathcal{O}))$ is the space of $H^{-1}(\mathcal{O})$-valued distributions on $(0,T)$. It remains to show that $\eta \in \widetilde{H}(x,z)$, a.e. in $(0,T)\times\mathcal{O}$. Because the map $z \to \widetilde{H}(x,z)$ is maximal monotone in $L^2((0,T)\times\mathcal{O}) \times L^2((0,T)\times\mathcal{O})$, to this end it suffices to show that

$$\limsup_{h\to 0} \int_0^T \int_{\mathcal{O}} z^h\eta^h dx\,dt \le \int_0^T \int_{\mathcal{O}} z\,\eta\, dx\,dt. \tag{4.24}$$

(See [26], p. 41.) By (4.19), we have

$$\limsup_{h\to 0} \int_0^T \int_{\mathcal{O}} z^h\eta^h dx\,dt + \frac{1}{2}\left(|z(T)|_{-1}^2 - |z_0|_{-1}^2\right) \le 0,$$

whereas (4.23) yields

$$\int_0^T \int_{\mathcal{O}} \eta z \, dx \, dt + \frac{1}{2} (|z(T)|_{-1}^2 - |z_0|_{-1}^2) = 0,$$

because $\frac{dz}{dt} \in L^2(0, T; H^{-1}(\mathcal{O}))$. (Here $|\cdot|_{-1}$ is, as usually, the norm of $H^{-1}(\mathcal{O})$.) This implies $\eta \in \widetilde{H}(x, z)$, a.e. in $(0, T) \times \mathcal{O}$, as claimed.

Hence, z is a strong solution to (4.7) in the sense of (4.17), (4.18). Moreover, by equation (4.23), it follows that the function $\varphi(t) = \int_{\mathcal{O}} j(x, z(x, t))dx$ is absolutely continuous on $[0, T]$ and (see [26], p. 158)

$$\frac{d}{dt} \varphi(t) = \left\langle \eta(t), \frac{dz}{dt}(t) \right\rangle_{-1}, \quad \text{a.e. } t \in (0, T).$$

Assume now that $z_0 \in L^\infty(\mathcal{O})$. To show that $z \in L^\infty((0, T) \times \mathcal{O})$, it suffices to check that

$$\|z^h\|_{L^\infty((0,T)\times\mathcal{O})} \leq C < \infty, \ \forall h > 0.$$

More precisely, we shall prove that the solution z^h to (4.15)–(4.16) satisfies the estimate

$$|z_{i+1}^h|_\infty \leq |z_i^h|_\infty \leq |z_0|_\infty, \ \forall i = 1, 2, \dots, N, \ h > 0. \tag{4.25}$$

It is readily seen that, for each i, $z_i^h = \lim\limits_{\lambda \to 0} (z_\lambda^h)_i$ strongly in $H(\mathcal{O})$, where $\lambda > 0$, and

$$\begin{aligned} &(z_\lambda^h)_{i+1} - h\Delta(\widetilde{H}(z_\lambda^h)_{i+1} + \lambda(z_\lambda^h)_{i+1}) \ni z_i^h \text{ in } \mathcal{O}, \\ &(z_\lambda^h)_{i+1} \in H_0^1(\mathcal{O}). \end{aligned} \tag{4.26}$$

Now, by (4.26), we see that, for $M_i = \operatorname{ess\,sup} z_i^h$, we have by Green's formula

$$\begin{aligned} &\int_{\mathcal{O}} (((z_\lambda^h)_{i+1} - M_i)^+)^2 dx + h \int_{\mathcal{O}} \nabla(\widetilde{H}(z_\lambda^h)_{i+1} + \lambda(z_\lambda^h)_{i+1}) \cdot \nabla((z_\lambda^h)_{i+1} - M_i)^+ dx \\ &\quad = \int_{\mathcal{O}} ((z_\lambda^h)_i - M_i)((z^h)_{i+1} - M_i)^+ dx \leq 0. \end{aligned}$$

This yields $\lim\limits_{\lambda \to 0} \lambda_i((z_\lambda^h)_{i+1} - M_i)^+ = 0$, a.e. in $\mathcal{O}$ and, therefore, $z_{i+1}^h \leq M_i$, a.e. in $\mathcal{O}$.

Similarly, we obtain, for $M_i^* = \operatorname{ess\,sup}\limits_{\mathcal{O}} \{-z_i^h\}$, that $(z_\lambda^h + M_i^*)^- = 0$, a.e. in $\mathcal{O}$. Hence, $z_{i+1}^h \geq -M_i^*$, a.e. in $\mathcal{O}$ and so (4.25) holds.

Finally, if $z_0 \geq 0$, a.e. in $\mathcal{O}$, we obtain in a similar way that $z^h_{i+1} \geq 0$, a.e. in $\mathcal{O}$, and so, by (4.22), we conclude that $z \geq 0$, a.e. in $(0, T) \times \mathcal{O}$.

This completes the proof.

Now, we come back to the controllability problem and assume that

$$y_0, y_1 \in L^\infty(\mathcal{O}), \ \ y_0 \geq y_1 \geq \rho > 0, \ \text{a.e. on } \mathcal{O}, \tag{4.27}$$

where ρ is a positive constant. We shall prove that, for α suitable chosen (dependent of T), the solution z to equation (4.7) satisfies

$$z(T, x) = 0, \quad \text{a.e. } x \in \mathcal{O}. \tag{4.28}$$

In other words, the feedback controller u defined by (4.4) steers in the time T the initial data y_0 in the state y_1.

To this end, let $p^* = \frac{2d}{d-2}$ if $d \geq 3$ and $p^* > 2$ if $d = 1, 2$. Then, by the Sobolev embedding theorem, $H^1_0(\mathcal{O}) \subset L^{p^*}(\mathcal{O})$ (see, e.g., [55], p. 278) and so

$$\gamma = \sup\{\|u\|_{L^{p^*}(\mathcal{O})};\ \|u\|_{H^1_0(\mathcal{O})} = 1\} < \infty. \tag{4.29}$$

We come back to (4.19) and note that, by (4.29), we have

$$\left(\int_{\mathcal{O}} |\eta^h_{i+1}|^{p^*} dx\right)^{\frac{1}{p^*}} \leq \gamma \left(\int_{\mathcal{O}} |\nabla \eta^h_{i+1}|^2 dx\right)^{\frac{1}{2}}, \ \forall i, \tag{4.30}$$

while, by (4.20) and (4.30), it follows that

$$\int_{\mathcal{O}} j(x, z^h_{i+1}(x))dx + h\gamma^{-2}\left(\int_{\mathcal{O}} |\eta^h_{i+1}(x)|^{p^*} dx\right)^{\frac{2}{p^*}} \leq \int_{\mathcal{O}} j(x, z^h_i(x))dx, \quad \forall i = 0, 1, \dots \tag{4.31}$$

Recalling that, by (4.19),

$$\eta^h_{i+1}(x) \in \widetilde{H}(z^h_{i+1}(x)) = \alpha(z^h_{i+1} + y_1(x))H(z^h_{i+1}(x)), \ \text{a.e. } x \in \mathcal{O},$$

we get

$$\left(\int_{\mathcal{O}} |\eta^h_{i+1}|^{p^*} dx\right)^{\frac{2}{p^*}} = \alpha^2 \left(\int_{[z^h_{i+1} > 0]} |z^h_{i+1}(x) + y_1(x)|^{p^*} dx\right)^{\frac{2}{p^*}}$$

and so, (4.31) yields

$$\begin{aligned}\int_{\mathcal{O}} j(x, z^h_{i+1}(x))dx + \alpha^2\gamma^{-2} \sum_{j=k}^{i+1} \left(\int_{[z^h_j(x) > 0]} |z^h_j(x) + y_1(x)|^{p^*} dx\right)^{\frac{2}{p^*}} \\ \leq \int_{\mathcal{O}} j(x, z^h_k(x))dx, \ \forall k \leq i+1.\end{aligned} \tag{4.32}$$

Then, by (4.27), we have

$$\int_{\mathcal{O}} j(x, z_{i+1}^h(x))dx + \alpha^2\gamma^{-2}\rho^2 \sum_{j=k}^{i+1} (m(x \in \mathcal{O}; z_{i+1}^h(x) > 0))^{\frac{2}{p^*}}$$
$$\leq \int_{\mathcal{O}} j(x, z_k^h(x))dx,$$

where m is the Lebesgue measure. Taking into account that by (4.25)

$$\int_{\mathcal{O}} z_{i+1}^h(x)dx \leq |z_0|_\infty m(x \in \mathcal{O}; z_{i+1}^h(x) > 0),$$

we get, for $1 \leq k \leq i+1$,

$$\int_{\mathcal{O}} j(x, z_{i+1}^h)dx + \alpha^2\gamma^{-2}\rho^2|z_0|_\infty^{-\frac{2}{p^*}} \sum_{j=k}^{i+1} \left(\int_{\mathcal{O}} z_j(x)dx\right)^{\frac{2}{p^*}}$$
$$\leq \int_{\mathcal{O}} j(x, z_k(x))dx.$$

Letting $h \to 0$, we obtain that

$$\int_{\mathcal{O}} j(x, z(x,t))dx + \alpha^2\gamma^{-2}|z_0|_\infty^{-\frac{2}{p^*}} \rho^2 \int_s^t \left(\int_{\mathcal{O}} (z(\tau, x))dx\right)^{\frac{2}{p^*}} d\tau$$
$$\leq \int_{\mathcal{O}} j(x, z(x,s))dx, \ 0 \leq s \leq t \leq T.$$

We set

$$\varphi(t) = \int_{\mathcal{O}} j(x, z(x,t))dx, \ t \in [0, T],$$

and obtain

$$\varphi(t) + \alpha^2\gamma^{-2}|z_0|_\infty^{-\frac{2}{p^*}} \rho^2 \int_s^t \left(\int_{\mathcal{O}} z(\tau, x)dx\right)^{\frac{2}{p^*}} d\tau \leq \varphi(s), 0 \leq s \leq t \leq T. \tag{4.33}$$

Since φ is absolutely continuous, we get by (4.33) the differential inequality

$$\begin{aligned} &\varphi'(t) + \alpha^2\gamma^{-2}|z_0|_\infty^{-\frac{2}{p^*}} \rho^2 \left(\int_{\mathcal{O}} z(x,t)dx\right)^{\frac{2}{p^*}} \leq 0, \ \text{a.e. } t > 0, \\ &\varphi(0) = \varphi_0 = \int_{\mathcal{O}} j(x, z_0(x))dx. \end{aligned} \tag{4.34}$$

Taking into account that $j(x, z) \geq \alpha\rho z, \forall z \geq 0$, we have that

$$\int_{\mathcal{O}} z(x,t)dx \geq \frac{1}{\alpha\rho} \varphi(t), \ \forall t > 0.$$

We set

$$\lambda_0 = \alpha^2\gamma^{-2}\rho^2|z_0|_\infty^{-\frac{2}{p^*}}.$$

Then, by (4.34), we get

$$\varphi'(t) + \lambda_0(\varphi(t))^{\frac{2}{p^*}} \le 0, \quad \text{a.e. } t > 0,$$

and, therefore, $\varphi(T) = 0$ if

$$\lambda_0 T = \frac{p^*}{p^*-2}\left(\int_{\mathcal{O}} j(x, z_0(x))dx\right)^{1-\frac{2}{p^*}} = \frac{p^*}{p^*-2}\left(\int_{\mathcal{O}} j(x, y_0(x)-y_1(x))dx\right)^{1-\frac{2}{p^*}}. \tag{4.35}$$

We have proved, therefore,

Theorem 4.2. *Let condition* (4.27) *hold. If α is chosen in such a way that* (4.35) *holds, then* (4.28) *follows.*

Now, coming back to equation (4.2), we have by Theorem 4.2

Theorem 4.3. *Let y_0 and y_1 be such that $y_0, y_1 \in L^\infty(\mathcal{O})$, $y_0 \ge y_1 \ge \rho > 0$, a.e. in $\mathcal{O}$. Then, the feedback controller*

$$u(t) = \alpha H(y(t) - y_1), \ t \in [0, T], \tag{4.36}$$

steers y_0 into y_1 in time T if $\alpha > 0$ is chosen as in equation (4.35).

In other words, equation (4.1) is exactly controllable in y_1 on $[0, T]$ by the feedback controller (4.36). It should be mentioned also that, if the solution y to the closed loop system (4.2) reaches y_1 in $t = T$, then, by the uniqueness of the solution z to (4.7), it follows that $y(t) = y_1$ for all $t \ge T$.

If we interpret y_1 as a critical state of diffusion system, the physical significance of Theorem 4.3 is that the supercritical region $\{x;\ y(x,t) > y_1(x)\}$ is absorbed in time T in the critical region $\{x;\ y(x,t) = y_1(x)\}$. In other words, the feedback controller (4.4) induces a nonlinear diffusion dynamic which steers the initial supercritical state y_0 in the critical state y_1.

Remark 4.1. The above exact controllability approach applies as well to the nonlinear parabolic equation

$$\begin{aligned}
&\frac{\partial y}{\partial t} - \Delta(uy) + f(y) = 0 \text{ in } (0, T) \times \mathcal{O},\\
&y(x, 0) = y_0(x) \text{ in } \mathcal{O},\\
&y = 0 \text{ on } (0, T) \times \partial\mathcal{O},
\end{aligned}$$

where $f : \mathbb{R} \to \mathbb{R}$ is a continuous and monotonically nondecreasing function such that $f(0) = 0$. We note that also in this case the operator $z \to -\alpha\Delta(z + y_1)H(z) +$

$f(z+y_1)$ is accretive in $L^1(\mathcal{O})$ and satisfies condition (4.10). Then the feedback law (4.4) has a similar exact controllability effect.

Note also that the above results extend to parabolic equations of the form

$$\frac{\partial y}{\partial t} - \sum_{i,j=1}^{d} a_{ij} \frac{\partial^2}{\partial x_i \partial x_j}(uy) = 0 \text{ in } (0,\infty)\times\mathcal{O},$$

where $a_{ij} = a_{ji}$ and $\sum_{i,j=1}^{d} a_{ij}\xi_i\xi_j \ge \gamma|\xi|_d^2$, $\forall \xi \in \mathbb{R}^d$, $\gamma > 0$. We omit the details.

4.2 Exact Controllability via Fast Diffusion Equation

We consider here system (4.1) with the feedback controller

$$u(t) = \alpha|y|^{m-1}\text{sign}\, y, \tag{4.37}$$

where $\alpha > 0$ and

$$m \in \left[\frac{d-2}{d+2}, 1\right),\ d \ge 2, \tag{4.38}$$

and $m = 0$ if $d = 1$.Here

$$\text{sign}\, y = \begin{cases} 1 & \text{for } y > 0, \\ [-1,1] & \text{for } y = 0, \\ -1 & \text{for } y < 0. \end{cases}$$

Theorem 4.4. *Let $y_0 \in H^{-1}(\mathcal{O}) \cap L^1(\mathcal{O})$. Then, equation* (4.1) *with the feedback controller* (4.37) *has a unique strong solution $y \in C([0,T]; L^1(\mathcal{O}) \cap H^{-1}(\mathcal{O}))$. Moreover, for*

$$\begin{aligned} \alpha &= \frac{1}{1-m}\|y_0\|_{H^{-1}(\mathcal{O})}^{1-m}(\gamma^{m+1}T)^{-1},\ d \ge 2, \\ \gamma &= \sup\{\|y\|_{H^{-1}(\mathcal{O})};\ \|u\|_{L^{m+1}(\mathcal{O})} = 1\}, \end{aligned} \tag{4.39}$$

one has

$$y(T) = 0. \tag{4.40}$$

If $d = 1$ and $m = 0$, then $y(x,T) = 0$ for $\alpha = \|y_0\|_{H^{-1}(\mathcal{O})}(\gamma T)^{-1}$.

Proof. By Theorem 5.3 in [26] (see, also, Theorem 1.3), it follows the existence of a strong solution $u \in C([0,T]; H^{-1}(\mathcal{O}) \cap L^1(\mathcal{O}))$ to the multivalued porous media equation

$$\begin{aligned}
&\frac{\partial y}{\partial t} - \frac{\alpha}{m}\Delta(|y|^m \eta) \ni 0 \ \text{ in } (0,T)\times\mathcal{O},\\
&y(x,0) = y_0(x), \ \ x \in \mathcal{O},\\
&y = 0 \ \text{ on } (0,T)\times\mathcal{O},\\
&\eta \in \operatorname{sign} y, \ \text{ a.e. in } (0,T)\times\mathcal{O}.
\end{aligned} \tag{4.41}$$

Moreover, one has

$$t^{\frac{1}{2}}\frac{d}{dt}y \in L^2(0,T;H^{-1}(\mathcal{O})),\ t^{\frac{1}{2}}(|y|^m\eta) \subset L^2(0,T;H_0^1(\mathcal{O})).$$

If we multiply scalarly in $H^{-1}(\mathcal{O})$ equation (4.41) by y, we get

$$\frac{1}{2}\frac{d}{dt}\|y(t)\|^2_{H^{-1}(\mathcal{O})} + \alpha\int_{\mathcal{O}} |y(x,t)|^{m+1}dx = 0,\ \forall t > 0,$$

and, since by the Sobolev embedding theorem,

$$\gamma\|y\|_{H^{-1}(\mathcal{O})} \le \|y\|_{L^{m+1}(\mathcal{O})} \text{ for } m \ge \frac{d-2}{d+2},$$

we obtain that

$$\frac{1}{2}\frac{d}{dt}\|y(t)\|^2_{H^{-1}(\mathcal{O})} + \alpha\gamma^{m+1}\|y(t)\|^{m+1}_{H^{-1}(\mathcal{O})} \le 0.$$

This yields

$$\frac{d}{dt}\|y(t)\|_{H^{-1}(\mathcal{O})} + \alpha\gamma^{m+1}\|y(t)\|^{m}_{H^{-1}(\mathcal{O})} \le 0,\ t > 0,$$

and so, by (4.39), (4.40) follows.

The case $d = 1$, $m = 0$ follows in a similar way from the equation

$$\frac{1}{2}\frac{d}{dt}\|y(t)\|^2_{-1} + \alpha\|y(t)\|_{L^1(\mathcal{O})} \le 0, \ \text{ a.e. } t > 0,$$

taking into account that $\|y\|_{L^1(\mathcal{O})} \ge \gamma\|y\|_{H^{-1}(\mathcal{O})}$. This completes the proof.

The case $m = 1$ in (4.37), which was ruled out by condition (4.38), leads to the feedback relay controller

$$u(t) = \alpha \operatorname{sign} y(t),\ t \in [0,T]. \tag{4.42}$$

The corresponding closed loop system is

$$
\begin{aligned}
&\frac{\partial y}{\partial t} - \alpha\Delta(\operatorname{sign} y) \ni 0 \text{ in } (0,T)\times\mathscr{O},\\
&y(x,0) = y_0(x), \qquad x\in\mathscr{O},\\
&y = 0 \qquad\qquad\qquad \text{on } (0,T)\times\partial\mathscr{O}.
\end{aligned}
\tag{4.43}
$$

As in the previous case, it follows that, for a suitable constant α, the controller (4.42) steers the initial data y_0 in origin in the time T. Namely, one has

Theorem 4.5. *Let $y_0 \in L^1(\mathscr{O}) \cap L^\infty(\mathscr{O})$ be such that $y_0 \geq 0$ and $T > 0$ be given. Then there is $\alpha = \alpha(y_0, T) > 0$ such that the solution $y \in C([0,T]; L^1(\mathscr{O}))$ to equation* (4.43) *satisfies $y(T) = 0$.*

Proof. Likewise all porous media equations of the form $\frac{\partial y}{\partial t} - \Delta\psi(y) \ni 0$ with the maximal monotone function $\psi : \mathbb{R} \to 2^{\mathbb{R}}$, problem (4.43) can be represented as the Cauchy problem in the space $L^1(\mathscr{O})$,

$$
\begin{aligned}
&\frac{dy}{dt} + Ay(t) \ni 0,\ t \geq 0,\\
&y(0) = y_0,
\end{aligned}
\tag{4.44}
$$

where the operator $A : D(A) \subset L^1(\mathscr{O}) \to L^1(\mathscr{O})$ is defined by

$$
Ay = \{-\alpha\Delta\eta;\ \eta\in W_0^{1,1}(\mathscr{O}),\ \eta(\xi)\in\operatorname{sign}(y(\xi)),\ \text{a.e. } \xi\in\mathscr{O}\},
\tag{4.45}
$$

and $D(A)$ consists of all $y \in L^1(\mathscr{O})$ for which such a section $\eta \in W_0^{1,1}(\mathscr{O})$ of sign $y(x)$ exists. We note that the multivalued operator A is m-accretive in $L^1(\mathscr{O})$ (see Theorem 1.3 in Chapter 1) and so, for each $y_0 \in \overline{D(A)} = L^1(\mathscr{O})$ and $T > 0$, the Cauchy problem (4.44) has a unique mild solution $y \in C([0,T]; L^1(\mathscr{O}))$ given by the exponential formula

$$
y(t) = \lim_{n\to\infty}\left(I + \frac{t}{n}A\right)^{-n} y_0 \text{ strongly in } L^1(\mathscr{O}),\ t\geq 0.
\tag{4.46}
$$

Equivalently,

$$
y(t) = \lim_{h\to 0} y_h(t) \ \text{ uniformly in } t,
\tag{4.47}
$$

where $y_h : [0,T] \to L^1(\mathscr{O})$ is the step function

$$
y_h(t) = y_i^h \ \text{ for } t\in[ih,(i+1)h),
\tag{4.48}
$$

$$
\begin{aligned}
&y_{i+1}^h + hAy_{i+1} \ni y_i^h,\ \ i = 0,1,2,\dots,N,\ \ N\in\left[\tfrac{T}{h}\right],\\
&y_0^h = y_0.
\end{aligned}
\tag{4.49}
$$

Moreover, we have

Proposition 4.1. *Assume that $y_0 \in L^2(\mathcal{O})$. Then (4.43) (equivalently, (4.44)) has a unique solution y satisfying*

$$y(t) \in C([0,T]; L^1(\mathcal{O}) \cap H^{-1}(\mathcal{O})) \cap W^{1,1}([0,T]; H^{-1}(\mathcal{O})),$$

$$\begin{aligned} &\frac{dy}{dt}(t) - \alpha\Delta\eta(t) = 0, \quad a.e.\ t \in [0,T], \\ &\eta(\xi,t) \in \operatorname{sign}(y(\xi,t)), \quad a.e.\ (\xi,t) \in (0,T) \times \mathcal{O}, \end{aligned} \tag{4.50}$$

where $\eta \in L^1(0,T; H_0^1(\mathcal{O}))$. Moreover, $t \to \int_{\mathcal{O}} |y(\xi,t)|d\xi$ is absolutely continuous. If $y_0 \in L^\infty(\mathcal{O})$, then $y \in L^\infty((0,T) \times \mathcal{O})$. If $y_0 \geq 0$, a.e. in $\mathcal{O}$, then $y \geq 0$, a.e. in $(0,T) \times \mathcal{O}$.

Proof. The proof is much similar to that of Theorem 4.1 and so it will be outlined only. Namely, we fix $h > 0$. By (4.46) and (4.48), we see that

$$y_{i+1}^h - h\alpha\Delta\eta_{i+1}^h = y_i^h \ \text{ in } \mathcal{O},\ i = 0, 1, \dots, N, \tag{4.51}$$

where $\eta_{i+1}^h \in \operatorname{sign} y_{i+1}^h$, a.e. in $\mathcal{O}$. This yields

$$\int_{\mathcal{O}} j(y_{i+1}^h(\xi))d\xi + h\alpha \int_{\mathcal{O}} |\nabla(\eta_i^h(\xi))|^2 d\xi \leq \int_{\mathcal{O}} j(y_i^h(\xi))d\xi,\ i = 0, 1, \dots, N,$$

where $j(r) = |r|$. Hence

$$\begin{aligned} \int_{\mathcal{O}} |y_h(\xi,t)|d\xi + \alpha \int_0^t \int_{\mathcal{O}} |\nabla\eta^h(s,\xi)|^2 ds\, d\xi \leq \int_{\mathcal{O}} |y_0(\xi)|d\xi, \\ t \in (0,T),\ \forall h > 0, \end{aligned} \tag{4.52}$$

where $\eta^h(\xi,t) = \eta_i^h(\xi)$ for $t \in (ih, (i+1)h),\ h \in \mathcal{O}$. We also have, by (4.49), that

$$\int_{\mathcal{O}} |y_{i+1}^h|^2 d\xi \leq \int_{\mathcal{O}} |y_i^h|^2 d\xi,\ \forall i,\ h > 0,$$

and, therefore,

$$\int_{\mathcal{O}} |y_i^h(t)|^2 d\xi \leq \int_{\mathcal{O}} |y_0|^2 d\xi,\ \forall t \in (0,T),\ h > 0. \tag{4.53}$$

Hence, we have by (4.46), (4.52), and (4.53) that, for $h \to 0$,

$$\begin{aligned} y_h \to y \text{ strongly on } &L^\infty(0,T; H^{-1}(\mathcal{O})), \\ &\text{weak-star in } L^\infty(0,T; L^2(\mathcal{O})), \\ \eta^h \to \eta \text{ weakly in } &L^2(0,T; H_0^1(\mathcal{O})). \end{aligned} \tag{4.54}$$

For any test function $\varphi \in C^1([0,T]; H^{-1}(\mathcal{O})) \cap C([0,T]; H_0^1(\mathcal{O}))$, $\varphi(0) = \varphi(T) = 0$, we have by (4.51) that

$$\int_h^T \left\langle y_h(t), \frac{d\varphi}{dt}(t) \right\rangle_{-1} dt + \alpha \int_0^T \int_{\mathcal{O}} \eta^h(t)\varphi(t) d\xi\, dt = 0, \ \forall h > 0,$$

and this yields $y \in W^{1,2}([0,T]; H^{-1}(\mathcal{O}))$, that is, $\frac{dy}{dt} \in L^2(0,T; H^{-1}(\mathcal{O}))$, and

$$\frac{dy}{dt} - \alpha \Delta \eta = 0 \ \text{ in } \mathcal{D}'(0,T). \tag{4.55}$$

It remains to show that $\eta \in \text{sign}(y)$, a.e. in $(0,T) \times \mathcal{O}$. This will be shown as above by checking that

$$\limsup_{h\to 0} \int_0^T \int_{\mathcal{O}} y_h \eta^h \, d\xi\, dt \le \int_0^T \int_{\mathcal{O}} y\, \eta\, d\xi\, dt. \tag{4.56}$$

In fact, by (4.51), (4.54), we have

$$\alpha \limsup_{h\to 0} \int_0^T \int_{\mathcal{O}} y_h \eta^h d\xi\, dt + \frac{1}{2}\left(|y(T)|_{-1}^2 - |y_0|_{-1}^2\right) \le 0,$$

while (4.55) yields

$$\alpha \int_0^T \int_{\mathcal{O}} \eta y\, d\xi\, dt + \frac{1}{2}\left(|y(T)|_{-1}^2 - |y_0|_{-1}^2\right) = 0,$$

because $\frac{dy}{dt} \in L^2(0,T; H^{-1}(\mathcal{O}))$. This implies (4.56), as claimed.

Hence, y is strong solution to (4.44) in sense of (4.50). Moreover, by (4.50) it follows that the function $\varphi(t) = \int_{\mathcal{O}} |y(\xi,t)| d\xi$ is absolutely continuous on $[0,T]$ and, as seen above,

$$\frac{d}{dt}\varphi(t) = \left\langle \eta(t), \frac{dy}{dt}(t) \right\rangle_{-1}, \quad \text{a.e. } t \in (0,T).$$

Assume now that $y_0 \in L^\infty(\mathcal{O})$. We have

$$\|y_h\|_{L^\infty((0,T)\times\mathcal{O})} \le C < \infty, \quad \forall h > 0, \tag{4.57}$$

as a consequence of the estimate

$$|y_{i+1}^h|_\infty \le |y_i|_\infty^h \le |y_0|_\infty, \ \ \forall i = 1, 2, \dots, N, \ h > 0. \tag{4.58}$$

Indeed, as in the previous case, we have, for each i, $y_{i+1}^h = \lim_{\lambda\to\infty} y_\lambda^h$ strongly in $H^1(\mathscr{O})$, where $\lambda > 0$, and

$$\begin{aligned} & y_\lambda^h - \varepsilon\alpha\Delta(\mathrm{sign}(y_\lambda^h) + \lambda y_\lambda^h) \ni y_i^h \text{ in } \mathscr{O}, \\ & y_\lambda^h \in H_0^1(\mathscr{O}). \end{aligned} \tag{4.59}$$

Now, by (4.59), we see that, for $M_i = \text{ess sup}_{\mathscr{O}}\, y_i^\varepsilon$, we have via Green's formula

$$\begin{aligned} \int_{\mathscr{O}} ((y_\lambda^h - M_i)^+)^2 d\xi + h\alpha \int_{\mathscr{O}} \nabla\widetilde{H}(y_\lambda^h) + \lambda y_\lambda^h) \cdot \nabla(y_\lambda^h - M_i)^+ d\xi \\ = \int_{\mathscr{O}} (y_i^h - M_i)(y_\lambda^h - M_i)^+ d\xi \le 0. \end{aligned}$$

This yields $(y_\lambda^h - M_i)^+ = 0$, a.e. in $\mathscr{O}$ and, therefore, $y_{i+1}^h \le M_i$, a.e. in $\mathscr{O}$. Similarly, we obtain, for $M_i^* = \text{ess sup}_{\mathscr{O}}\{-y_i^h\}$ that $(y_\lambda^h + M_i^*)^- = 0$, a.e. in $\mathscr{O}$. Hence, $y_{i+1}^h \ge -M_i^*$, a.e. in $\mathscr{O}$, and so (4.58) holds.

Finally, if $y_0 \ge 0$, a.e. in $\mathscr{O}$, we obtain in a similar way that $y_{i+1}^h \ge 0$, a.e. in $\mathscr{O}$, and so by (4.54) we conclude that $y \ge 0$, a.e. in $(0, T) \times \mathscr{O}$. This completes the proof of Proposition 4.1.

To complete the proof of Theorem 4.5, we shall prove that

$$y(x, t) \equiv 0, \text{ a.e. } x \in \mathscr{O}, \ \forall t \in T^*, \tag{4.60}$$

where

$$T^* = \frac{p^*}{p^* - \alpha}\, \gamma^2 |y_0|_\infty^{\frac{2}{p^*}} \left(\int_{\mathscr{O}} y_0 d\xi\right)^{t - \frac{2}{p^*}}.$$

Here, $p^* = \frac{2d}{d-2}$ for $d \ge 3$, $p^* > 2$ for $d = 1, 2$, and $\gamma = \sup\left\{|u|_{p^*} \|u\|_{H_0^1(\mathscr{O})}^{-1}\right\}$.

We start with the difference scheme (4.49) (or (4.51)), that is,

$$\begin{aligned} & y_{i+1}^h - h\alpha\Delta\eta_{i+1}^h = y_i^h \text{ in } \mathscr{O}, \ i = 0, 1, \ldots, N, \ N = \left[\tfrac{T}{N}\right], \\ & \eta_{i+1}^h \in H_0^1(\mathscr{O}), \ \eta_{i+1}^h \in \mathrm{sign}(y_{i+1}^h), \text{ a.e. in } \mathscr{O}. \end{aligned} \tag{4.61}$$

Taking into account that $y_i^h \ge 0$, a.e. in $\mathscr{O}$, and $\langle \eta_{i+1}^h, y_{i+1}^h - y_i^h\rangle \ge \int_{\mathscr{O}} (y_{i+1}^h - y_i^h) d\xi$, we obtain as above that

$$\int_{\mathscr{O}} y_{i+1}^h(\xi) d\xi + h\alpha \int_{\mathscr{O}} |\nabla\eta_{i+1}^h(\xi)|^2 d\xi \le \int_{\mathscr{O}} y_i^h(\xi) d\xi, \ \forall i = 0, 1. \tag{4.62}$$

On the other hand, by the Sobolev embedding theorem, we have (see (4.30))

$$\left(\int_{\mathscr{O}} |\eta_{i+1}^h|^{p^*} d\xi\right)^{\frac{1}{p^*}} \leq \gamma \left(\int_{\mathscr{O}} |\nabla \eta_{i+1}^h|^2 d\xi\right)^{\frac{p^*}{2}}, \ \forall i = 0, 1, \ldots \tag{4.63}$$

Then, by (4.62) we obtain

$$\int_{\mathscr{O}} y_{i+1}^h d\xi + h\alpha\gamma^{-2}\left(\int_{\mathscr{O}} |\eta_{i+1}^h|^{p^*} d\xi\right)^{\frac{2}{p^*}} \leq \int_{\mathscr{O}} y_i^h d\xi, \ \forall i = 0, 1, \ldots \tag{4.64}$$

and recalling that $\eta_{i+1}^h \in \text{sign}(y_{i+1}^h)$, a.e. in $\mathscr{O}$, we have

$$\left(\int_{\mathscr{O}} |\eta_{i+1}^h|^{p^*} d\xi\right)^{\frac{2}{p^*}} \geq (m[\xi \in \mathscr{O};\ y_{i+1}^h(\xi) > 0])^{\frac{2}{p^*}}, \tag{4.65}$$

where m stands for the Lebesgue measure. On the other hand, we have by (4.58)

$$0 \leq y_{i+1}^h(\xi) \leq |y_0|_\infty, \ \text{ a.e. } \xi \in \mathscr{O},\ i = 1, \ldots,$$

and so, we obtain that

$$|y_0|_\infty m[\xi \in \mathscr{O};\ y_{i+1}^h(\xi) > 0] \geq \int_{\mathscr{O}} y_{i+1}^h(\xi) d\xi, \ \forall i. \tag{4.66}$$

By (4.62)–(4.63), we have that

$$\int_{\mathscr{O}} y_{i+1}^h(\xi) d\xi + h\alpha\gamma^{-2} \sum_{j=k}^{i+1} \left(\int_{\mathscr{O}} |\eta_{j+1}^h|^{p^*} d\xi\right)^{\frac{2}{p^*}} \leq \int_{\mathscr{O}} y_k^h(\xi) d\xi,$$

for all $i > k$.

Summing up and letting ε tend to zero and keeping in mind (4.48), we obtain by virtue of (4.44) that, for all $0 \leq s < t < \infty$, we have

$$\int_{\mathscr{O}} y(\xi, t)\xi + \alpha\gamma^{-2}|y_0|_\infty^{-\frac{2}{p^*}} \int_s^t \left(\int_{\mathscr{O}} y(\tau, \xi) d\xi\right)^{\frac{2}{p^*}} d\tau \leq \int_{\mathscr{O}} y(s, \xi) d\xi. \tag{4.67}$$

We set

$$\varphi(t) = \int_{\mathscr{O}} y(\xi, t) d\xi, \ \ t \geq 0,$$

and rewrite (4.67) as

$$\varphi(t) + \alpha\gamma^{-2}|y_0|_\infty^{-\frac{2}{p^*}} \int_s^t (\varphi(\tau))^{\frac{2}{p^*}} d\tau \leq \varphi(s), \ 0 \leq s < \infty.$$

Recalling that φ is absolutely continuous on $[0, T]$ and $p^* > 2$, we obtain that

$$\frac{p^*}{p^*-2}\frac{d}{dt}(\varphi(t))^{1-\frac{2}{p^*}} + \alpha\gamma^{-2}|y_0|_\infty^{-\frac{2}{p^*}} \le 0, \quad \text{a.e. } t > 0,$$

and, therefore,

$$\varphi(t) = 0 \text{ for } t > T^* = \frac{p^*}{(p^*-2)}\alpha^{-1}\gamma^2|y_0|_\infty^{\frac{2}{p^*}}\left(\int_{\mathscr{O}} y_0 d\xi\right)^{1-\frac{2}{p^*}}.$$

Hence, $y(\xi, t) = 0$, a.e. $\xi \in \mathscr{O}$ for $t \ge T^*$, and taking

$$\alpha = \frac{p^*}{(p^*-2)T}\gamma^2|y_0|_\infty^{\frac{2}{p^*}}\left(\int_{\mathscr{O}} y_0 dx\right)^{1-\frac{2}{p^*}}, \tag{4.68}$$

we get $T = T^*$, as claimed.

If the control parameter u is taken the diffusivity of medium, then by the Fick diffusion law we have for the concentration y the equation

$$\begin{aligned} &\tfrac{\partial y}{\partial t} - \operatorname{div}(u\nabla y) = 0 \text{ in } (0, T) \times \mathscr{O}, \\ &y(x, 0) = y_0(x). \end{aligned} \tag{4.69}$$

The feedback controller $u(t) = \frac{\alpha}{m}|y(t)|^{m-1}$, $t \ge 0$, inserted in (4.69) yields a closed loop system of the form (4.41), that is,

$$\begin{aligned} &\frac{\partial y}{\partial t} - \frac{\alpha}{m}\Delta(|y|^m \operatorname{sign} y) = 0 \text{ in } (0, T) \times \mathscr{O}, \\ &y(x, 0) = y_0(x) \text{ in } \mathscr{O}, \\ &y(x, t) = 0, \quad \text{on } (0, T) \times \partial\mathscr{O}, \end{aligned} \tag{4.70}$$

while, for $m = 0$, one gets a system of the form (4.43). Then, by Theorem 4.4 and Theorem 4.7, it follows that $y(T) = 0$ if $m \in \left[\frac{d-2}{d+2}, 1\right)$ and a condition of the form (4.39) (respectively, (4.68)) holds. Moreover, as easily seen, if $y_0 \ge 0$, a.e. in $\mathscr{O}$, then $y \ge 0$ on $(0, T) \times \mathscr{O}$.

Remark 4.2. It is easily seen that the linear system

$$\begin{aligned} &\frac{\partial y}{\partial t} - \Delta y + u = 0 \text{ in } (0, \infty) \times \mathscr{O}, \\ &y(x, 0) = y_0(x), \ x \in \mathscr{O}, \\ &y = 0 \text{ on } (0, \infty) \times \partial\mathscr{O}, \end{aligned}$$

can be controlled to y_1 in a finite time via the feedback controller $u = -\alpha H(y-y_1)$ if $y_1 \geq \rho > 0$ in $\mathcal{O}$. However, it should be mentioned that in this case the physical significance of controller (4.4) (as well as of (4.37)) is different since it intervenes within the flow of the diffusion system.

4.3 Exact Controllability via Total Variation Flow

For the linear parabolic equation (4.69), it turns out that a controller u of the form

$$u(x,t) = \alpha|\nabla y(x,t)|_d^{-1}, \ t \geq 0, \ x \in \mathcal{O}, \tag{4.71}$$

steers y_0 into origin in time T for a suitable positive constant $\alpha = \alpha(y_0, T)$. Indeed, substituting (4.71) in (4.69), we get the nonlinear parabolic equation

$$\begin{aligned}
&\frac{\partial y}{\partial t} - \alpha \operatorname{div}\left(\frac{\nabla y}{|\nabla y|_d}\right) = 0, \ t \geq 0, \ x \in \mathcal{O},\\
&y(x,0) = y_0(x), \qquad\qquad x \in \mathcal{O},\\
&y(x,t) = 0, \qquad\qquad\quad x \in \partial\mathcal{O}.
\end{aligned} \tag{4.72}$$

It should be said, however, that problem (4.72) (as well as controller (4.71)) is not well posed in the Sobolev space $W_0^{1,1}(\mathcal{O})$, but in the space $BV(\mathcal{O})$ of functions with bounded variations on $\mathcal{O}$. Namely, it has a natural formulation as the infinite dimensional Cauchy problem

$$\begin{aligned}
&\frac{dy}{dt} + Ay \ni 0 \text{ on } (0,T),\\
&y(0) = y_0,
\end{aligned} \tag{4.73}$$

where A is the subdifferential $\partial\varphi : L^2(\mathcal{O}) \to L^2(\mathcal{O})$ of the convex function

$$\varphi(u) = \alpha|Du| + \alpha \int_{\partial\mathcal{O}} \gamma_0(u)\, d\mathcal{H}^{d-1}, \ u \in BV(\mathcal{O}),$$

where $|Du|$ is the total variation of $u \in BV(\mathcal{O})$, that is,

$$|Du| = \sup\left\{\int_{\mathcal{O}} u \operatorname{div}\varphi\, dx;\ \varphi \in C_0^\infty(\mathcal{O}),\ |\varphi|_\infty \leq 1\right\}.$$

$\gamma_0(u)$ is the trace of u to $\partial\mathcal{O}$ and $d\mathcal{H}^{d-1}$ is the $d-1$ dimensional Hausdorff measure on $\partial\mathcal{O}$ (see, e.g., [4, 41]).

By Theorem 1.1, the Cauchy problem (4.73) in the space $X = L^2(\mathcal{O})$ has, for each $y_0 \in L^2(\mathcal{O})$, a unique strong solution $y \in C([0,T]; L^2(\mathcal{O})) \cap$

$L^1(0,T;BV(\mathcal{O}))$ and $y \in L^\infty(0,T;BV(\mathcal{O}))$ if $y_0 \in BV(\mathcal{O})$. As a matter of fact, $t \to y(t)$ is $L^2(\mathcal{O})$-absolutely continuous if $y_0 \in D(\partial\varphi)$. In this sense, problem (4.72) is well posed for each $y_0 \in L^2(\mathcal{O})$. For simplicity, we assume $1 \le d \le 2$ and so $BV(\mathcal{O}) \subset L^2(\mathcal{O})$, but this condition can be removed by redefining $\varphi(u) = +\infty$ for $u \in L^2(\mathcal{O} \setminus BV(\mathcal{O})$. We also have

Theorem 4.6. *Assume that $\mathcal{O}$ is bounded and convex, and let $y_0 \in L^2(\mathcal{O})$ and $T > 0$ be arbitrary but fixed. Then, there is $\alpha > 0$ such that $y(T) = 0$ and $y(t) = 0$ for all $t > T$.*

Proof. To fix the idea, we give first a formal argument which may work if $y(t)$ is a strong solution to (4.72) in the Sobolev space $W_0^{1,1}(\mathcal{O})$. If we multiply equation (4.72) by y and integrate on $(0,t) \times \mathcal{O}$, we get

$$\frac{1}{2}\,|y(t)|_2^2 + \alpha \int_0^t \int_{\mathcal{O}} |\nabla y(x,s)|_d dx\, ds \le \frac{1}{2}\,|y_0|_2^2,\ \forall t \ge 0.$$

On the other hand, by the Sobolev embedding theorem, we have

$$\int_{\mathcal{O}} |\nabla y(x,s)|_d dx \ge \gamma |y(s)|_2,\ \forall s \ge 0.$$

This yields

$$|y(t)|_2 + \alpha\gamma t \le |y_0|_2,\ \forall t \in (0,T), \tag{4.74}$$

and, therefore, $y(t) = 0$ for $t \ge T = (\alpha\gamma)^{-1}|y_0|_2$. This argument can be made rigorous if we take into account that equation (4.73) can be approximated by

$$\begin{aligned} &\frac{dy}{\partial t} - \varepsilon\Delta y - \alpha\,\mathrm{div}\left(\frac{\nabla y}{|\nabla y|_d}\right) = 0 \text{ in } (0,T) \times \mathcal{O},\\ &y(x,0) = y_0(x),\ x \in \mathcal{O},\\ &y(x,t) = 0,\ x \in \partial\mathcal{O}. \end{aligned} \tag{4.75}$$

This follows via the Kato-Trotter theorem (see [26], p. 168) because, under our assumptions, the operator $A_\varepsilon u = -\varepsilon\Delta u - \alpha\,\mathrm{div}\left(\frac{\nabla u}{|\nabla u|_d}\right)$ with the domain $D(A_\varepsilon) = H_0^1(\mathcal{O}) \cap H^2(\mathcal{O})$ is m-accretive in $L^2(\mathcal{O})$ and, for each $f \in L^2$, the solution $u_\varepsilon \in H_0^1(\mathcal{O}) \cap H^2(\mathcal{O})$ to the equation

$$u_\varepsilon - \varepsilon\Delta u_\varepsilon - \alpha\,\mathrm{div}\left(\frac{\nabla u_\varepsilon}{|\nabla u_\varepsilon|}\right) = f \text{ in } \mathcal{O}, \tag{4.76}$$

is strongly convergent, for $\varepsilon \to 0$, to $(I - \partial\varphi)^{-1} f$. To get the m-accretivity of A_ε and so the existence in (4.76), we approximate in this equation the mapping $u \to \frac{u}{|u|_d}$

by $\psi_\lambda(u) = \frac{1}{\lambda} u$ for $|u|_d \le \lambda$, $\psi_\lambda(u) = \frac{u}{|u|_d}$ for $|u|_d \ge \lambda$, and let $\lambda \to 0$ in the resulting equation, by taking into account that (see Corollary (8.2) in [41])

$$\int_{\mathscr{O}} \Delta y \cdot \operatorname{div} \psi_\lambda(\nabla y) dx \ge 0, \ \forall\, y \in H_0^1(\mathscr{O}) \cap H^2(\mathscr{O}), \ \lambda > 0.$$

Then, by Theorem 4.1, equation (4.75) has a unique solution

$$y_\varepsilon \in L^2(0, T; H_0^1(\mathscr{O})) \cap C([0, T]; L^2(\mathscr{O})), \ \sqrt{t}\, \frac{\partial y_\varepsilon}{\partial t} \in L^2(0, T; L^2(\mathscr{O})),$$

and, arguing as above, we get the estimate

$$\frac{1}{2}\, |y_\varepsilon(t)|_2^2 + \int_0^t \int_{\mathscr{O}} (\varepsilon |\nabla y_\varepsilon|^2 + |\nabla y_\varepsilon|) ds\, dx = \frac{1}{2}\, |y_0|_2^2, \ \forall\, t \in (0, T), \ \forall\, \varepsilon > 0,$$

which implies that $|y_\varepsilon(t)|_2 + \alpha\gamma t \le |y_0|_2$, $\forall\, t \in [0, T]$. Letting $\varepsilon \to 0$, we get, for the solution $y = \lim\limits_{\varepsilon \to 0} y_\varepsilon$ to (4.72), estimate (4.74), as claimed.

The physical significance of Theorem 4.6 is obvious: the concentration y of the diffusion system (4.69) can be driven at zero in a finite time by a diffusion controller y proportional with the inverse $|\nabla y(t)|_d^{-1}$ of flux magnitude.

4.4 Exact Null Controllability in $\mathbb{R}^d$

Consider here the control system

$$\begin{aligned} &\frac{\partial y}{\partial t} + (\lambda - \Delta)(uy) = 0 \text{ in } (0, \infty) \times \mathbb{R}^d, \\ &y(x, 0) = y_0(x), \end{aligned} \tag{4.77}$$

where $\lambda > 0$ and u is the feedback controller (4.37).

The corresponding closed loop equation

$$\begin{aligned} &\frac{\partial y}{\partial t} + \alpha(\lambda - \Delta)(y^m \operatorname{sign} y) = 0 \text{ in } (0, \infty) \times \mathbb{R}^d, \\ &y(x, 0) = y_0(x), \end{aligned} \tag{4.78}$$

has, for $y_0 \in L^1(\mathbb{R}^d) \cap H^{-1}(\mathbb{R}^n)$, a unique strong solution

$$y \in C([0, T]; L^1(\mathbb{R}^d) \cap H^{-1}(\mathbb{R}^d)), \forall\, T > 0.$$

(see, e.g., [26], p. 233.) We have

Theorem 4.7. *Let $y_0 \in L^1(\mathbb{R}^d) \cap H^{-1}(\mathbb{R}^d)$, $d \geq 3$, and $m = \frac{d-2}{d+2}$. Then the solution y to* (4.78) *satisfies*

$$y(T) = 0 \text{ for } \alpha = \frac{1}{\gamma^{m+1} T(1-m)} \|y_0\|^{1-m}_{H^{-1}(\mathbb{R}^d)}, \tag{4.79}$$

where $\gamma = \sup\{\|y\|_{H^{-1}(\mathbb{R}^d)};\ \|y\|_{L^{m+1}(\mathbb{R}^d)} = 1\}$.

Here, the space $H^{-1}(\mathbb{R}^d)$ is endowed with the norm

$$\|y\|_{H^{-1}(\mathbb{R}^d)} = \left(\int_{\mathbb{R}^d} ((I-\Delta)^{-1} y, y)\, dx\right)^{\frac{1}{2}}.$$

Proof. By multiplying (4.78) scalarly in $H^{-1}(\mathbb{R}^d)$ with y and integrating on $\mathbb{R}^d$, we get

$$\frac{1}{2} \frac{d}{dt} \|y(t)\|^2_{H^{-1}(\mathbb{R}^d)} + \alpha \int_{\mathbb{R}^d} |y(x,t)|^{m+1} dx = 0, \text{ a.e. } t > 0.$$

By the Sobolev embedding theorem, we have $\|y\|_{L^{m+1}(\mathbb{R}^d)} \geq \gamma \|y\|_{H^{-1}(\mathbb{R}^d)}$ for $m = \frac{d-2}{d+2}$ (see, e.g., [55], p. 278). This yields

$$\frac{d}{dt} \|y(t)\|_{H^{-1}(\mathbb{R}^d)} + \alpha \gamma^{m+1} \|y(t)\|^m_{H^{-1}(\mathbb{R}^d)} \leq 0, \text{ a.e. } t > 0.$$

We set $\varphi(t) = \|y(t)\|_{H^{-1}(\mathbb{R}^d)}$ and get the differential inequality

$$\frac{\varphi'(t)}{(\varphi(t))^m} + \alpha \gamma^{m+1} \leq 0, \text{ a.e. } t > 0.$$

This yields

$$\frac{1}{1-m} (\varphi(t))^{1-m} + \alpha \gamma^{m+1} t \leq \frac{1}{1-m} \|y_0\|^{1-m}_{H^{-1}(\mathbb{R}^d)}.$$

Hence, $\varphi(T) = 0$ for $\alpha = \frac{1}{T\gamma^{m+1}(1-m)} \|y_0\|^{1-m}_{H^{-1}(\mathbb{R}^d)}$, as claimed.

4.5 Exact Controllability of Linear Stochastic Parabolic Equations

Consider the linear stochastic equation

$$\begin{aligned} & dX - \Delta(uX)dt = XdW \text{ in } (0,T) \times \mathcal{O}, \\ & X(x,0) = X_0(x), \\ & X(x,t) = 0, \ (x,t) \in [0,T] \times \partial\mathcal{O}, \end{aligned} \tag{4.80}$$

where $\mathscr{O} \subset \mathbb{R}^d$ is bounded and open, while W is a cylindrical Gaussian process in a probability space $(\Omega, \mathscr{F}, \mathbb{P})$. More precisely, W is given by $W(t) = \sum_{j=1}^{\infty} \mu_j e_j \beta_j(t)$, where $\{\beta_j\}_{j=1}^{\infty}$ are independent Brownian motions, $\{e_j\}_{j=1}^{\infty} \subset H^{-1}(\mathscr{O})$ is an orthonormal basis in $H^{-1}(\mathscr{O})$, and μ_j are suitable chosen. (See Section 3.6.)

The controllability problem considered here is to design a feedback controller $u = \Phi(X)$ which steers X_0 in origin in time T. The controller is of the form (4.37), where $m \in [0, 1)$ and the corresponding closed loop stochastic differential system is

$$\begin{aligned} &dX - \alpha\Delta(|X|^m \operatorname{sign} X)dt = X\,dW \text{ in } (0, \infty) \times \mathscr{O}, \\ &X(0) = X_0 \text{ in } \mathscr{O}, \\ &X = 0 \text{ on } (0, \infty) \times \partial\mathscr{O}. \end{aligned} \tag{4.81}$$

Theorem 4.8. *Assume that* $m \in [0, 1)$, $1 \le d < \frac{2(m+1)}{1-m}$, *and* $X_0 \in H^{-1}(\mathscr{O})$. *Then equation* (4.81) *has a unique solution* $X \in L^2(\Omega; C([0, T]; H^{-1}(\mathscr{O}))$, $\forall T > 0$, *and, if* $\tau = \tau(\omega)$ *is the stopping time* $\tau(\omega) = \inf\{t > 0;\ X(t, \omega) = 0\}$, $\omega \in \Omega$, *then we have*

$$\mathbb{P}(\tau > t) \le C\alpha^{-1}\|X_0\|_{H^{-1}(\mathscr{O})}^{1-m}, \tag{4.82}$$

where $C > 0$ *is independent of* ω.

Theorem 4.4 amounts to saying that $\mathbb{P}$-a.s. system (4.80) is exactly controlled in origin in a random time $T = T(\omega)$ with a probability $\mathbb{P}$ estimated by formula (4.82).

The existence of the solution X as well as estimate (4.82) for the stopping time τ was proved in [48], p. 68 (Theorem 3.73) and we omit here the details. We only note that (4.82) follows by (4.77) via Itô's formula applied to (4.81) with the Lyapunov function $\varphi(X) = \|X\|_{H^{-1}(\mathscr{O})}^{1-m}$, which yields after some calculation

$$\begin{aligned} \|X(t)\|_{H^{-1}(\mathscr{O})}^{1-m} &+ \alpha\frac{1-m}{1+m}\int_r^t \|X(s)\|_{H^{-1}(\mathscr{O})}^{-m-1}\|X(s)\|_{L^{m+1}(\mathscr{O})}^{m+1}\mathbf{1}_{[\|X(s)\|_{H^{-1}(\mathscr{O})}>0]}ds \\ &\le \|X(r)\|_{H^{-1}(\mathscr{O})}^{1-m} + C\int_r^t \|X(s)\|_{H^{-1}(\mathscr{O})}^{1-m}ds \end{aligned}$$

for $0 < r < t < \infty$. This implies (4.82) via the stochastic Gronwall's lemma.

4.6 Notes on Chapter 4

The results of this chapter are new in the controllability theory context. The feedback controllers constructed here transform the original equation in a nonlinear closed loop system of phase transition type which is at the origin of the exact controllability mechanism. Roughly speaking, this approach relies on the finite time extinction property of solutions to nonlinear fast diffusion equations (the main references about this mechanism are the works [27, 33, 40, 46, 48]). As a matter of fact, this property is true for an entire class of nonlinear infinite dimensional equations and is expressed in the following abstract result: *Let X be a Banach space, $J : X \to X^*$ the duality mapping of X and $A : D(A) \subset X \to X$ be an m-accretive (multivalued) operator such that ${}_X(v, J(u))_{X^*} \geq \rho \|u\|_X$, $\forall u \in D(A)$, $v \in Au$, where $\rho > 0$. Then every mild solution $u \in C([0, T]; X)$ to the Cauchy problem $\frac{du}{dt} + Au \ni 0$, $t \geq 0$, $u(0) = u_0$, has the property $u(t) = 0$ for $t \geq \rho^{-1}\|u_0\|_X$.* The proof is quite immediate if u is a strong solution and follows by the convergence of the finite difference scheme $u_{i+1} + hAu_{i+1} \ni u_i$ to the mild solution u in the general case. One might expect to find such nonlinear feedback controllers for more general parabolic equations and also for controllers with internal or boundary support. A nice feature of this method is that it provides exact controllability by feedback and not by open loop controllers as usually happens by the linearization technique presented in Chapter 3.

Chapter 5
Feedback Stabilization of Semilinear Parabolic Equations

We shall discuss here the internal and boundary feedback stabilization of equilibrium solutions to semilinear parabolic equations. The main conclusion is that such an equation is stabilizable by a feedback controller with finite dimensional structure dependent of the unstable spectrum of the corresponding linearized system around the equilibrium solution.

5.1 Riccati-based Internal Stabilization

An important objective of the control theory is to design a feedback controller that stabilizes the differential system around a given orbit which in most situations is a steady-state or periodic orbit of the system. Here we shall study this problem for semilinear parabolic equations of the form (1.9). As in the exact controllability problem, it is desirable to design stabilizable feedback controllers which are supported by accessible small subdomains or on the boundary.

Compared with the exact null controllability property discussed in Chapter 3, the stabilization which is a weaker one has two important advantages very appreciated in engineering applications: it is stable to structural modifications of the system and may be achieved by feedback controllers, that is, in real time.

Let $\mathscr{O} \in \mathbb{R}^d$, $d = 1, 2, 3$, be an open and bounded subset with smooth boundary $\partial\mathscr{O}$, $\omega \subset \mathscr{O}$ be an open subset and let $m = \mathbf{1}_\omega$ be the characteristic function of ω. Let $Q = \mathscr{O} \times (0, \infty)$ and $\Sigma = \partial\mathscr{O} \times (0, \infty)$.

Consider the quasilinear controlled reaction-diffusion equation

$$\begin{array}{ll} y_t(x,t) - \Delta y(x,t) + f(x, y(x,t), \nabla y(x,t)) = m(x)u(x,t) & \text{in } Q, \\ y(x,0) = y_0(x), & \text{in } \mathscr{O}, \\ y(x,t) = 0, & \text{on } \Sigma, \end{array} \tag{5.1}$$

V. Barbu, *Controllability and Stabilization of Parabolic Equations*,
Progress in Nonlinear Differential Equations and Their Applications 90,
https://doi.org/10.1007/978-3-319-76666-9_5

and a steady-state (equilibrium) solution y_e to (5.1), that is,

$$\begin{aligned} -\Delta y_e(x) + f(x, y_e(x), \nabla y_e(x)) &= 0, \text{ in } \mathcal{O}, \\ y_e(x) &= 0, \qquad\qquad \text{on } \partial\mathcal{O}. \end{aligned} \tag{5.2}$$

Let $H = L^2(\mathcal{O})$ with the norm $|\cdot|_2$ and $V = H_0^1(\mathcal{O})$ with the norm $\|\cdot\|$. We shall denote by $|\cdot|_s$ the norm of $L^s(\mathcal{O})$ and by $\|\cdot\|_r$ the norm of $H^r(\mathcal{O})$. We use $(\cdot,\cdot)$ to denote the inner product in H and the paring between V and V' and between $H^2(\mathcal{O})$ and $(H^2(\mathcal{O}))'$, respectively. Use $|\cdot|_\omega$ and $(\cdot,\cdot)_\omega$ to denote the norm and the inner product of $L^2(\omega)$, respectively. The same notations will be used for the complex space $L^2(\mathcal{O})$ and $L^2(\omega)$, respectively. We shall consider now a set of technical conditions on the steady-state solution y_e and the nonlinear function f.

(H_1) $y_e \in C(\overline{\mathcal{O}})$, $\nabla y_e \in (C(\overline{\mathcal{O}}))^d$.
(H_2) $f(x, y, \theta) \equiv g(y) + a(x)y + b(x) \cdot \theta$, where $g : \mathbb{R} \to \mathbb{R}$ is of class C^2, $a \in C(\overline{\mathcal{O}})$, $b \in C(\overline{\mathcal{O}}; \mathbb{R}^d)$ and

$$g'(r) \ge -\gamma, \quad \forall r \in \mathbb{R}, \tag{5.3}$$

$$|g''(r)| \le \gamma_1(|r|^{\eta-2} + 1), \quad \forall r \in \mathbb{R}, \tag{5.4}$$

where $\gamma \in \mathbb{R},\ \gamma_1 \ge 0$, and

$$2 \le \eta \le \frac{d+3}{d-1}\cdot \tag{5.5}$$

The standard example is the polynomial function

$$g(r) \equiv a_1|r|^{\eta-1}r + a_2|r|r,$$

where $a_i, a_2 \ge 0$ and η satisfies condition (5.5).

Here, we shall not discuss the existence of a solution y to (5.1) under assumptions (H_1)–(H_2) though this problem can be treated as in the case of Theorem 1.4. The problem to be addressed here is that to design a feedback controller $u = \Phi(t, y)$ such that the solution y to (5.1) where $u = \Phi(t, y)$ satisfies

$$\lim_{t\to\infty} y(t) = y_e \text{ strongly in } L^2(\mathcal{O})$$

for all y_0 in a suitable chosen neighborhood of y_e.

By the substitution $y \to y_e + y$, we reduce the problem of stabilization of y_e to that of stabilization of the null solution to the equation

$$\begin{aligned} & y_t - \Delta y + f(x, y + y_e, \nabla y + \nabla y_e) - f(x, y_e, \nabla y_e) = mu && \text{in } Q, \\ & y(x, 0) = y^0(x) = y_0(x) - y_e(x) && \text{in } \mathcal{O}, \\ & y = 0 && \text{on } \Sigma. \end{aligned} \tag{5.6}$$

We consider the corresponding linearized equation about zero solution

$$\begin{aligned} &y_t - \Delta y + g'(y_e)y + ay + b \cdot \nabla y = mu \text{ in } Q,\\ &y(x,0) = y^0(x) \qquad \text{in } \mathscr{O},\\ &y = 0 \qquad \text{on } \Sigma. \end{aligned} \tag{5.7}$$

It should be said that, as seen earlier in Chapter 3, under the above assumptions for each T there is an open loop controller $u^* : [0,T] \to L^2(\mathscr{O})$ such that the corresponding solution y^{u^*} to (5.1) satisfies $y^{u^*}(T) = y_e$. If we extend this controller with 0 on (T,∞), we may view u^* as a stabilizable controller for y_e. However, being obtained by fixed point theoretical arguments the above controller u^* is not in feedback form and its structure is somehow unclear. We shall develop in the following a constructive procedure to obtain a feedback stabilizable controller u based on Riccati-based stabilization of linear system (5.7) and which is expressed in function of a finite system of unstable modes of the system.

Let $A = -\Delta$ with the domain $D(A) = H^2(\mathscr{O}) \cap H_0^1(\mathscr{O})$, and

$$A_0 y = g'(y_e)y + a(x)y + b(x) \cdot \nabla y.$$

We denote by $A^s, s \in (0,1)$ the fractional power of operator A. Let $W = D(A^{\frac{1}{4}})$ with the graph norm $\|y\|_W = |A^{\frac{1}{4}}y|_2$. Recall that (see [84], pp. 66),

$$D(A^{\frac{1}{4}}) = \{y \in H^{\frac{1}{2}}(\mathscr{O});\ (\mathrm{dist}(x, \partial\mathscr{O}))^{-\frac{1}{2}} y \in L^2(\mathscr{O})\}.$$

In terms of A and A_0, equation (5.7) can be rewritten as

$$\begin{aligned} &y' + Ay + A_0 y = mu,\ t \in (0,T)\\ &y(0) = y_0. \end{aligned} \tag{5.8}$$

Consider the operator $\mathscr{F}_0 = -(A + A_0)$ with the domain $D(\mathscr{F}_0) = D(A) = H_0^1(\mathscr{O}) \cap H^2(\mathscr{O})$ and denote by "$\mathscr{F}$ the extension of $\mathscr{F}_0$ on the complexified space $H = L^2(\mathscr{O}) \oplus iL^2(\mathscr{O})$. This means that

$$\mathscr{F}(y_1 + iy_2) = \mathscr{F}_0 y_1 + i\mathscr{F}_0 y_2,\ \forall y_1, y_2 \in D(A).$$

We shall denote again $(\cdot,\cdot)$ the scalar product in the space H. It is readily seen that $\mathscr{F}$ has a compact resolvent and generates an analytic C_0-semigroup on $H = L^2(\mathscr{O})$. Then, by the Fredholm–Riesz theorem, the operator $\mathscr{F}$ has a countable set of complex eigenvalues λ_j and corresponding *proper eigenvectors* φ_j, that is, $\mathscr{F}\varphi_j = \lambda_j\varphi_j$. For each λ_j there is a finite number m_j of linear independent vectors $\{\varphi_{ij};\ i = 1, \dots, m_j\}$ such that $(\mathscr{F} - \lambda_j)^\ell \varphi_{ij} = 0$, $1 \le \ell \le m_j$. These vectors are called *generalized eigenvectors* or, simply, eigenvectors (eigenfunctions). The *algebraic multiplicity* ℓ_j of λ_j is the number of generalized eigenvectors, while the *geometric multiplicity* of λ_j is the number of proper vectors corresponding to λ_j. We note also that the operator $\mathscr{F}$ has a finite number N of eigenvalues λ_j with $\operatorname{Re}\lambda_j \geq 0$ (unstable eigenvalues). In the following, the eigenvalues are repeated according to their algebraic multiplicity ℓ_j. An eigenvalue λ_j is said to be *simple* if $m_j = 1$ and *semisimple* if the algebraic and geometric multiplicity coincide. This means that λ_j is a pole of order m_j. Let $\{\varphi_j\}_{j=1}^N$ be the corresponding system of eigenfunctions, $\varphi_j = \varphi_j^1 + i\varphi_j^2$ and let

$$Y_N^m = \operatorname{span}\{\varphi_j^m\}_{j=1}^N, \ m = 1, 2. \tag{5.9}$$

Denote by M the number of distinct unstable eigenvalues, that is, $\ell_1 + \ell_2 + \cdots + \ell_M = N$, and let

$$K = \max\{\ell_j\}_{j=1}^M. \tag{5.10}$$

We shall prove first the following stabilization result for equation (5.7) or, equivalently, (5.8).

Theorem 5.1. *Under assumption* (H_1), *there are the functions* $\{\psi_i\}_{i=1}^K \subset Y_N^1$, $\{\psi_i\}_{i=K+1}^{2K} \subset Y_N^2$ *and a linear self-adjoint operator* $G : D(G) \subset H \to H$ *such that for some* $0 < \gamma_1 < \gamma_2$, $C_1 > 0$,

$$\gamma_1 |A^{\frac{1}{4}} y|_2^2 \leq (Gy, y) \leq \gamma_2 |A^{\frac{1}{4}} y|_2^2, \ \forall y \in D(A^{\frac{1}{4}}), \tag{5.11}$$

$$|Gy|_2 \leq C_1 \|y\|, \ \forall y \in V, \tag{5.12}$$

$$(\mathscr{F} y, Gy) + \frac{1}{2}\sum_{i=1}^{2K} (\psi_i, Gy)_\omega^2 = \frac{1}{2} \, |A^{\frac{3}{4}} y|_2^2, \ \forall y \in D(A). \tag{5.13}$$

Moreover, the feedback controller

$$u = -\sum_{i=1}^{2K} (Gy, \psi_i)_\omega \psi_i \tag{5.14}$$

stabilizes exponentially system (5.8).

Equation (5.13) can be, equivalently, written as the algebraic operational Riccati equation

$$G\mathscr{F} + \mathscr{F}^* G + \mathscr{N} = A^{\frac{3}{2}}, \tag{5.15}$$

where the linear operator $\mathscr{N} : H \to H$ is defined by

$$\mathscr{N} y = \sum_{i=1}^{2K} \int_\omega (Gy)(\xi)\psi_i(\xi) d\xi \, G(m\psi_i), \ \forall y \in H.$$

It turns out that the stabilizing feedback controller (5.14) also stabilizes the semilinear system (5.6). Namely, we have

Theorem 5.2. *Under assumptions* (H_1), (H_2), *the feedback controller*

$$u = -\sum_{i=1}^{2K} (G(y - y_e), \psi_i)_\omega \psi_i \tag{5.16}$$

exponentially stabilizes the steady-state solution y_e *to* (5.1) *in a neighborhood*

$$\mathscr{U}_\rho = \{y_0 \in D(A^{\frac{1}{4}});\ \|y_0 - y_e\|_W < \rho\}$$

of y_e *for a suitable* $\rho > 0$. *More precisely, if* $\rho > 0$ *is sufficiently small, then for each* $y_0 \in \mathscr{U}_\rho$ *there is a solution* $y \in C(\mathbb{R}^+; H) \cap L^2_{\rm loc}(\mathbb{R}^+; V)$ *to the closed loop system*

$$\begin{aligned} & y_t - \Delta y + f(x, y, \nabla y) + \sum_{i=1}^{2K} m\psi_i (G(y - y_e), \psi_i)_\omega = 0 \ \text{in } Q, \\ & y(x, 0) = y_0(x) \qquad \text{in } \mathscr{O}, \\ & y = 0 \qquad \text{on } \Sigma, \end{aligned} \tag{5.17}$$

such that

$$\int_0^\infty |A^{\frac{3}{4}}(y(t) - y_e)|_2^2 dt \le C\|y_0 - y_e\|_W^2 \tag{5.18}$$

$$|y(t) - y_e|_2 \le Ce^{-\gamma_0 t}\|y_0 - y_e\|_W, \ \forall t \ge 0, \tag{5.19}$$

for some $\gamma_0 > 0$.

Remark 5.1. In particular, it follows by Theorem 5.2 that, if all unstable eigenvalues λ_j are simple, that is, $K = 1$, then the feedback controller (5.16) is of the form

$$u = -(G(y - y_e), \psi)_\omega,$$

where $\psi \in \text{span}\{\text{Re}\,\varphi_j, \text{Im}\,\varphi_j\}_{j=1}^N$ is arbitrary.

Remark 5.2. As mentioned earlier, by controllability theorems established in Chapter 3, we know that equation (5.7) (or (5.8)) is exactly null controllable in every finite time which implies by a Riccati-based device, mentioned in Section 1.5, the feedback stabilization (see (1.92)–(1.93)). The novelty of Theorem 5.1 and 5.2 is that the stabilizable controllers (5.14), (5.16) found here have a simple finite dimensional structure which is not the case with the controllers arising in Chapter 3. As seen in Theorems 3.1–3.3 for exact null controllability, it was necessary a Carleman-type inequality for the solutions to dual parabolic equations, while as seen later on, for stabilization it is sufficient, and in a certain sense necessary, a weaker property, i.e., the unique continuation property of eigenfunctions the operator $A + A_0$.

As in the case of the exact controllability, the stabilization result given by Theorem 5.2 is local and works only for nonlinear functions f with *mild* polynomial growth. This can be explained by the fact that a large nondissipative reaction term f is a source of instability, which cannot be compensated by the linear dissipative operator $-\Delta$. It should be mentioned also that the stabilization property, obtained via this approach, is only local for initial data in a neighborhood of an equilibrium solution.

Proof of Theorem 5.1. We shall proceed as in [37] (see, also, [45]), via the pole location technique. In a few words, the idea is to decompose the system in an unstable finite-dimensional system generated by unstable eigenvectors and an infinite-dimensional exponentially stable system. Then the finite-dimensional unstable system is stabilized via the standard pole allocation technique. We shall consider the extension of system (5.8) on the complexified space $L^2(\mathcal{O}) \oplus i L^2(\mathcal{O}) = H$, that is,

$$\begin{aligned} &\frac{dz}{dt} - \mathcal{F} z = mv, \quad t > 0, \\ &z(0) = z_0 = y^0 + i\widetilde{y}^0, \end{aligned} \tag{5.20}$$

where $z = y + i\widetilde{y}$, $v = u + i\widetilde{u}$, $z_0 = y^0 + i\widetilde{y}^0$. As noted below, the operator $\mathcal{F}$ has a compact resolvent in H, and generates an analytic C_0-semigroup on H, so that only a finite number of eigenvalues of $\mathcal{F}$ are in the right complex half-plane $\{\lambda \in \mathbb{C};\ \mathrm{Re}\,\lambda \geq 0\}$. We have already denoted by $\lambda_1, \lambda_2, \ldots, \lambda_N$ these (unstable) eigenvalues repeated according to their algebraic multiplicity ℓ_i so that

$$\mathrm{Re}\,\lambda_{N+1} < 0 \leq \mathrm{Re}\,\lambda_N \leq \cdots \leq \mathrm{Re}\,\lambda_1.$$

Let Γ_j be a closed curve enclosing λ_j but no other point of the spectrum of $\mathcal{F}$. Define the eigenprojection $P_{N,j}$,

$$P_{N,j} = \frac{1}{2\pi i} \int_{\Gamma_j} (\lambda I - \mathcal{F})^{-1} d\lambda, \ j = 1, \ldots, M.$$

The space $Z_{N,j} = R(P_{N,j})$ (the range of $P_{N,j}$) is called the *algebraic eigenspace* for the eigenvalue λ_j and $\ell_j = \dim Z_{N,j}$ is just the *algebraic multiplicity* for λ_j. We recall (see [82], Theorem 6.17) that

$$Z_{N,j} = \{z \in H;\ (\lambda_j - \mathcal{F})^{\ell_j} z = 0\}, \ j = 1, \ldots, M.$$

We now denote by $\{\varphi_{ij}\}_{j=1}^{\ell_i}$ the (normalized) linearly independent system of generalized eigenfunctions corresponding to each unstable distinct eigenvalue λ_i of $\mathcal{F}$. Then the space H can be decomposed as (see [82], p. 178)

$$H = Z_N^u \oplus Z_N^s;\ Z_N^u = \text{span}\{\varphi_{ij},\ i = 1, \ldots, M;\ j = 1, \ldots, \ell_i\};\ \dim Z_N^u = N,$$

where each of the spaces Z_N^u and Z_N^s is invariant under $\mathcal{F}$.

Denote by P_N the projection, explicitly identified as a contour integral, $P_N = \frac{1}{2\pi i}\int_\Gamma(\lambda I - \mathcal{F})^{-1}d\lambda$ where Γ separates the unstable spectrum from the stable one and similarly define $P_N^* = \frac{1}{2\pi i}\int_{\overline{\Gamma}}(\lambda I - \mathcal{F}^*)^{-1}d\lambda$. We set

$$\mathcal{F}_N^u = P_N\mathcal{F}|_{Z_N^u};\ \ \mathcal{F}_N^s = (I - P_N)\mathcal{F} = \mathcal{F}|_{Z_N^s}.$$

We then have that the spectra of $\mathcal{F}$ on Z_N^u and Z_N^s coincide with $\{\lambda_j\}_{j=1}^N$ and $\{\lambda_j\}_{j=N+1}^\infty$, respectively:

$$\sigma(\mathcal{F}_N^u) = \{\lambda_j\}_{j=1}^N;\ \ \sigma(\mathcal{F}_N^s) = \{\lambda_j\}_{j=N+1}^\infty.$$

Moreover, since $\mathcal{F}$ generates a C_0-analytic semigroup on H, then its restriction $\mathcal{F}_N^s$ to Z_N^s generates likewise a C_0-analytic semigroup on Z_N^s. This implies that it satisfies the spectrum determined growth condition on Z_N^s (see [114]), and so we have

$$\|e^{\mathcal{F}_N^s t}\|_{\mathcal{L}(H;H)} \leq C_{\gamma_0}e^{-\gamma_0 t},\ \ \forall t \geq 0, \tag{5.21}$$

for any $\gamma_0 > |\text{Re}\,\lambda_{N+1}|$, where C_{γ_0} depends on γ_0.

If $(\mathcal{F}_N^u)^*$ is the adjoint operator of $\mathcal{F}_N^u$ on Z_N^u, then the eigenvalues of $(\mathcal{F}_N^u)^*$ are just the complex conjugate $\overline{\lambda}_j$ of eigenvalues λ_j of $\mathcal{F}_N$, $j = 1, \ldots, N$.

The eigenvalue λ_j of $\mathcal{F}_N^u$ is called *semisimple* if the algebraic and geometric multiplicity of λ_j coincide. This means that $\mathcal{F}_N^u\varphi_{jk} = \lambda_j\varphi_{jk}$, $j = 1, \ldots, M$, $k = 1, \ldots, \ell_j$, and we have also $(\mathcal{F}_N^u)^*\varphi_{jk}^* = \overline{\lambda}_j\varphi_{jk}^*$ for an eigenvalue $\overline{\lambda}_j$ of $(\mathcal{F}_N^u)^*$. If all the eigenvalues of $\mathcal{F}_N^u$ are semisimple, that is, the operator $\mathcal{F}_N^u$ is diagonlizable (see [82], p. 41), then the systems $\{\varphi_{jk}\}_{j=1}^{M}{}_{k=1}^{\ell_j}$, $\{\varphi_{jk}^*\}_{j=1}^{M}{}_{k=1}^{\ell_j}$ can be chosen in such a way to form a bi-orthogonal system, that is,

$$(\varphi_{jk}, \varphi_{im}^*) = 0\ \text{ if } j \neq 1,\ j = i,\ k \neq m,$$

and

$$(\varphi_{jk}, \varphi_{jk}^*) = 0\ \text{ for } k = 1, \ldots, \ell_j,\ j = 1, \ldots, M.$$

(See [42] for the proof.)

Then system (5.20) can be, accordingly, decomposed as

$$z = z_N + \zeta_N, \quad z_n = P_N z, \quad \zeta_N = (I - P_N)z, \tag{5.22}$$

$$\text{on } Z_N^u: \; z_N' - \mathscr{F}_N^u z_N = P_N(mv), \; z_N(0) = P_N z_0, \tag{5.23}$$

$$\text{on } Z_N^s: \; \zeta_N' - \mathscr{F}_N^s \zeta_N = (I - P_N)(mv), \; \zeta_N(0) = (I - P_N)z_0. \tag{5.24}$$

We shall prove now that the finite dimensional unstable system (5.23) is stabilizable by a controller v with finite structure (Lemma 5.1) and show afterwards that the infinite dimensional system (5.24) is stable.

Lemma 5.1. *Given $\gamma_1 > 0$ arbitrarily large, there is a controller $v = v_N = \sum_{i=1}^{K} v_N^i(t)\psi_i$, $\psi_i \in Z_N^u$, such that once inserted in* (5.23), *yields the estimate*

$$|z_N(t)| + |v_N(t)| \le C_{\gamma_1} e^{-\gamma_1 t} |P_N z_0|, \quad t \ge 0. \tag{5.25}$$

Proof. Let $w_1, w_2, \ldots, w_k \in Z_N^u$, $\lambda_i, i = 1, \ldots, M$, be the distinct unstable eigenvalue of $\mathscr{F}$, and $\{\{\varphi_{ij}\}_{j=1}^{\ell_i}\}_{i=1}^M = \{\varphi_j\}_{j=1}^N$ its corresponding system of generalized eigenfunctions linearly, independent on $L^2(\mathscr{O})$. By the classical procedure, we may choose the system $\{\{\varphi_{ij}\}_{j=1}^{\ell_i}\}_{i=1}^M$ such that the matrix Λ corresponding to $\mathscr{F}_N^u$ in this basis has the Jordan canonical form. More precisely,

$$\mathscr{F}_N^u \{\varphi_{ij}\}_{j=1}^{\ell_i} = J_i \{\varphi_{ij}\}_{j=1}^{\ell_i}, \quad \forall i = 1, \ldots, M, \tag{5.26}$$

where J_i is the Jordan block for λ_i.

In the following, we shall simply write the system $\{\varphi_{ij}\}_{j=1}^{\ell_i}\,{}_{i=1}^{M}$ as $\{\varphi_k\}_{k=1}^N$. (We recall that $\ell_1 + \ell_2 + \ell_M = N$.)

Consider the orthonormal system $\{\phi_j\}_{j=1}^N \in Z_N^u$ obtained from $\{\varphi_k\}_{k=1}^N$ by Schmidt's orthogonalization procedure. Let us represent ϕ_j as

$$\phi_j = \sum_{k=1}^{N} a_{jk}\varphi_k, \quad j = 1, 2, \ldots, N$$

and set

$$\chi_\ell = \sum_{k=1}^{N} b_{k\ell}\varphi_k, \;\; b_{k\ell} = \sum_{j=1}^{N} a_{jk}a_{j\ell}, \;\; k, \ell = 1, \ldots, N.$$

We shall regroup the vectors χ_ℓ from χ_1 to χ_{ℓ_1}, then from χ_{ℓ_1+1} to $\chi_{\ell_1+\ell_2}$, and so on. We rename them as

$$\begin{aligned} \chi_j &= \chi_{1j}, \;\; j = 1, \ldots, \ell_1, \\ \chi_{\ell_1+j} &= \chi_{2j}, \;\; j = 1, \ldots., \ell_2, \\ &\ldots\ldots\ldots\ldots\ldots\ldots\ldots\ldots \\ \chi_{\ell_1+\ldots+\ell_{M-1}+j} &= \chi_{M_j}, \; j = 1, \ldots., \ell_M. \end{aligned}$$

For $k = K$ and any fixed vector $\{\psi_1, \dots, \psi_k\} \in Z^u_N$, consider the $\ell_i \times k$ matrix D_i defined by

$$D_i = \left\| \begin{matrix} (\psi_1, P^*_N \chi_{i_1})_\omega & (\psi_2, P^*_N \chi_{i_1})_\omega & \cdots & (\psi_k, P^*_N \chi_{i_1})_\omega \\ (\psi_1, P^*_N \chi_{i_2})_\omega & (\psi_2, P^*_N \chi_{i_2})_\omega & \cdots & (\psi_k, P^*_N \chi_{i_2})_\omega \\ \cdots & \cdots & \cdots & \cdots \\ (\psi_1, P^*_N \chi_{i\ell_1})_\omega & (\psi_2, P^*_N \chi_{i\ell_1})_\omega & \cdots & (\psi_k, P^*_N \chi_{i\ell_1})_\omega \end{matrix} \right\| \tag{5.27}$$

for $i = 1, 2, \dots, M$. Then we shall use the following finite-dimensional feedback stabilization result which will play a key role in the proof of Lemma 5.1.

Lemma 5.2. *Let* $K = \max\{\ell_i;\ i = 1, \dots, M\}$ *and assume that*

$$\operatorname{rank} D_i = \ell_i, \quad \forall i = 1, \dots, M. \tag{5.28}$$

Then there is a controller $v = v_N$ *such that* (5.25) *holds.*

Proof of Lemma 5.2. In order to illustrate better the proof technique, we shall assume first that all eigenvalues λ_j, $j = 1, \dots, N$, are simple. Hence, $K = \ell_i = 1$ for all $i = 1, \dots, N = M$ and

$$\varphi_{ij} \equiv \phi_i,\ i = 1, \dots, N,\ j = 1,$$

the matrix Λ is diagonal, $J_i = \{\varphi_i\}$ and $\phi_j = \varphi_j$, $b_{jk} = a_{jk} = \delta_{jk}$ (Kronecker's symbol) $\chi_\ell = \varphi_\ell$, $\forall \ell = 1, \dots, N$. Then, for $\psi_1 \in Z^u_N$, the matrix (5.27) is

$$D_i = (\psi_1, P^*_N \varphi_i)_\omega,\ i = 1, \dots, N,$$

and condition (5.28) reduces to

$$(\psi_1, P^*_N \varphi_j)_\omega \neq 0,\ j = 1, \dots, N. \tag{5.29}$$

Let us show that, under this assumptions, system (5.23), that is,

$$z'_N - \mathscr{F}^u_N z_N = P_N(m\psi_1)v \tag{5.30}$$

is exponentially stable, that is, there is $v \in C([0, T])$ such that (5.25) holds.

To prove this, we set $z_N = \sum\limits_{j=1}^N z^j_N \varphi_j$ and so rewrite (5.30) as

$$\sum_{j=1}^N (z^j_N)' \varphi_j = \sum_{j=1}^N \mathscr{F}^u_N(\varphi_j) z^j_N(t) + v_1(t) P_N(m\psi_1).$$

Equivalently,

$$\sum_{j=1}^{N}((z_N^j)'(t) - \lambda_j z_N^j(t))\varphi_j = v_1(t) P_N(m\psi_1).$$

By using the Schmidt orthogonalization technique, we may replace the system $\{\varphi_j\}_{j=1}^N$ by an orthonormal one in H, and so we get

$$(z_N^j)' - \lambda_j z_N^j = (P_N(m\psi_1), \varphi_j)v_1 = (w_1, P_N^*\varphi_j)_\omega v_1 = D_j v_1, \ j = 1, \dots, N. \tag{5.31}$$

Recall that, by the Kalman controllability theorem (see condition (1.90) in Section 1.5), the linear system (5.31) is exactly null controllable if and only if

$$\text{rank}\|D, DA, \dots, DA^{N-1}\| = N,$$

where $A = \text{diag}\|\lambda_j\|_{j=1}^N$, $D = \text{col}\|D_j\|_{j=1}^N$. Equivalently,

$$\det\|\text{col}(D_j)_{j=1}^N, \ \lambda_1 \text{col}(D_j)_{j=1}^N, \dots, \lambda_1^{N_1}\text{col}(D_j)_{j=1}^N\| \neq 0.$$

By condition (5.29), this condition is obviously satisfied for system (5.31).

The above argument extends to the case where all unstable eigenvalues λ_j are semisimple. That means that the Jordan block J_i is a diagonal matrix of dimension $\ell_i \times \ell_i$, that is,

$$J_i = \left\| \begin{matrix} \lambda_i & & 0 \\ & \ddots & \\ 0 & & \lambda_i \end{matrix} \right\|, \ i = 1, 2, \dots, M. \tag{5.32}$$

Indeed, assume that $\{\psi_i\}$ satisfy condition (5.28) and consider the controlled system

$$z_N' - \mathscr{F}_N^u z_N = P_N\left(\sum_{i=1}^{K} v_N^i(t) m\psi_i\right) = \sum_{i=1}^{K} v_N^i(t) P_N(m\psi_i) \tag{5.33}$$

obtained from (5.23), with the controller

$$v = \sum_{i=1}^{K} v_N^i(t) w_i, \tag{5.34}$$

where $v_N = \{v_N^i\}_{i=1}^K \subset \mathbb{C}^K$. In this case, we have by (5.27)

$$\mathscr{F}_N^u(\varphi_{ij} = \lambda_i \varphi_{ij}, \ i = 1, \dots, M; \ j = 1, \dots, \ell_i,$$
$$(\mathscr{F}_N^u)^*(\varphi_{ij}^*) = \overline{\lambda}_i \varphi_{ij}^*, \ i = 1, \dots, M; \ j = 1, \dots, \ell_i.$$

If we rename the sequences $\{\varphi_{ij}\}_{i=1,j=1}^{M\ \ \ell_i}$, $\{\varphi_{ij}^*\}_{i=1,j=1}^{M\ \ \ell_i}$ as $\{\phi_i\}_{i-1}^N$ and $\{\phi_i^*\}_{j=1}^N$, respectively, we may assume that

$$(\phi_i, \phi_j^*) = \delta_{ij},\ i, j = 1, \dots, N. \tag{5.35}$$

If we insert, as above, $z_N = \sum_{j=1}^{N} z_N^j \phi_j$ in system (5.33), we get

$$\sum_{j=1}^{N} (z_N^j)' \phi_j = \sum_{j=1}^{N} \mathscr{F}_N^u(\phi_j) z_N^j + \sum_{i=1}^{K} v_N^i P_N(m\psi_i).$$

Taking into account (5.35) and that $\mathscr{F}_N^u \phi_j = \lambda_j$, we get

$$(z_N^j)' = \lambda_j z_N^j + \sum_{i=1}^{K} (\psi_i, P_N^*(\phi_j^*))_\omega v_N^i,\ j = 1, \dots, N.$$

Equivalently,

$$\widetilde{z}_N' - \Lambda \widetilde{z}_N = D v_N, \tag{5.36}$$

where $\widetilde{z}_N = \mathrm{col}\|z_N^j\|_{j=1}^N$, $D = \|d_{ij}\|_{i,j=1}^{K,N}$, $d_{ij} = (\psi_i, P_N^*(\phi_j^*))_\omega$ and Λ is the $N \times N$ matrix

$$\Lambda = \begin{Vmatrix} J_1 & & 0 \\ & \ddots & \\ 0 & & J_M \end{Vmatrix}, \tag{5.37}$$

where J_i, $i = 1, \dots, M$, are given by (5.32). It should be noted that D is just the matrix $\mathrm{col}\|D_i\|_{i=1}^M$, where $\chi_\ell = \phi_\ell$, $\ell = 1, \dots, N$. Taking into account that

$$\Lambda^m = \begin{Vmatrix} J_1^m & & 0 \\ & \ddots & \\ 0 & & J_M^m \end{Vmatrix}, \ \forall m \in \mathbb{N},$$

it is easily seen by assumption (5.28) that, also in this case, the Kalman condition

$$\mathrm{rank}\|D, D\Lambda, \dots, D\Lambda^{N-1}\| = N$$

holds, which implies the exact null controllability of system (5.36), as desired.

The general case of not semisimple unstable eigenvalues λ_j follows similarly, but with some technical details. Namely, also in this case, system (5.32) reduces to

(5.35), but the matrices Λ and D are given by (5.37) and, respectively,

$$D = \|(w_i, P_N^* \chi_j)_\omega\|_{i=1}^{K} {}_{j=1}^{N} = \begin{bmatrix} D_1 \\ D_2 \\ \vdots \\ D_M \end{bmatrix}.$$

Here J_i is the Jordan block of dimension ℓ_i corresponding to the eigenvalue λ_i of $\mathscr{F}$, that is, of $\mathscr{F}_N^u$.

We shall test the controllability of the pair $\{\Lambda, D\}$ via Kalman's theorem recalled above. To this end, we introduce for $n = 1, 2, \ldots$ the matrices

$$\Lambda^n D = \begin{bmatrix} J_1^n & & & \\ & J_2^n & & \\ & & \ddots & \\ & & & J_M^n \end{bmatrix} \begin{bmatrix} D_1 \\ D_2 \\ \vdots \\ D_M \end{bmatrix} = \begin{bmatrix} J_1^n D_1 \\ J_2^n D_2 \\ \vdots \\ J_M^n D_M \end{bmatrix}$$

The Kalman controllability matrix

$$[D, \Lambda D, \Lambda^2 D, \ldots, \Lambda^{N-1} D] = \begin{bmatrix} D_1 & J_1 D_1 & \lambda_1^2 D_1 & \cdots & J_1^{N-1} D_1 \\ D_2 & J_2 D_2 & \lambda_2^2 D_1 & \cdots & J_2^{N-1} D_2 \\ \vdots & \vdots & \vdots & \vdots & \vdots \\ D_M & J_M D_M & \lambda_M^2 D_1 & \cdots & J_M^{N-1} D_M \end{bmatrix}$$

of size $N \times KN$, $N = \dim Z_N^u$, is of full rank that satisfies Kalman's controllability condition, if and only if rank $W_i = \ell_i$, $i = 1, \ldots, M$, which is true by assumption (5.28). Thus, there is a controller v_N such that $v_N(T) = 0$.

In conclusion, given any $\gamma_1 > 0$, there exist $\{p_1, \ldots, p_K\} \subset Z_N^u$ such that the solution z_N to (5.30) satisfies the estimate

$$|v_N(t)| + |z_N(t)| = \left| e^{\mathscr{F}_N^u t} z_N(0) \right| \le C_{\gamma_1} e^{-\gamma_1 t} |z_N(0)|.$$

Now, to complete the proof of Lemma 5.1, we need to check condition (5.28). The proof relies on the following basic properties of the generalized eigenfunctions $\{\varphi_{ij}\}_{i=1}^{M} {}_{j=1}^{\ell_i} = \{\varphi_j\}_{j=1}^{N}$, of the operator $\mathscr{F}_N^u$.

1° The eigenfunction system $\{\varphi_{ij}, 1 \le i \le M, 1 \le j \le \ell_i\}$ is linearly independent in $L^2(\omega)$.

2° The system $\{P_N^* \varphi_{ij}, 1 \le i \le M, 1 \le j \le \ell_i\}$ is linearly independent in $L^2(\mathscr{O})$ and also in $L^2(\omega)$.

3° If $\varphi \in Z_N^u$ and $P_N^* \varphi = 0$ on ω then $\varphi \equiv 0$.

We shall see that all these properties are consequences of the unique continuation property for the elliptic equation

$$\begin{aligned}&-\Delta\varphi+f_y(x,y_e,\nabla y_e)\varphi+f_\theta(x,y_e,\nabla y_e)\cdot\nabla\varphi=\lambda\varphi\\&\varphi=0\quad\text{on }\partial\mathscr{O}\end{aligned}\tag{5.38}$$

and for the corresponding dual equation. More precisely, if $\varphi=0$ on ω, then $\varphi\equiv 0$. To begin with, we treat first where the eigenvalues $\{\lambda_j\}$, $j=1,\dots,M$, are semisimple, that is, the case of proper eigenvectors (eigenfunctions) φ_j.

1°. We note first that, if $\{\varphi_j, 1\le j\le\mu\}$ is a system of eigenfunctions corresponding to the same eigenvalue λ_j, it is linearly independent on $\mathscr{O}$, and so, by the unique continuation property for solutions φ_{ij} to (5.38), it is linearly independent on ω, too.

Consider now the case of two distinct eigenvalues λ_1,λ_2. Assume by contradiction that the corresponding system of eigenfunctions $\{\varphi_{1j},\varphi_{2k};\ j=1,2,\dots,\ell_1,$ $k=1,2,\dots,\ell_2\}$ is linearly dependent on ω. We may assume, therefore, that

$$\varphi_{2,\ell_2}=\sum_{j=1}^{\ell_1}\alpha_j\varphi_{1j}+\sum_{k=1}^{\ell_2-1}\beta_k\varphi_{2k}\text{ on }\omega,$$

where $\sum\limits_{j=1}^{\ell_1}\alpha_j>0$. Taking into account that $\mathscr{F}\varphi_{1j}=\lambda_1\varphi_{1j}$, $\mathscr{F}\varphi_{2k}=\lambda_2\varphi_{2k}$ and $\mathscr{F}\varphi_{2\ell_2}=\lambda_2\varphi_{2\ell_2}$ on $\mathscr{O}$, it follows after some calculation that

$$(\lambda_2-\lambda_1)\sum_{j=1}^{\ell_1}\alpha_j\varphi_{1j}=0\text{ on }\omega.$$

Since, by the first step, the system $\{\varphi_{1j}\}$ is linearly independent on ω, and so $\alpha_j=0$ for all j, which is a contradiction. By induction with respect to number q of distinct eigenvalues λ_j, 1° follows in general. Indeed, assume that 1° is true for $q-1$ and consider a system $\{\varphi_{ij}, 1\le i\le q,\ 1\le j\le\ell_i\}$. Assume that this system is linearly dependent on ω and we shall argue from this to a contradiction. As above, we may assume that

$$\varphi_{q\ell_q}=\sum_{j=1}^{\ell_1}\alpha_{j1}\varphi_{j1}+\ldots+\sum_{j=1}^{\ell_q-1}\alpha_{jq}\varphi_{jq}\text{ on }\omega,$$

where $\sum\limits_{j=1}^{q-1}\sum\limits_{k=1}^{\ell_k}\alpha_{jk}>0$. Taking in account that $\mathscr{F}\varphi_{ij}=\lambda_i\varphi_{ij}$ on $\mathscr{O}$, and consequently on ω too, this implies that

$$(\lambda_1(\lambda_q)^{-1}-1)\sum_{j=1}^{\ell_1}\alpha_{j1}\varphi_{j1}+\ldots+(\lambda_{q-1}(\lambda_q)^{-1}-1)\sum_{j=1}^{\ell_{q-1}}\alpha_{j(q-1)}\varphi_{j(q-1)}=0\text{ on }\omega.$$

By the inductive hypothesis, the latter implies that all α_{jk} are zero. The contradiction we arrived at completes the proof of 1°.

We note that property 1° remains true, by the same argument, for the eigenfunctions φ^*_{ij}, $i = 1, \dots, M$, $j = 1, \dots, \ell_i$ of the dual operator $\mathscr{F}^{u\,*}_N$. In particular, it follows by 1° that, if $\varphi \in Z^u_N$ is zero on ω, then it is zero on all of $\mathscr{O}$.

To prove 2°, we assume that $\sum_{i=1,j=1}^{M,\ell_i} \alpha_{ij} P^*_N \varphi_{ij} = 0$ on $\mathscr{O}$ and we will show that $\alpha_{ij} = 0$. Let $h \in H$ be arbitrary but fixed. We have

$$\left(h, \sum_{i=1,j=1}^{M,\ell_i} \alpha_{ij} P^*_N \varphi_{ij} \right) = \left(P_N h, \sum_{i=1,j=1}^{M,\ell_i} \alpha_{ij} \varphi_{ij} \right) = 0.$$

Taking into account that, as h runs H, $P_N h$ fills all of Z^u_N, we conclude that $\sum \alpha_{ij} \varphi_{ij} = 0$ on $\mathscr{O}$. Since the system $\{\varphi_{ij}\}$ is linearly independent, we conclude that $\alpha_{ij} = 0$, as claimed. Now, assume that

$$\varphi = \sum_{i=1,j=1}^{M,\ell_i} \alpha_{ij} P^*_N \varphi_{ij} = 0 \text{ on } \omega.$$

We may write $\varphi = \sum_{k=1}^{N} \beta_k \varphi^*_k$, where φ^*_k are eigenfunctions of the dual operator $\mathscr{F}^*_N$. By property 1° applied to the dual operator $\mathscr{F}^*_N$, we infer that system $\{\varphi^*_j\}$ is linearly independent on ω and so all β_k are zero. Hence $\varphi = 0$ on $\mathscr{O}$, as claimed.

As regards 3°, it is an immediate consequence of 1° and 2°. Indeed, if $P^*_N \varphi = 0$ on ω and $\varphi = \sum_{j=1}^{N} \alpha_j \varphi_j$, then by 2° it follows that all α_j are zero. Hence, by 1°, φ is identically zero, which is absurd.

Now, let us prove properties 1°, 2°, and 3° in the general case of generalized eigenfunctions. We note first that, if $\varphi \in Z_{N,j}$, that is, $(\lambda_j - \mathscr{F})^{\ell_j} \varphi = 0$ and $\varphi = 0$ on ω, then according to first step $(\lambda_j - \mathscr{F})^{\ell_j - 1} \varphi \equiv 0$ (because $(\lambda_j - \mathscr{F})^{\ell_j} \varphi = 0$ on ω and, finally, it follows that $\varphi \equiv 0$.

Assume, now, that $\varphi_1 \in Z_{N,1}$, $\varphi_2 \in Z_{N,2}$ are such that $\varphi_1 = \alpha \varphi_2$ on ω. We may take $\alpha = 1$ and assume $\ell_1 < \ell_2$. We have $(\lambda_j I - \mathscr{F})^k \varphi_j = 0$, $j = 1, 2$, where $k = \ell_1$, and so $\sum_{j=1}^{k} \binom{j}{k} \lambda_1^j (-1)^j \mathscr{F}^{k-j} \varphi_1 = \sum_{j=1}^{k} \binom{j}{k} \lambda_2^j (-1)^j \mathscr{F}^{k-j} \varphi_1 = 0$ on ω. This yields $(\lambda_1 - \mathscr{F})^k \varphi_1 = (\lambda_2 - \mathscr{F})^k \varphi_1$ on ω and so we have, for $k = \ell_1 < \ell_2$,

$$\sum_{j=1}^{k} \binom{j}{k} (\lambda_1)^j (-1)^{k-j} \mathscr{F}^{k-j} \varphi_1 = \sum_{j=1}^{k} \binom{j}{k} (\lambda_2)^j (-1)^{k-j} \mathscr{F}^{k-j} \varphi_1.$$

This yields

$$\sum_{j=1}^{k} \binom{j}{k} (\lambda_1^j - \lambda_2^j)(-1)^{k-j} \mathscr{F}^{k-j} \varphi_1 = 0 \text{ on } \omega.$$

Keeping in mind that $(\lambda_1 - \mathscr{F})^{\ell_1} \varphi_1 = 0$, we get, for

$$z = \sum_{j=1}^{k} \frac{j}{k} (\lambda_1^j - \lambda_2^j)(-1)^{k-j} \mathscr{F}^{k-j} \varphi_1,$$

$$(\lambda_1 - \mathscr{F})^{\ell_1} z = 0, \quad z = 0 \text{ on } \omega.$$

Hence $z \equiv 0$ on $\mathscr{O}$ and this implies that

$$(\lambda_1 - \mathscr{F})^{\ell_1} \varphi_1 = (\lambda_2 - \mathscr{F})^{\ell_1} \varphi_1 = 0 \text{ on } \mathscr{O}.$$

Therefore, φ_1 is the eigenfunction of $\mathscr{F}$ for the eigenvalue λ_2. The contradiction we arrived at shows that φ_1 and φ_2 are linearly independent on ω. By induction with respect to $m = 1, \dots, N$, it follows that system $\{\varphi_j\}_{j=1}^M$ is linearly independent on ω, as claimed.

Similarly, it follows that $\{P_N^* \varphi_j\}_{j=1}^N$ is linearly independent on ω.

Taking into account that the orthogonalization matrix $\{a_{jk}\}$ as well as $\{b_{jk}\}$ are nonsingular, it follows that properties 1°, 2°, *and* 3° remain true for the systems $\{\phi_j\}_{j=1}^N$ and $\{\chi_j\}_{j=1}^N$ as well.

Now, let us turn to the proof of existence of $\{w_i\}_{i=1}^K$ satisfying (5.28). For simplicity, we shall assume first that λ_j, $j = 1, \dots, N$, are all distinct eigenvalues. Then we must prove that there is $w \in Z_N^u$ such that $(w, P_N^* \varphi_i)_\omega \neq 0$ for all $i = 1, \dots, N$. We choose w of the form $w = \sum_{j=1}^{N} \alpha_j \varphi_j$. Taking into account (5.27) and the fact that the matrix $\{b_{k\ell}\}$ is nonsingular (it is symmetric and positive definite), we conclude that, for existence of w, it suffices to show that $\det \|(\varphi_j, P_N^* \varphi_i)_\omega\| \neq 0$. But the latter is obviously true because otherwise, there is $\eta = \sum_{i=1}^{N} \beta_i P_N^* \varphi_i$ such that $(\eta, \varphi_j)_\omega = 0$ for all $j = 1, \dots, N$. Since, as follows by 1° and 2°, both systems $\{\varphi_j\}$ and $\{P_N^* \varphi_j\}$ are linearly independent on $(L^2(\omega))^d$, we arrived to a contradiction. It should be mentioned in this context also that there are infinitely many solutions $w_1, w_2, \dots, w_K$ to problem (5.28).

Let us consider now the general case of multiple eigenvalues.

We set $\Phi_i = \{\chi_k\}_{k=\ell_{i-1}+1}^{\ell_{i-1}+\ell_i} \subset (Z_N^u)^{\ell_i}$, $i = 1, \dots, M$ and $D^0 = [w_1, \dots, w_K] \in (Z_N^u)^K$. Choose the $K \times \ell_i$ matrices D_i such that $\operatorname{rank} D_i = \ell_i$ and consider the system

$$(P_N^* \Phi_i, D^0)_\omega = D_i, \ i = 1, \dots, M. \tag{5.39}$$

Here $P_N^*\Phi_i = \text{col}[P_N^*\chi_{\ell_{i-1}+1}, \dots P_N^*\chi_{\ell_{i-1}+\ell_i}]$. We set

$$\widetilde{D} = \text{col}[D_1, \dots, D_M]$$

and denote by $d_{kj},\ k = 1, \dots, N,\ j = 1, \dots, K$ the elements of the matrix $\widetilde{D}$. We look for w_k of the form

$$w_k = \sum_{j=1}^{N} \alpha_{kj}\varphi_j,\ k = 1, \dots, K.$$

Then system (5.39) reduces to

$$\sum_{j=1}^{N} \alpha_{\ell j}(\varphi_j, P_N^*\chi_i)_\omega = d_{\ell i},\ i = 1, \dots, N,\ \ell = 1, \dots, k.$$

Thus system (5.39) has a solution if $\det(\varphi_j, P_N^*\chi_i)_\omega$ is not zero.

But the letter is an immediate consequence of properties 1°, 2°. This completes the proof of Lemma 5.1. ■

Now, we shall return back to system (5.8) and prove that

Lemma 5.3. *For each $y^0 \in H = L^2(\mathcal{O})$, there is a real valued controller $u = u_N(x,t)$,*

$$u_N(x,t) = \sum_{i=1}^{2K} u_N^i(t)\psi_i(x), \tag{5.40}$$

where $\psi_i \in Y_N^1$ for $1 \le i \le K$, $\psi_i \in Y_N^2$ for $K+1 \le i \le 2K$, such that

$$y \in L^\infty(0,T; L^2(\mathcal{O})) \cap L^2(0,T; H_0^1(\mathcal{O})) \cap L^2(\delta,T; H^2(\mathcal{O})),\ \forall\, 0 < \delta < T, \tag{5.41}$$

$$|y(t)|_2 + |u_N^i(t)| \le Ce^{-\gamma_0 t}|y^0|_2,\ \forall\, t \ge 0, \tag{5.42}$$

where $0 < \gamma_0 < |\text{Re}\,\lambda_{N+1}|$.

Proof. If we substitute the feedback controller v_N given by Lemma 5.1 in system (5.24), we get by (5.21) that

$$|\zeta_N(t)| \le Ce^{-\gamma_1 t}|(I - P_N)z_0|$$

and, therefore, the solution z to system (5.20) satisfies

$$|z(t)| \le Ce^{-\gamma_1 t}|z_0|,\ \forall\, t \ge 0. \tag{5.43}$$

Hence there is a controller $v = v_N$ of the form (5.34) with $w_i \in Z_N^u$ such that the solution z to (5.20) satisfies estimate (5.43). Hence

$$v_N = \sum_{j=1}^{K} (v_j^1 + i v_j^2)(\phi_j^1 + i\phi_j^2),$$

where $v_j^1(t), v_j^2, \phi_j^1, \phi_j^2$ are real valued and $\phi_j^1 \in Y_N^1, \phi_j^2 \in Y_N$, for $j = 1, \dots, K$. Coming to system (5.8), we infer that there is a controller u_N of the form (5.40) such that (5.41), (5.42) hold. This completes the proof.

Remark 5.3. The constant $2K$ arising in Lemma 5.3 and defined by (5.10) is the minimal dimension of the stabilizing controller u. As mentioned earlier, if all unstable eigenvalues are simple, then $K = 1$ and, if all unstable eigenvalues are real and simple, then the dimension of the controller u is 1. However, the proof of Lemma 5.3, that is, the existence of a controller u which stabilizes the linear system (2.8) becomes much simpler for $K = N$. In fact, if in system (5.23) we look for a stabilizing controller v of the form $v_N = \sum\limits_{j=1}^{N} v_j^N w_j$, then the corresponding system is exactly null controllable because one can find $\{w_j\}_{j=1}^N \subset Z_N^u$ such that $\det D \neq 0$. The details are omitted.

Proof of Theorem 5.1. The idea of the proof was already outlined in Section 1.5 (see (1.92), (1.93)). Namely, consider the optimal control problem

$$\varphi(y^0) = \text{Min}\left\{\frac{1}{2}\int_0^\infty \left(|A^{\frac{3}{4}} y(t)|_2^2 + |u(t)|_{2K}^2\right) dt\right\} \tag{5.44}$$

subject to $u \in L^2(0, \infty; \mathbb{R}^{2K})$ and to the state system

$$\begin{aligned} y' + Ay + A_0 y &= m \sum_{i=1}^{2K} u_i \psi_i \\ y(0) &= y^0, \ t \in (0, \infty). \end{aligned} \tag{5.45}$$

Here $|u|_{2K} = \left(\sum\limits_{i=1}^{2K} u_i^2\right)^{1/2}$ and the functions ψ_i are as in Lemma 5.3.

We set $D(\varphi) = \{y^0 \in H;\ \varphi(y^0) < \infty\}$. By Lemma 5.3, we know that for each y^0 there is $u \in L^2(0, \infty; R^{2K})$ such that $|y(t)|_2 \leq Ce^{-\gamma t}, \forall t > 0$. If we multiply equation (5.45) by $A^{\frac{1}{2}} y$, we obtain

$$\frac{1}{2}\frac{d}{dt}|A^{\frac{1}{4}} y(t)|_2^2 + |A^{\frac{3}{4}} y(t)|_2^3 \leq |(A_0 y, A^{\frac{1}{2}} y)| + C|u(t)|_{2K}|A^{\frac{1}{2}} y(t)|_2.$$

Taking into account that $|A^{\frac{1}{2}}y|_2^2 \le |A^{\frac{1}{4}}y|\,|A^{\frac{3}{4}}y|_2$, $|(A_0y, A^{\frac{1}{2}}y)|_2 \le C|A^{\frac{1}{2}}y|_2^2$ and $|y|_2 \in L^\infty(0,\infty)$, we see that $A^{\frac{3}{4}}y \in L^2(0,\infty;H)$ if $y^0 \in D(A^{\frac{1}{4}})$. Hence

$$\varphi(y^0) \le C|A^{\frac{1}{4}}y^0|_2^2,\quad \forall\, y^0 \in D(A^{\frac{1}{4}}). \tag{5.46}$$

On the other hand, we have

$$\varphi(y^0) \ge C|A^{\frac{1}{4}}y^0|_2^2,\quad \forall\, y^0 \in D(A^{\frac{1}{4}}). \tag{5.47}$$

Indeed, it is easy to see that for each $y^0 \in D(\varphi)$, problem (5.44) has a unique solution $(y^*, u^*) \in L^2(\mathbb{R}^+; D(A^{\frac{3}{4}}))\times L^2(\mathbb{R}^+; \mathbb{R}^{2K})$. Moreover, $y^* \in C(\mathbb{R}^+; D(A^{\frac{1}{4}}))$. If we multiply the equation (5.45), where $y = y^*$, $u = u^*$, by $A^{\frac{1}{2}}y^*$ and integrate it on $(0,\infty)$, we obtain

$$\begin{aligned}\frac{1}{2}\,|A^{\frac{1}{4}}y^0|_2^2 &\le \int_0^\infty ((Ay^*, A^{\frac{1}{2}}y^*) + (A_0y^*, A^{\frac{1}{2}}y^*) + |u^*|_{2K}|A^{\frac{1}{2}}y^*|_2)dt\\ &\le C\int_0^\infty (|A^{\frac{3}{4}}y^*|_2^2 + |u|_{2K}^2)dt = C\varphi(y^0)\end{aligned}$$

because

$$|(A_0y^*, A^{\frac{1}{2}}y^*)| \le C\|g'(y_e)\|_{C(\overline{\mathcal{O}})}(|y^*|_2|A^{\frac{1}{2}}y^*|_2 + |A^{\frac{1}{2}}y^*|_2^2) \le C\|y^*\|^2.$$

This implies that $D(\varphi) = D\left(A^{\frac{1}{4}}\right) = W$. We note that, by (5.44), φ is of the form

$$\varphi(y^0) = \frac{1}{2}\int_0^\infty (|A^{\frac{3}{4}}L_1(y^0)|_2^2 + |L_2(y^0)|_{2K}^2)dt, \tag{5.48}$$

where $L_1 : W \to D(A^{\frac{3}{2}})$, $L_2 : W \to \mathbb{R}^{2k}$, are linear continuous operators. Hence, there exists a linear self-adjoint positive operator $G : H \to H$ with domain $D(G)\subset W$ such that

$$\frac{1}{2}\,(Gy^0, y^0) = \varphi(y^0),\quad \forall\, y^0 \in D(G). \tag{5.49}$$

Moreover, the operator G extends to all of W and $G \in L(W, W')$. ■

Lemma 5.4. *Let* $(y^*, u^*) \in L^2(0,\infty; D(A^{\frac{3}{4}}))\times L^2(0,\infty;\mathbb{R}^{2K})$ *be the optimal pair for problem* (5.44) *corresponding to* $y^0 \in D(A^{\frac{1}{4}})$. *Then*

$$u^*(t) = -\{(Gy^*(t), \psi_i)_\omega\}_{i=1}^{2K},\quad \forall\, t \ge 0. \tag{5.50}$$

Moreover, $V \subset D(G)$, that is,

$$|Gy|_2 \leq C\|y\|, \quad \forall\, y \in V \tag{5.51}$$

and there exist $C_i > 0$, $i = 1, 2$, such that

$$C_1\|y\|_W^2 \leq (Gy, y) \leq C_2\|y\|_W^2, \quad \forall\, y \in W. \tag{5.52}$$

The operator G is the solution to the algebraic Riccati equation

$$(Ay + A_0y, Gy) + \frac{1}{2}\sum_{i=1}^{2K}(Gy(t), \psi_i)_\omega^2 = \frac{1}{2}|A^{\frac{3}{4}}y|_2^2, \quad \forall\, y \in D(A). \tag{5.53}$$

Proof. Estimate (5.52) follows immediately from (5.46) and (5.47). By the dynamic programming principle, it follows that, for each $T > 0$, (y^*, u^*) is the solution to optimal control problem

$$\min\left\{\frac{1}{2}\int_0^T(|A^{\frac{3}{4}}y(t)|^2 + |u(t)|_{2K}^2)dt + \varphi(y(T));\ (y, u) \text{ subject to (5.45)}\right\}.$$

By the maximum principle, we obtain that

$$u^*(t) = \{(q^T(t), \psi_i)_\omega\}_{i=1}^{2K}, \tag{5.54}$$

where q^T is the solution to adjoint equation

$$\begin{aligned} &\frac{d}{dt}q^T - (A + A_0)^*q^T = A^{\frac{3}{2}}y^*, \quad t \in (0, T), \\ &q^T(T) = -Gy^*(T). \end{aligned} \tag{5.55}$$

Since $q^T(t) \in W' \subset V'$, it follows from the standard existence theory for linear evolution equations that $q^T \in L^2(0, T; H) \cap C([0, T], V')$. Moreover, if $y^0 \in V$, we have $q^T \in C([0, T]; H)$. Indeed, $y^* \in L^2(0, T; D(A))$ if $y^0 \in V$. If we set $z = A^{-\frac{1}{2}}q^T$, it follows from (5.47) that

$$z' - Az - A^{-\frac{1}{2}}A_0^*A^{\frac{1}{2}}z = Ay^*.$$

It is easy to check that the estimate

$$\frac{1}{2}\frac{d}{dt}\|z(t)\|^2 \geq |Az(t)|_2^2 - |Ay^*(t)|_2\,|Az(t)|_2 - C|Az(t)|_2\|z(t)\|, \text{ a.e. } t \in (0, T),$$

which, together with (5.55), shows that $z \in C([0, T]; V)$. Hence $q^T \in C([0, T]; H)$, as claimed.

By (5.55) and by the unique continuous property for the backward linear parabolic equation

$$q_t - (A + A_0)^* q = 0 \text{ in } Q,$$
$$q = 0 \text{ on } \partial\mathscr{O},$$

it follows that $q^T = q^{T'}$ on $(0, T)$ for $0 < T < T'$. Hence $q^T = q$ is independent of T and so (5.54) and (5.55) extend to all T of $\mathbb{R}^+$. Moreover, we have

$$Gy^0 = -q^T(0). \tag{5.56}$$

Here is the argument. For all $z^0 \in D\left(A^{\frac{1}{4}}\right)$, we have

$$\varphi(y^0) - \varphi(z^0) \le \int_0^T ((A^{\frac{3}{4}} y^*(t), A^{\frac{3}{4}}(y^*(t) - z^*(t)) + (u^*(t), u^*(t) - v^*(t))_{2K})dt$$
$$+(Gy^*(T), y^*(T) - z^*(T)),$$

where (z^*, v^*) is the optimal pair of problem (5.45) corresponding to z^0. On the other hand, it follows from (5.55) that

$$\frac{d}{dt}(q^T(t), y^*(t) - z^*(t)) = (A^{\frac{3}{4}} y^*(t), A^{\frac{3}{4}}(y^*(t) - z^*(t))) + (u^*(t), u^*(t) - v^*(t))_{2K}.$$

Integrating it on $(0, T)$ and then substituting the result into the previous inequality, we obtain that

$$\varphi(y^0) - \varphi(z^0) \le -(q^T(0), y^0 - z^0)$$

which yields (5.56), as desired.

By (5.55) and (5.56), we infer that $q(t) = -Gy^*(t)$ for all $t \ge 0$, which shows (5.50), as desired.

Now, let $y^0 \in V$. Then, by the previous argument, we see that $q \in C([0, T]; H)$ which, together with (5.60), show that $G : V \to H$. By the closed graph theorem, one obtains (5.51), as desired.

Next, we have

$$\varphi(y^*(t)) = \frac{1}{2}\int_t^\infty (|A^{\frac{3}{4}} y^*(s)|_2^2 + |u^*(s)|_{2K}^2)ds, \quad \forall t \ge 0$$

and, therefore,

$$\left(Gy^*(t), \frac{d}{dt} y^*(t)\right) + \frac{1}{2}|A^{\frac{3}{4}} y^*(t)|_2^2 + \frac{1}{2}\sum_{i=1}^{2K} (Gy^*(t), \psi_i)_\omega^2 = 0.$$

Hence we have, for all $t \geq 0$,

$$-(Gy^*(t), Ay^*(t) + A_0y^*(t)) - \frac{1}{2}\sum_{i=1}^{2K}(Gy^*(t), \psi_i)_\omega^2 + \frac{1}{2}|A^{\frac{3}{4}}y^*(t)|_2^2 = 0$$

which implies (5.53), thereby completing the proof.

Proof of Theorem 5.2 (Continued). Let G be the operator defined by (5.49). Then, putting the feedback controller (5.50) in system (5.45), we are led to the closed loop system

$$\begin{aligned} &y_t + Ay + f(x, y, \nabla y) + \sum_{i=1}^{2K} m(G(y - y_e), \psi_i)_\omega \psi_i = 0, \quad t > 0, \\ &y(0) = y_0. \end{aligned} \tag{5.57}$$

By Theorem 1.5, it follows that, under hypotheses (H_1) and (H_2), for $y_0 \in L^2(\mathscr{O})$, equation (5.57) has a unique solution $y \in L^2(0, T; V) \cap C([0, T]; H)$.

Moreover, $t^{\frac{1}{2}}y \in L^2(0, T; D(A)) \cap W^{1,2}([0, T]; H)$. Also, for $\varepsilon \to 0$, we have

$$y_\varepsilon \to y \text{ strongly in } C([0, T]; L^2(\mathscr{O})), \text{ weakly in } L^2(0, T; V), \tag{5.58}$$

where y_ε is the solution to the approximating equation (see (1.46))

$$\begin{aligned} &(y_\varepsilon)_t + Ay_\varepsilon + g_\varepsilon(y_\varepsilon) + ay_\varepsilon + b \cdot \nabla y_\varepsilon + \sum_{i=1}^{2K} m(G(y_\varepsilon - y_e)\psi_i)_\omega \psi_i = 0 \\ &\qquad \text{in } (0, \infty) \times \mathscr{O}, \\ &y_\varepsilon(0) = y_0 \text{ in } \mathscr{O}, \end{aligned} \tag{5.59}$$

where $g_\varepsilon = g((1 + \varepsilon g)^{-1})$ is the Yosida approximation of g.

We shall show that, if $y_0 \in \mathscr{U}_\rho$ for ρ sufficiently small, then, for $t \to \infty$, the solution y to (5.57) goes exponentially to y_e. To this end, we substitute y by $y + y_e$ into (5.57) and reduce the problem to that of stability of the null solution to equation

$$\begin{aligned} &y_t + Ay + A_0y + R(y) + \sum_{i=1}^{2K} m(Gy, \psi_i)_\omega \psi_i = 0, \ t > 0, \\ &y(0) = y^0 \equiv y_0 - y_e, \end{aligned} \tag{5.60}$$

where

$$R(y) = f(x, y+y_e, \nabla(y+y_e)) - f(x, y_e, \nabla y_e) - A_0y \equiv g(y+y_e) - g(y_e) - g'(y_e)y.$$

As seen above, we may approximate (5.60) by (see (5.59))

$$(y_\varepsilon)_t + Ay_\varepsilon + A_0 y_\varepsilon + R_\varepsilon(y_\varepsilon) + \sum_{i=1}^{2K} m(Gy_\varepsilon, \psi_i)_\omega \psi_i = 0,$$
$$y_\varepsilon(0) = y^0 = y_0 - y_e, \tag{5.61}$$

where $R_\varepsilon(y) \equiv g_\varepsilon(y+y_e) - g_\varepsilon(y_e) - g'_\varepsilon(y_e)y$. Taking into account that

$$g'_\varepsilon(y) \equiv \frac{g'(J(y))}{1+\varepsilon g'(J(y))}, \quad J(y) = (I+\varepsilon g)^{-1}(y),$$
$$g''_\varepsilon(y) \equiv \frac{g''(J(y))}{(1+\varepsilon g'(J(y)))^3}, \ y \in \mathbb{R},\ \varepsilon > 0,$$

by (5.4) and (5.61), we see that

$$|R_\varepsilon(y)| \le C(|y|^\eta + |y|^2),\ \forall\, y \in \mathbb{R},\ \varepsilon > 0, \tag{5.62}$$

where C is independent of ε. We multiply (5.61) by Gy and use (5.51) and (5.53) to get after some calculation that

$$\frac{d}{dt}(Gy_\varepsilon(t), y_\varepsilon(t)) + \sum_{i=1}^{2K}(Gy_\varepsilon, \psi_i)^2_\omega + |A^{\frac{3}{4}} y_\varepsilon(t)|_2^2 \le 2|(Gy_\varepsilon(t), R_\varepsilon y_\varepsilon(t))|. \tag{5.63}$$

We shall show that there exist $C > 0$ independent of ε, such that

$$|(Gy, R(y))| \le C(|A^{\frac{1}{4}} y|_2 + |A^{\frac{1}{4}} y|_2^{\eta-1})|A^{\frac{3}{4}} y|_2^2,\ \ \forall\, y \in D(A^{\frac{3}{4}}). \tag{5.64}$$

To this end, by virtue of (5.62), we prove that

$$|(Gy, |y|^\eta)| \le C|A^{\frac{1}{4}} y|_2^{\eta-1} |A^{\frac{3}{4}} y|_2^2,\ \ \forall\, y \in D(A^{\frac{3}{4}}) \tag{5.65}$$

and

$$|(G(y), |y|^2)| \le C|A^{\frac{1}{4}} y|_2 |A^{\frac{3}{4}} y|_2^2,\ \ \forall\, y \in D(A^{\frac{3}{4}}). \tag{5.66}$$

We recall that $D(A^s) = H_0^{2s}(\mathcal{O})$ for $s > \frac{1}{4}$ and $D(A^{\frac{1}{4}}) \subset H^{\frac{1}{2}}(\mathcal{O})$. Thus, the norm $|\cdot|_{D(A^s)}$ and $\|\cdot\|_{2s}$ are equivalent for $s > \frac{1}{4}$.

By (5.51) and by the interpolation inequality

$$\|y\| \le |A^{\frac{3}{4}} y|_2^{\frac{1}{2}} |A^{\frac{1}{4}} y|_2^{\frac{1}{2}},\ \forall\, y \in D(A^{\frac{3}{4}}),$$

it follows that

$$|(Gy,|y|^\eta)| \le C\|y\||y|^\eta_{2\eta} \le C|A^{\frac14}y|_2^{\frac12}|A^{\frac34}y|_2^{\frac12}|y|^\eta_{2\eta} \tag{5.67}$$

while, by Sobolev's embedding theorem (see, e.g., [2], p 217),

$$|y|_{2\eta} \le C\|y\|_\alpha, \tag{5.68}$$

for $\alpha = \frac{d(\eta-1)}{2\eta}$ · Then, again by the interpolation inequality, we obtain that

$$\|y\|_\alpha \le C|A^{\frac14}y|_2^{\frac32-\alpha}|A^{\frac34}y|_2^{\alpha-\frac12},\ \alpha = \frac{d(\eta-1)}{2\eta}\cdot$$

This, together with (5.67), implies that

$$|(Gy,|y|^\eta)| \le C|A^{\frac14}y|^{\frac12(3\eta+1-(\eta-1)d)}|A^{\frac34}y|_2^{\frac12(\eta-1)(d-1)} \le C|A^{\frac14}y|_2^{\eta-1}|A^{\frac34}y|_2^2, \tag{5.69}$$

and so, (5.67) follows. Similarly, we have

$$\begin{aligned}|(G(y),|y|^2)| &\le C\|y\|\,|y|_4^2 \le C|A^{\frac14}y|_2^{\frac12}|A^{\frac34}y|_2^{\frac12}\|y\|^2_{\frac d4}\\ &\le C|A^{\frac14}y|_2^{\frac12(7-d)}|A^{\frac34}y|^{\frac{d-1}{2}} \le C|A^{\frac14}y|_2|A^{\frac34}y|_2^2,\end{aligned}$$

and (5.66) follows.

Now, we come back to (5.60). Then, by (5.52)and (5.63)–(5.64), we see that

$$\begin{aligned}&\frac{d}{dt}(Gy_\varepsilon(t),y_\varepsilon(t)) + \frac12|A^{\frac34}y_\varepsilon(t)|_2^2 \le C(\|y_\varepsilon(t)\|_W^{\eta-1} + \|y_\varepsilon(t)\|_W)|A^{\frac34}y_\varepsilon(t)|_2^2\\ &\le \frac14|A^{\frac34}y_\varepsilon(t)|_2^2 + C\left(\varphi(\|y_\varepsilon(t)\|_W) - \frac{1}{4C}\right)|A^{\frac34}y_\varepsilon(t)|_2^2,\end{aligned} \tag{5.70}$$

where $\varphi(r) = r^{\eta-1} + r$. Taking into account (5.50), we get by (5.70) that

$$\begin{aligned}&\tfrac{d}{dt}\,(Gy_\varepsilon(t),y_\varepsilon(t)) + \tfrac12\,A^{\frac34}y_\varepsilon(t)|_2^2\\ &\le C\left(\varphi\left(\left(\tfrac{(Gy_\varepsilon(t),y_\varepsilon(t))}{C_1}\right)^{\frac12}\right) - \tfrac{1}{2C}\right)|A^{\frac34}y_\varepsilon(t)|_2^2,\end{aligned} \tag{5.71}$$

where C, C_1 are independent of ε.

If we multiply (5.61) by $A^{\frac12}y_\varepsilon$ and take into account that, since g_ε is Lipschitz,

$$|R_\varepsilon(y)| \le C_\varepsilon(|y|+1),\ \forall\, y\in\mathbb{R},$$

we get

$$\frac{1}{2}\frac{d}{dt}\|y_\varepsilon(t)\|_W^2 + \frac{1}{2}A^{\frac{3}{4}}y_\varepsilon(t)|^2 \le C_\varepsilon^1\|y_\varepsilon(t)\|_W(1+\|y_\varepsilon(t)\|), \text{ a.e. } t>0.$$

This yields

$$\|y_\varepsilon(t)\|_W \le \|y^0\|_W + C_\varepsilon^1 t, \ \forall t>0. \tag{5.72}$$

We denote by $(0, T_\varepsilon)$ the maximal interval with the property that

$$\varphi\left(\left(\frac{(Gy_\varepsilon(t), y_\varepsilon(t))}{C_1}\right)^{\frac{1}{2}}\right) - \frac{1}{2C} \le 0, \ t\in[0, T_\varepsilon).$$

By (5.72), it follows that, for $\|y^0\|_W \le \rho$ sufficiently small, $T_\varepsilon > 0$. On the other hand, by (5.71) we have

$$\frac{d}{dt}(Gy_\varepsilon(t), y_\varepsilon(t)) + \frac{1}{2}|A^{\frac{3}{4}}y_\varepsilon(t)|^2 \le 0, \text{ a.e. } t\in(0, T_\varepsilon).$$

Since φ is monotonically decreasing and $t \to (Gy_\varepsilon(t), y_\varepsilon(t))$ is decreasing on $(0, T_\varepsilon)$, we infer that $T_\varepsilon = \infty$ and

$$\frac{d}{dt}(Gy_\varepsilon(t), y_\varepsilon(t)) + \frac{1}{2}|A^{\frac{3}{4}}y_\varepsilon(t)|_2^2 \le 0, \text{ a.e. } t>0.$$

Letting $\varepsilon \to 0$, this yields, for some $\gamma > 0$,

$$\frac{d}{dt}(Gy(t), y(t)) + \gamma(Gy(t), y(t)) \le 0, \quad \text{a.e. } t>0 \tag{5.73}$$

and

$$\int_0^\infty |A^{\frac{3}{4}}y(t)|_2^2 dt \le 2(Gy^0, y^0).$$

Moreover, by (5.50) and (5.73), we have

$$|y(t)|_2 \le \|y(t)\|_W \le C|y^0|e^{-\gamma_0 t}, \quad \forall t \ge 0.$$

This completes the proof. ■

Remark 5.4. It should be mentioned that the above argument shows that the feedback controller (5.16) is robust with respect to smooth perturbations. More precisely, if $Q_\varepsilon \in L(V, H)$ is such that $|Q_\varepsilon y| \le \delta(\varepsilon)|Gy|$, $\forall y \in V$, where

$\delta(\varepsilon) \to 0$ as $\varepsilon \to 0$, then, for all $\varepsilon > 0$ sufficiently small, the feedback controller

$$u = -\sum_{i=1}^{2K} ((G + Q_\varepsilon)(y - y_e), \psi_i)_\omega \psi_i$$

still exponentially stabilizes y_e in the neighborhood $\mathscr{U}_\rho$ of y_e.

This implies that, if G_N is a finite dimensional approximation of the Riccati equation (5.13) (equivalently, (5.15)), that is

$$G_N \mathscr{F}_N + \mathscr{F}_N^* G_N + \mathscr{N}_N = (A^{\frac{3}{2}})_N,$$

then, for N large enough, the corresponding feedback controller (5.16) is stabilizable.

Example 5.1. The steady-state solution $y_e = 0$ to the nonlinear heat equation

$$\begin{aligned} & y_t - y_{xx} - y \quad y^3 = mu && \text{in } (0,\pi)\times(0,\infty),\\ & y(0,t) = y(\pi,t) = 0 && \text{in } (0,\infty),\\ & y(x,0) = y_0(x) && \text{in } (0,\pi) \end{aligned}$$

where $m = \mathbf{1}_{[0,\frac{\pi}{2}]}$ is unstable. Clearly, $A = -\Delta$, $D(A) = H_0^1(0,\pi) \cap H^2(0,\pi)$, $A_0 y = -y$ and $\psi_j(x) = \sin(jx)$, $\lambda_j = 1 - j^2$. The stabilizable feedback controller (5.16) is, in this case, $u = -\left(\int_0^{\frac{\pi}{2}} G(y)(x) \sin x \, dx\right) \sin x$, where $G : L^2(0,\pi) \to L^2(0,\pi)$ is the solution to the Riccati equation (5.13) (or (5.15)),

$$\begin{aligned} -G(\Delta y + y) - (\Delta + 1)G(y) + \int_0^{\frac{\pi}{2}} G(y)(\xi) \sin \xi \, d\xi \; G(m(x) \sin x) \\ = \sum_{j=1}^{\infty} j^{\frac{3}{2}} \left(\int_0^{\pi} y(\xi) \sin(j\xi) d\xi \right) \sin jx. \end{aligned} \tag{5.74}$$

If we define

$$G^N(z) = \sum_{i,j=1}^{N} g_{ij} z_i \sin jx, \; z(x) = \sum_{i=1}^{N} z_i \sin ix,$$

we approximate (5.74) by the finite dimensional Riccati equation

$$2G^N A_N + G^N B_N B_N^* G^N = F^N, \tag{5.75}$$

where

$$F^N = \text{diag}\|j^{\frac{3}{2}}\|_{j=1}^N \text{ and } \qquad A_N = \text{diag}\|j^2 - 1\|_{j=1}^N,$$

$$B_N = \text{column}\left\| \int_0^{\frac{\pi}{2}} \sin x \sin jx\, dx \right\|_{j=1}^N, \ F^N = \text{column}\|j^{\frac{3}{2}}\|_{j=1}^N.$$

If $G^N = \|g_{ij}^N\|_{i,j=1}^N$ is the solution to (5.75), then the feedback controller

$$u(x,t) = -\sum_{i,j=1}^N g_{ij}^N y_i(t) b_j \sin x,$$
$$b_j = \int_0^{\frac{\pi}{2}} \sin jx \sin x\, dx, \ y_i(t) = \int_0^{\pi} y(x,t) \sin ix\, dx,$$

is stabilizable for N sufficiently large.

Remark 5.5. Theorem 5.1 extends along the above lines to the boundary stabilization of the semilinear heat equation

$$\begin{aligned} & y_t - \Delta y + ay + g(y) = 0 \ \text{ in } Q, \\ & y = u \qquad\qquad\qquad\quad\ \text{ on } \Sigma, \\ & y(x,0) = y_0(x) \qquad\quad\ \ \text{ in } \mathcal{O}, \end{aligned}$$

where $g \in C^2(\mathbb{R})$ satisfies conditions (5.3) and (5.4) and $a \in \mathbb{R}$.

Remark 5.6. Theorems 5.1 and 5.2 extend to more general parabolic equations (5.1) where Δ is replaced by the second order elliptic operator

$$Ly \equiv \sum_{i,j=1}^n \frac{\partial}{\partial x_i}\left(a_{ij}(x)\frac{\partial}{\partial x_j}\right),$$

where a_{ij} satisfy condition (1.11). Moreover, the Dirichlet homogeneous condition in (5.1) can be replaced by the Neumann boundary condition $\frac{\partial y}{\partial \nu} = 0$. The proofs are exactly the same.

It should be said also that, taking into account the Sobolev embedding theorem, condition $1 \le d \le 3$ can be relaxed.

Stabilization of Semilinear Parabolic Systems
Consider here the reaction-diffusion system

$$\begin{aligned}
&y_t(x,t) - \Delta y(x,t) + f(y(x,t), z(x,t)) = m(x)u(x,t) \text{ in } Q \equiv \mathcal{O}\times R^+,\\
&z_t(x,t) - \alpha\Delta z(x,t) + g(y(x,t), z(x,t)) = 0, \text{ in } Q,\\
&y(x,t) = z(x,t) = 0 \text{ on } \Sigma \equiv \partial\mathcal{O}\times\mathbb{R}^+,\\
&y(x,0) = y_0(x),\ z(x,0) = z_0(x) \text{ in } \mathcal{O},
\end{aligned} \tag{5.76}$$

where α is a positive constant and m is the characteristic function of an open subset $\omega \subset \mathcal{O} \subset \mathbb{R}^d$, $d = 1, 2, 3$. This reaction diffusion system is relevant in mathematical description of several physical processes including chemical reactions, semiconductor theory, nuclear reactor dynamics, and population dynamics (see, e.g., [80] and references given there).

Let (y_e, z_e) be a steady-state (equilibrium) solution to (5.76), that is,

$$\begin{aligned}
-\Delta y_e(x) + f(y_e(x), z_e(x)) &= 0 \ \text{ in } \mathcal{O},\\
-\alpha\Delta z_e(x) + g(y_e(x), z_e(x)) &= 0 \ \text{ in } \mathcal{O},\\
y_e(x) = z_e(x) &= 0 \ \text{ on } \partial\mathcal{O}.
\end{aligned} \tag{5.77}$$

We shall prove here with the methods developed above that, if f and g are of quadratic growth in (y, z) (the exact conditions will be specified later), then the steady-state solution (y_e, z_e) to (5.76) is locally exponentially stabilizable by a feedback controller provided by a linear quadratic LQ control problem associated with the linearized systems. In the present situation it seems that system (5.76) is not locally controllable (anyway this is still an open problem) and so the stabilization cannot be established via local controllability. Then, proceeding as in the previous case, we first prove the exponential stabilization of the linearized system of (5.76) in a direct way and then introduce an appropriate infinite horizon LQ problem associated with the linearized system with unbounded cost functional, from which we find a solution R to an algebraic Riccati equation associated with the LQ problem, such that the feedback controller provided by R locally stabilizes the steady-state solution (y_e, z_e).

In a similar way will be treated the local stabilization of the phase-field system with one internal controller on the temperature field.

The following assumptions will be in effect everywhere in the following:

(K_1) The steady-state solution y_e is in $C(\overline{\mathcal{O}})$ with $\nabla y_e \in (C(\overline{\mathcal{O}}))^d$.
(K_2) $f, g \in C^1(\mathbb{R}\times\mathbb{R})$ satisfy the growth condition

$$\|\nabla f(y,z)\|_{\mathrm{Lip}} + \|\nabla g(y,z)\|_{\mathrm{Lip}} \le C,\ \forall\, y, z \in \mathbb{R}, \tag{5.78}$$

where $\|\cdot\|_{\mathrm{Lip}}$ is a Lipschitz norm and

$$\begin{aligned}
(f(y,z) - f(\bar{y},\bar{z}))(y-\bar{y}) + (g(y,z) - g(\bar{y},\bar{z}))(z-\bar{z})&\\
\ge -\gamma((y-\bar{y})^2 + (z-\bar{z})^2)&,
\end{aligned} \tag{5.79}$$

for some $\gamma \in \mathbb{R}$.

(K_3) $f_z(y_e, z_e) \equiv b,\ g_y(y_e, z_e) \equiv b_1,\ g_z(y_e, z_e) - \alpha f_y(y_e, z_e) \equiv c_0$, where c_0, b, and b_1 are constants, $b_1 \neq 0$ and α is a positive constant.

Let $A = -\Delta$ and $A_1 = -\Delta + aI$ with domain $D(A) = D(A_1) = H^2(\mathscr{O}) \cap H_0^1(\mathscr{O})$, where I is the identity operator on $H = L^2(\mathscr{O})$, $a = f_y(y_e, z_e) \in C(\overline{\mathscr{O}})$, and let $\mathscr{A}, \mathscr{A}_0 : H \times H \to H \times H$ be defined by

$$\mathscr{A} = \begin{bmatrix} A & 0 \\ 0 & \alpha A \end{bmatrix}, \ \mathscr{A}_0 = \begin{bmatrix} aI & bI \\ b_1 I & (a\alpha + c_0)I \end{bmatrix} = \begin{bmatrix} f_y(y_e, z_e) & f_z(y_e, z_e) \\ g_y(y_e, z_e) & g_z(y_e, z_e) \end{bmatrix},$$

respectively, with the domain $D(\mathscr{A}) = D(A) \times D(A)$. It is clear that $\mathscr{A}$ is linear positive and self-adjoint operator on $H \times H$. We denote by $\mathscr{A}^s$ and A^s, $s \in (0, 1)$, the fractional powers of operator the $\mathscr{A}$ and A, respectively. If $W = D(A^{\frac{1}{4}})$ with the graph norm $\|y\|_W = |A^{\frac{1}{4}} y|_2$, then $W \times W = (D(A^{\frac{1}{4}}))^2$. Let $B : H \to H \times H$ be the operator defined by

$$B = \begin{bmatrix} mI \\ 0 \end{bmatrix},$$

and let $B^* : H \times H \to H$ be the adjoint operator of B, that is,

$$B^* \begin{pmatrix} p \\ q \end{pmatrix} = mp, \ \forall (p, q) \in H \times H.$$

We have

$$\mathscr{A}(y, z) \equiv \mathscr{A} \begin{pmatrix} y \\ z \end{pmatrix} = (Ay, \alpha Az), \ \forall (y, z) \in D(\mathscr{A}).$$

In the following, the vectors

$$\begin{pmatrix} f \\ g \end{pmatrix} \text{ and } \begin{pmatrix} y \\ z \end{pmatrix}$$

will be also denoted by (f, g) and (y, z), respectively.

We denote by $\langle \cdot, \cdot \rangle$ and $\langle\langle \cdot, \cdot \rangle\rangle$ the scalar products of H and $H \times H$, respectively.

Let λ_j and φ_j be the eigenvalues and corresponding orthonormal system of eigenfunctions of A_1, that is,

$$A_1 \varphi_j = \lambda_j \varphi_j, \ j = 1, \dots, \infty.$$

(We note that A_1 is self-adjoint and so all the eigenvalues λ_j are real.)

From now on, we shall omit all x, t in the functions of x, t if there is no ambiguity, and we shall use the same symbol C to denote several positive constants.

Theorem 5.3. *Suppose that* (K_1), (K_2) *and* (K_3) *hold. Then there exist* N *and a linear positive self-adjoint operator* $R : D(R) \subset H\times H \to H\times H$ *such that the feedback controller*

$$u = -\sum_{i=1}^{N}(R_{11}(y-y_e)+R_{12}(z-z_e),\varphi_i)_\omega\,\varphi_i$$

exponentially stabilizes (y_e, z_e) *in a neighborhood* $E_\rho = \{(y_0, z_0) \in W\times W;\ \|y_0 - y_e\|_W + \|z_0 - z_e\|_W < \rho\}$ *of* (y_e, z_e). *More precisely, for each pair* $(y_0, z_0) \in E_\rho$ *there is a solution* $(y, z) \in C(\mathbb{R}^+; H) \cap L^2_{loc}(\mathbb{R}^+; V)\times C(\mathbb{R}^+; H) \cap L^2_{loc}(R^+; V)$ *to closed loop system*

$$\begin{aligned}&(y_t, z_t) + \mathscr{A}(y, z) + (f(y, z), g(y, z))\\ &\qquad + B\left(\sum_{i=1}^{N}(R_{11}(y-y_e)+R_{12}(z-z_e),\varphi_i)_\omega\,\varphi_i\right) = 0,\ t>0,\\ &(y(0), z(0)) = (y_0, z_0),\end{aligned} \tag{5.80}$$

such that

$$\int_0^\infty |\mathscr{A}^{\frac34}(y(t), z(t)) - \mathscr{A}^{\frac34}(y_e, z_e)|^2_{H\times H}dt \le C(\|y_0 - y_e\|^2_W + \|z_0 - z_e\|^2_W) \tag{5.81}$$

and

$$|y(t) - y_e|_2 + |z(t) - z_e|_2 \le Ce^{-\gamma t}(\|y_0 - y_e\|_W + \|z_0 - z_e\|_W),\ \forall t > 0, \tag{5.82}$$

for some $\gamma > 0$, *where* $(\cdot,\cdot)_\omega$ *denotes the inner product in* $L^2(\omega)$ *and*

$$R = \begin{bmatrix} R_{11} & R_{12}\\ R_{12} & R_{22}\end{bmatrix},$$

where $R_{ij} \in L(H; H),\ i, j = 1, 2$. *Moreover,* R *is the solution to the algebraic Riccati equation*

$$\begin{aligned}&\langle\langle R(y, z), (\mathscr{A} + \mathscr{A}_0)(y, z)\rangle\rangle\\ &\quad +\frac12\sum_{i=1}^{N}(R_{11}(y-y_e)+R_{12}(z-z_e),\varphi_i)^2_\omega = \frac12\,|\mathscr{A}^{\frac34}(y, z)|^2_{H\times H}\end{aligned} \tag{5.83}$$

for all $(y, z) \in D(\mathscr{A})$.

Here $\{\varphi_i\}$ are eigenvectors of A_1 corresponding to λ_i.

Theorem 5.3 can be understood better in the framework of stability theory for nonlinear parabolic systems. If the spectrum $\sigma(L)$ of the linearized operator

$$L = \begin{bmatrix} -\Delta & 0 \\ 0 & -\alpha\Delta \end{bmatrix} + \begin{bmatrix} f_y(y_e, z_e) & f_z(y_e, z_e) \\ g_y(y_e, z_e) & g_z(y_e, z_e) \end{bmatrix}$$

has a nonempty intersection with $\{\lambda; \operatorname{Re}\lambda < 0\}$, then the equilibrium solution (y_e, z_e) is unstable. However, under assumptions (K_1), (K_2) the nonlinear system (5.76) is locally stabilizable by a linear feedback controller u with the support in an arbitrary open subset ω of $\mathcal{O}$.

An Example
The parabolic system

$$\begin{aligned} &y_t - y_{xx} + ay + bz + f(y, z) = 0,\ t > 0,\ x \in (0, \pi), \\ &z_t - z_{xx} + cy + dz + g(y, z) = 0,\ t > 0,\ x \in (0, \pi), \\ &y(0, t) = y(\pi, t) = 0,\ z(0, t) = z(\pi, t) = 0 \end{aligned} \tag{5.84}$$

arises in chemical reactor theory. (See, e.g., [11].) Here f, g are smooth functions such that $f(0, 0) = g(0, 0) = 0$, $\nabla f(0, 0) = \nabla g(0, 0) = 0$ and a, b, c, d are constants. It is easily seen that the eigenvalues $\lambda = \lambda_j$ of the corresponding linearized system should satisfy the equation

$$\lambda^2 + \lambda(d + 2j^2 + a) + (a + j^2)(d + j^2) - bc = 0.$$

(To get this, we look for eigenfunctions (φ_j, ψ_j) of the form $\sum_{n=1}^{\infty} (\varphi_j^n, \psi_j^n) \sin nx$.)
This implies that, for $bc - (a + j^2)(d + j^2) > 0$, there are positive eigenvalues λ_j, that is, the zero solution is unstable. However, according to Theorem 5.3, for each open subset $\omega \subset (0, \pi)$ there is a controller u with the support in $\omega \times (0, \infty)$ which stabilizes the system

$$\begin{aligned} &y_t - y_{xx} + ay + bz + f(y, z) = mu,\ t > 0,\ x \in (0, \pi), \\ &z_t - z_{xx} + cy + dz + g(y, z) = 0,\ t > 0,\ x \in (0, \pi), \\ &y(0, t) = y(\pi, t) = 0,\ z(0, t) = z(\pi, t) = 0. \end{aligned}$$

The above example reveals another interesting phenomenon which has deep implications in biology and the dynamics of chemical reactions.

We have seen that the constant solution (0, 0) to system (5.84) is unstable for a, b, c, d chosen as above but, as easily seen, for corresponding ordinary differential system

$$\begin{aligned} &\frac{dy}{dt} + ay + bz + f(y, z) = 0, \\ &\frac{dz}{dt} + cy + dz + g(y, z) = 0, \end{aligned}$$

the solution (0, 0) is asymptotically stable because the eigenvalues of this system satisfy

$$\lambda^2 + (a + d)\lambda + ad - bc = 0$$

and so, for $a+d > 0$, $(a+d)^2 - 4(ad-bc) < 0$, all have negative real parts. This is the so-called Turing's instability (see Murray [101]), which roughly speaking means that the diffusion may destabilize a $2D$ ordinary differential system. More precisely, the stable constant solutions to ordinary differential systems could become unstable in the presence of diffusions and this is at the origin of spatial pattern which can be experimentally seen in chemical reaction or biological evolution.

Proof of Theorem 5.3. Proceeding as in Theorem 5.2, we shall linearize system (5.76) in (y_e, z_e). Namely,

$$\begin{aligned} &y_t + A_1 y + bz = mu, \ t > 0, \\ &z_t + \alpha A_1 z + c_0 z + b_1 y = 0, \ t > 0, \\ &y(0) = y^0 \equiv y_0 - y_e, \ z(0) = z^0 \equiv z_0 - z_e, \end{aligned} \tag{5.85}$$

where $A_1 = A + aI$. We shall prove first

Lemma 5.5. *There exist N and $u_j \in L^\infty(\mathbb{R}^+)$, $j = 1, \dots, N$, such that the controller*

$$u(x,t) = \sum_{i=1}^{N} u_j(t)\varphi_i(x)$$

stabilizes exponentially system (5.85). *More precisely, we have*

$$|y(t)| + |z(t)| + \sum_{j=1}^{N} |u_j(t)| \le Ce^{-\gamma t}(|y^0| + |z^0|), \ \forall t > 0,$$

for some $\gamma, C > 0$, where (y, z) is the solution to (5.85) *corresponding to $u = \sum\limits_{j=1}^{N} u_j(t)\varphi_j(x)$.*

Proof. Let $\{\varphi_i\}_{i=1}^\infty$ be an orthonormal basis formed with the eigenfunctions of A_1 and let $\{\lambda_i\}_{i=1}^\infty$ be the corresponding eigenvalues of A_1. Since by hypotheses (K_1) and (K_2), $a = f_y(y_e, z_e) \in C(\overline{\mathcal{O}})$, it follows that $\lambda_i \to \infty$ as $i \to \infty$. Let $X_N = \text{span}\{\varphi_i\}_{i=1}^N$, $P_N : L^2(\mathcal{O}) \to X_N$ be the orthonormal projection on X_N and $Q_N = I - P_N$. Set $y_N = P_N y$, $z_N = P_N z$, $\widetilde{y}_N = Q_N y$ and $\widetilde{z}_N = Q_N z$. Applying P_N and Q_N to system (5.85), respectively, and taking into account that $y = y_N + \widetilde{y}_N$, $z = z_N + \widetilde{z}_N$, we obtain that

$$
\begin{aligned}
&\frac{dy_N^i}{dt} + \lambda_i y_N^i + bz_N^i = P_N(mu, \varphi_i), \quad t > 0,\ i = 1, \dots, N, \\
&\frac{dz_N^i}{dt} + \lambda_i \alpha z_N^i + c_0 z_N^i + b_1 y_N^i = 0, \quad t > 0,\ i = 1, \dots, N, \\
&y_N^i(0) = (P_N y^0)^i,\ z_N^i(0) = (P_N z^0)^i,
\end{aligned}
\tag{5.86}
$$

where $y_N = (y_N^1, \dots, y_N^N)$, $z_N = (z_N^1, \dots, z_N^N)$, $P_N y^0 = ((P_N y^0)^1, \dots, (P_N y^0)^N)$ and $P_N z^0 = ((P_N z^0)^1, \dots, (P_N z^0)^N)$, and

$$
\begin{aligned}
&\frac{d}{dt}\widetilde{y}_N + Q_N A_1 \widetilde{y}_N + b\widetilde{z}_N = Q_N(mu) \qquad t > 0, \\
&\frac{d}{dt}\widetilde{z}_N + \alpha Q_N A_1 \widetilde{z}_N + c_0 \widetilde{z}_N + b_1 \widetilde{y}_N = 0 \ \ t > 0, \\
&\widetilde{y}_N(0) = Q_N y^0,\ \widetilde{z}_N(0) = Q_N z^0.
\end{aligned}
\tag{5.87}
$$

We prove first for N large enough the exact null controllability of the finite–dimensional system (5.86), where u is given as above, that is,

$$
\begin{aligned}
&\frac{d}{dt} y_N^i + \lambda_i y_N^i + bz_N^i = \sum_{j=1}^{N} u_j(t)(\varphi_j, \varphi_i)_\omega,\ t > 0,\ i = 1, \dots, N, \\
&\frac{d}{dt} z_N^i + \lambda_i \alpha z_N^i + c_0 z_N^i + b_1 y_N^i = 0, \qquad t > 0,\ i = 1, \dots, N, \\
&y_N(0) = P_N y^0,\ z_N(0) = P_N z^0.
\end{aligned}
\tag{5.88}
$$

The backward dual system corresponding to (5.88) is the following:

$$
\begin{aligned}
&\frac{d}{dt} p_N^i - \lambda_i p_N^i - b_1 q_N^i = 0, \qquad t > 0,\ i = 1, \dots, N, \\
&\frac{d}{dt} q_N^i - \lambda_i \alpha q_N^i - c_0 q_N^i - bp_N^i = 0,\ t > 0,\ i = 1, \dots, N.
\end{aligned}
\tag{5.89}
$$

We set

$$
B_N = \|(\varphi_j, \varphi_i)_\omega\|_{i,j=1}^N, \tag{5.90}
$$

where $\|(\varphi_j, \varphi_i)_\omega\|_{i,j=1}^N$ denotes the $N \times N$ matrix whose components are $(\varphi_j, \varphi_i)_\omega$, $j, i = 1, \dots, N$. We note that the right-hand side of system (5.105) is $\mathscr{B}u = (B_N u, 0)$, $u = \{u_j\}_{j=1}^N$ and the adjoint $\mathscr{B}^* : \mathbb{R}^N \times \mathbb{R}^N \to \mathbb{R}^N$ of $\mathscr{B} : \mathbb{R}^N \to \mathbb{R}^N \times \mathbb{R}^N$ is given by

$$
B^*(p, q) = B_N p,\ \forall\, (p, q) \in \mathbb{R}^N \times \mathbb{R}^N,\ i = 1, \dots, N.
$$

Recall that (see Section 1.5) system (5.88) is exactly null controllable on $[0, T]$, $T > 0$, if and only if

$$B_N(p_N(t)) = 0, \quad \forall t \in (0, T) \tag{5.91}$$

implies that $p_N(t) \equiv 0$ and $q_N(t) \equiv 0$, $\forall t \in (0, T)$, where

$$p_N(t) = \left(p_N^1(t) \dots p_N^N(t)\right), \quad q_N(t) = \left(q_N^1(t) \dots q_N^N(t)\right). \tag{5.92}$$

By (5.91) and (5.92), it follows that

$$\sum_{i=1}^{N} (\varphi_j, \varphi_i)_\omega p_N^i(t) \equiv 0, \quad t \in (0, T), \; j = 1, \dots, N. \tag{5.93}$$

On the other hand, we have

$$\det \|(\varphi_j, \varphi_i)_\omega\|_{i,j=1}^N \neq 0. \tag{5.94}$$

To prove this, we give an argument by reduction to absurdity. If (5.94) does not hold, then the system $\{\varphi_j\}_{j=1}^N$ is linearly dependent in $L^2(\omega)$, and so, we might assume without loss of generality that

$$\varphi_N = \sum_{i=1}^{N-1} \gamma_i \varphi_i \quad \text{on } \omega, \text{ where } \gamma_i \neq 0, \text{ for some } i \in \{1, \dots, N-1\}.$$

Then, we have

$$(-\Delta + a)\varphi_N = \sum_{i=1}^{N-1} \gamma_i(-\Delta + a)\varphi_i = \sum_{i=1}^{N-1} \gamma_i \lambda_i \varphi_i = \lambda_N \varphi_N = \sum_{i=1}^{N-1} \lambda_N \gamma_i \varphi_i \text{ on } \omega,$$

which implies that

$$\sum_{i=1}^{N-1} (\lambda_N - \lambda_i)\gamma_i \varphi_i = 0 \text{ on } \omega.$$

Since, by hypotheses (K_1) and (K_2), $a = f_y(y_e, z_e) \in C(\overline{\mathcal{O}})$, it follows that $\lambda_i \to \infty$ as $i \to \infty$. So we may take N large enough such that $\lambda_N > \lambda_i$ for $i = 1, \dots, N-1$. Thus in this way one arrives to conclusion that there is at least one φ_i such that $\varphi_i = 0$ on ω. By virtue of the unique continuation property of the solutions to elliptic equations, this implies that $\varphi_i = 0$ on $\mathcal{O}$. This is a contradiction and so we obtain (5.94).

It follows that $p_N(x,t) = 0$, $\forall t \geq 0$. Then we obtain that $q_N(x,t) = 0$, $\forall t \geq 0$, as claimed. Thus system (5.88) is exactly null controllable and this implies that there are $\{u_j(t)\}_{j=1}^N$ (given in feedback form) such that system (5.86) is exponentially stable with arbitrary exponent γ_0. More precisely, we have

$$|y_N^i(t)| + |z_N^i(t)| + |u_i(t)| \leq C_{\gamma_0} e^{-\gamma_0 t}(|y^0| + |z^0|), \quad i = 1, \dots, N, \tag{5.95}$$

where C_{γ_0} is a positive constant independent of i and t but dependent on γ_0. Then, by (5.87), we obtain

$$\frac{d}{dt}((\widetilde{y}_N^i)^2 + (\widetilde{z}_N^i)^2) + 2\lambda_i(\widetilde{y}_N^i)^2 + (b+b_1)\widetilde{y}_N^i\widetilde{z}_N^i + 2(\alpha\lambda_i + c_0)(\widetilde{z}_N^i)^2$$
$$= \left(\sum_{j=1}^N (Q_N(mu_j), \varphi_i)\right)\widetilde{y}_N^i, \ i = 1+N, \dots,$$

which together with (5.95) implies that, for N large enough, we have

$$|\widetilde{y}_N^i(t)|^2 + |\widetilde{z}_N^i(t)|^2 \leq Ce^{-\gamma_N t}(|y^0|^2 + |z^0|^2) + \int_0^t e^{-\gamma_N(t-s)}\sum_{j=1}^N |u_j(s)|^2 ds$$
$$\leq Ce^{\gamma t}(|y^0|^2 + |z^0|^2), \quad \forall t \geq 0, \ i = N+1, \dots,$$

where γ, γ_N are positive constants. This completes the proof of the lemma.

Proof of Theorem 5.3 (Continued). Now, we rewrite system (5.85) as

$$\begin{aligned} &(y_t, z_t) + \mathscr{A}(y,z) + \mathscr{A}_0(y,z) = Bu, \ t > 0 \\ &(y(0), z(0)) = (y^0, z^0), \end{aligned} \tag{5.96}$$

and consider the LQ (linear quadratic) optimal control problem

$$\Psi(y^0, z^0) = \text{Min}\left\{\frac{1}{2}\int_0^\infty (|\mathscr{A}^{\frac{3}{4}}(y,z)|_{H\times H}^2 + |u|_2^2)dt \text{ subject to (5.96)}\right\}. \tag{5.97}$$

Let $D(\Psi)$ be the set of all $(y^0, z^0) \in H \times H$ such that $\Psi(y^0, z^0) < \infty$. We observe first that for each $(y^0, z^0) \in D(\mathscr{A}^{\frac{1}{4}})$, there exists $u \in L^2(\mathbb{R}^+, H)$ such that the system (5.96) has a unique solution $(y,z) \in L^2(R^+, D(\mathscr{A}^{\frac{3}{4}}))$. Indeed, it follows from Lemma 5.5 that, for each pair $(y^0, z^0) \in D(\mathscr{A}^{\frac{1}{4}})$, there exists a $u \in L^2(\mathbb{R}^+; H)$ such that the system (5.96) has a solution (y,z) satisfying

$$|(y(t), z(t)|_{H\times H}^2 \leq Ce^{-\gamma t}\left(|y^0|_2^2 + |z^0|_2^2\right), \quad \forall t > 0.$$

Multiplying scalarly in $H \times H$ equation (5.96) by $\mathscr{A}^{\frac{1}{2}}(y, z)$, we obtain after some calculation that

$$\begin{aligned}
&\frac{1}{2}\frac{d}{dt}|\mathscr{A}^{\frac{1}{4}}(y(t), z(t))|^2_{H\times H} + |\mathscr{A}^{\frac{3}{4}}(y(t), z(t))|^2_{H\times H} \\
&\leq \left|\left\langle\!\left\langle Bu, \mathscr{A}^{\frac{1}{2}}(y, x)\right\rangle\!\right\rangle\right| + \left|\left\langle\!\left\langle \mathscr{A}_0(y, z), \mathscr{A}^{\frac{1}{2}}(y, z)\right\rangle\!\right\rangle\right| \\
&\leq C\left(|mu|_2 + |y|_2 + |z|_2\right)|\mathscr{A}^{\frac{1}{2}}(y, z)| \\
&\leq C\left(|mu|_2 + |y|_2 + |z|_2\right)|\mathscr{A}^{\frac{3}{4}}(y, z)|_{H\times H} \\
&\leq C\left(|mu|_2^2 + |y|_2^2 + |z|_2^2 + \frac{1}{2}|\mathscr{A}^{\frac{3}{4}}(y, z)|^2_{H\times H}\right)
\end{aligned}$$

which implies that

$$\begin{aligned}
&\frac{d}{dt}|\mathscr{A}^{\frac{1}{4}}(y(t), z(t))|^2_{H\times H} + |\mathscr{A}^{\frac{3}{4}}(y(t), z(t))|^2_{H\times H} \\
&\quad\leq C\left(|y(t)|_2^2 + |z(t)|_2^2 + |mu(t)|_2^2\right) \\
&\quad\leq C\left(e^{-2\gamma t}\left(|y^0|_2^2 + |z^0|_2^2\right) + |mu(t)|_2^2\right).
\end{aligned}$$

Integrating it on $(0, \infty)$, we obtain that

$$\int_0^\infty |\mathscr{A}^{\frac{3}{4}}(y(t), z(t))|^2_{H\times H} dt \leq C < \infty$$

as desired.

Thus, for each pair $(y^0, z^0) \in D\left(\mathscr{A}^{\frac{3}{4}}\right)$, $\Psi(y^0, z^0) < \infty$ and, therefore,

$$\Psi(y^0, z^0) \leq C|\mathscr{A}^{\frac{1}{4}}(y^0, z^0)|^2_{H\times H}, \quad \forall (y^0, z^0) \in D\left(\mathscr{A}^{\frac{1}{4}}\right). \tag{5.98}$$

On the other hand, we have

$$\Psi(y^0, z^0) \geq C|\mathscr{A}^{\frac{1}{4}}(y^0, z^0)|^2_{H\times H}, \quad \forall (y^0, z^0) \in D\left(\mathscr{A}^{\frac{1}{4}}\right). \tag{5.99}$$

Indeed, it is readily seen for each pair $(y^0, z^0) \in D(\mathscr{A}^{\frac{1}{4}})$, problem (5.97) has a unique optimal solution

$$(y^*, z^*, u^*) \in L^2(\mathbb{R}^+; D(\mathscr{A}^{\frac{3}{4}}))\times L^2(\mathbb{R}^+; D(\mathscr{A}^{\frac{3}{4}}))\times L^2(\mathbb{R}^+; L^2(\mathscr{O})).$$

Multiplying, once again, equation (5.96) by $\mathscr{A}^{\frac{1}{2}}(y^*, z^*)$ and integrating on $\mathbb{R}^+$, we obtain that

$$\frac{1}{2}\left|\mathscr{A}^{\frac{1}{4}}(y^0,z^0)\right|^2_{H\times H}\le\int_0^\infty\left(\left|\left\langle\!\left\langle\mathscr{A}(y^*,z^*),\mathscr{A}^{\frac{1}{2}}(y^*,z^*)\right\rangle\!\right\rangle\right|\right.$$
$$\left.+\left|\left\langle\!\left\langle\mathscr{A}_0(y^*,u^*),\mathscr{A}^{\frac{1}{2}}(y^*,u^*)\right\rangle\!\right\rangle\right|+|mu^*|_2\left|\mathscr{A}^{\frac{1}{2}}y^*\right|_2\right)dt$$
$$\le C\int_0^\infty\left(\left|\mathscr{A}^{\frac{3}{4}}(y^*,u^*)\right|^2_{H\times H}+|u^*|_2^2\right)dt=C\Psi(y^0,z^0),$$

because

$$\left|\left\langle\!\left\langle\mathscr{A}_0(y^*,z^*),\mathscr{A}^{\frac{1}{2}}(y^*,z^*)\right\rangle\!\right\rangle\right|\le C\left(\left|\left\langle(1+a)y^*,A^{\frac{1}{2}}y^*\right\rangle\right|\right.$$
$$\left.+\left|\left\langle y^*,A^{\frac{1}{2}}z^*\right\rangle\right|+\left|\left\langle(1+a)z^*,A^{\frac{1}{2}}z^*\right\rangle\right|\right)\le C\left|\mathscr{A}^{\frac{3}{4}}(y^*,z^*)\right|^2_{H\times H}.$$

By (5.98) and (5.99), we see that

$$D(\Psi)=D(\mathscr{A}^{\frac{1}{4}})=W\times W.$$

Since Ψ likewise φ (see (5.48)) is a quadratic functional, there exists a linear positive and self-adjoint operator $R:H\times H\to H$ with the domain $D(R)\subset W\times W$ such that

$$\frac{1}{2}\left\langle\!\left\langle R(y^0,z^0),(y^0,z^0)\right\rangle\!\right\rangle=\Psi(y^0,z^0),\quad\forall\,(y^0,z^0)\in D(\Psi).\tag{5.100}$$

Moreover, R extends to all $W\times W$ and $R\in L(W\times W;\,W'\times W')$.

Lemma 5.6. *Let* (y^*,z^*,u^*) *be optimal for problem* (5.97) *corresponding to* $(y^0,z^0)\in W\times W$. *Then*

$$u^*(t)=-\sum_{i=1}^N(R_{11}y^*(t)+R_{12}z^*(t),\varphi_i)_\omega\varphi_i,\quad\forall t>0,\tag{5.101}$$

Moreover, $V\times V\subset D(R)$, *that is,*

$$|R(y,z)|^2_{H\times H}\le C\|(y,z)\|^2_{V\times V},\quad\forall\,(y,z)\in V\times V,\tag{5.102}$$

and there are constants $c_1>0,\ c_2>0$ *such that*

$$c_1\|(y,z)\|^2_{W\times W}\le\langle\langle R(y,z),(y,z)\rangle\rangle\le c_2\|(y,z)\|^2_{W\times W}.\tag{5.103}$$

The operator R *is the solution to algebraic Riccati equation*

$$\langle\langle(\mathscr{A}+\mathscr{A}_0)(y,z),R(y,z)\rangle\rangle+\frac{1}{2}\sum_{i=1}^N(R_{11}y+R_{12}z,\varphi_i)^2_\omega=\frac{1}{2}|\mathscr{A}^{\frac{3}{4}}(y,z)|^2_{H\times H}.\tag{5.104}$$

Proof. The proof follows closely that of Lemma 5.4. In fact, estimate (5.103) follows from the previous estimates. By the dynamic programming principle, it follows that $\forall\, T > 0$, (y^*, z^*, u^*) is the solution to optimal control problem

$$\text{Min}\left\{\frac{1}{2}\int_0^T \left(|\mathscr{A}^{\frac{3}{4}}(y, z)|_{H\times H}^2 + |u|^2\right) dt + \Psi(y(T), z(T)),\ \text{subject to (5.96)}\right\}.$$

By the maximum principle it follows that

$$u^*(t) = B^*(p^T(t), q^T(t)),\ u_i^*(t) = \left\langle p^T(t), \varphi_i\right\rangle_\omega,\ \forall\, t \in [0, T), \tag{5.105}$$

where (p^T, q^T) is the solution to the adjoint system

$$\begin{aligned} &\frac{d}{dt}(p^T, q^T) - (\mathscr{A} + \mathscr{A}_0^*)(p^T, q^T) = \mathscr{A}^{\frac{3}{2}}(y^*, z^*),\ \ t > 0, \\ &(p^T(T), q^T(T)) = -R(y^*(T), z^*(T)). \end{aligned} \tag{5.106}$$

Since $R(y^*(T), z^*(T)) \in W'\times W' \subset V'\times V'$, it follows from the standard existence theory for linear evolution equations that $(p^T, q^T) \in L^2(0, T; H\times H) \cap C([0, T]; V'\times V')$. Moreover, if $(y^0, z^0) \in V\times V$, then we have that $(p^T, q^T) \in C([0, T]; H\times H)$. Indeed, we have that $(y^*, z^*) \in L^2(0, T; D(\mathscr{A}))$ if $(y^0, z^0) \in V\times V$. Let $(\widetilde{p}, \widetilde{q}) = \mathscr{A}^{-\frac{1}{2}}(p^T, q^T)$. It follows from (5.106) that

$$\frac{d}{dt}(\widetilde{p}, \widetilde{q}) - \mathscr{A}(\widetilde{p}, \widetilde{q}) - \mathscr{A}^{-\frac{1}{2}}\mathscr{A}_0^*\mathscr{A}^{\frac{1}{2}}(\widetilde{p}, \widetilde{q}) = \mathscr{A}(y^*, z^*).$$

One can check easily that

$$\left|\left\langle\!\left\langle \mathscr{A}^{-\frac{1}{2}}\mathscr{A}_0^*\mathscr{A}^{\frac{1}{2}}(\widetilde{p}, \widetilde{q})\right\rangle\!\right\rangle\right| \le C|(\widetilde{p}, \widetilde{q})|_{V\times V}$$

which, combined with (5.106), implies that $(\widetilde{p}, \widetilde{q}) \in C([0, T]; V\times V)$. Hence $(p^T, q^T) \in C([0, T]; H\times H)$. By (5.105) and by the unique continuation property for the linear parabolic system

$$\frac{d}{dt}(p, q) - (\mathscr{A} + \mathscr{A}_0)^*(p, q) = 0,$$

it follows that $(p^T, q^T) = (p^{T'}, q^{T'})$ on $(0, T)$ for $0 < T < T'$. Hence $(p^T, q^T) = (p, q)$ is independent of T and so (5.105) and (5.106) extends to all of $\mathbb{R}^+$. Moreover, we have

$$R(y^0, z^0) = -(p(0), q(0)). \tag{5.107}$$

Indeed, for all $(y^1, z^1) \in D\left(\mathscr{A}^{\frac{1}{4}}\right)$, we have

$$
\begin{aligned}
&\Psi(y^0, z^0) - \Psi(y^1, z^1) \\
&\leq \int_0^T \left(\left\langle\!\left\langle \mathscr{A}^{\frac{3}{4}}(y^*, z^*), \mathscr{A}^{\frac{3}{4}}(y^* - y_1^*, z^* - z_1^*)\right\rangle\!\right\rangle + \left\langle u^*, u^* - u_1^*\right\rangle\right) dt \qquad (5.108) \\
&+ \langle\!\langle R(y^*(T), z^*(T)), (y^*(T) - y_1^*(T) - z_1^*(T))\rangle\!\rangle,
\end{aligned}
$$

where (y_1^*, z_1^*, u_1^*) is the optimal solution to problem (5.96) corresponding to (y^1, z^1). On the other hand, by (5.107), we have

$$
\begin{aligned}
&\frac{d}{dt}\left\langle\!\left\langle (p^T(t), q^T(t)), (y^*(t) - y_1^*(t), z^*(t) - z_1^*(t))\right\rangle\!\right\rangle \\
&\quad = \left\langle\!\left\langle \mathscr{A}^{\frac{3}{4}}(y^*, z^*), \mathscr{A}^{\frac{3}{4}}(y^* - y_1^*, z^* - z_1^*)\right\rangle\!\right\rangle + \left\langle u^*, u^* - u_1^*\right\rangle.
\end{aligned}
$$

Integrating it over $(0, T)$ and then substituting the result into (5.108), we obtain that

$$
\Psi(y^0, z^0) - \Psi(y^1, z^1) \leq -\left\langle\!\left\langle (p(0), q(0)), (y^0 - y^1, z^0 - z^1)\right\rangle\!\right\rangle
$$

which implies (5.107), as desired. By (5.106) and (5.107), we conclude that

$$
(p(t), q(t)) = -R(y^*(t), z^*(t)), \quad \forall t \geq 0,
$$

which along with (5.105) implies (5.107), as desired.

Now, let $(y^0, z^0) \in V \times V$. Then, it follows from the previous argument that $(p, q) \in C([0, T]; H \times H)$ which together with (5.107) implies that $R : V \times V \to H \times H$. By the closed graph theorem, we obtain (5.102), as desired.

Next, we observe that

$$
\Psi(y^*(t), z^*(t)) = \frac{1}{2}\int_t^\infty \left(|\mathscr{A}^{\frac{3}{4}}(y^*(s), z^*(s))|_{H\times H}^2 + |u^*(s)|_2^2\right) ds, \quad \forall t \geq 0,
$$

and, therefore,

$$
\begin{aligned}
&\left\langle\!\left\langle R(y^*(t), z^*(t)), \frac{d}{dt}(y^*(t), z^*(t))\right\rangle\!\right\rangle + \frac{1}{2}\left|\mathscr{A}^{\frac{3}{4}}(y^*(t), z^*(t))\right|_{H\times H}^2 \\
&\qquad + \frac{1}{2}|B^* R(y^*(t), z^*(t))|_H^2 = 0, \quad \text{a.e. } t > 0.
\end{aligned} \qquad (5.109)
$$

Since $|B^* R(y, z)|_H \leq C|(y, z)|_{V\times V}$, $\forall (y, z) \in V \times V$, we see that the operator $\mathscr{A} + \mathscr{A}_0 + BB^* R$ with the domain $D(\mathscr{A})$ generates a C_0-semigroup in $H \times H$. This implies that $\mathscr{A}(y^*, z^*)$, $\mathscr{A}_0(y^*, z^*)$, $BB^*R(y^*, z^*) \in C([0, \infty); H \times H)$. Then it follows from (5.109) that, for $(y^0, z^0) \in D(\mathscr{A})$,

$$-\langle\langle R(y^*(t), z^*(t)), (\mathscr{A} + \mathscr{A}_0)(y^*(t), z^*(t))\rangle\rangle - \frac{1}{2}|B^*R(y^*(t), z^*(t))|^2$$
$$+\frac{1}{2}\left|A^{\frac{3}{4}}(y^*(t), z^*(t))\right|^2_{H\times H} = 0, \ \forall t \geq 0,$$

which implies (5.104), thereby completing the proof of Lemma 5.6.

Now, let $F(y, z) = (f(y, z), g(y, z))$ and consider the closed loop system

$$(y_t, z_t) + \mathscr{A}(y, z) + F(y, z) + B\sum_{i=1}^{N}(R_{11}(y - y_e) + R_{12}(z - z_e), \varphi_i)_\omega \varphi_i = 0,$$
$$t > 0,$$
$$y(0), z(0) = (y_0, z_0). \tag{5.110}$$

It follows that, under hypotheses (K_1) and (K_2), for each pair $(y_0, z_0) \in W \times W$, systems (5.110) have a unique solution $(y, z) \in L^2(0, T; V \times V) \cap C([0, T]; H \times H)$ for all $T > 0$. Moreover, $t^{\frac{1}{2}}(y, z) \in L^2(0, T; D(\mathscr{A})) \cap W^{1,2}([0, T]; H \times H)$. (This follows as in Theorem 1.5 by proving first that the operator

$$(y, z) \to \mathscr{A}(y, z) + F(y, z) + B\sum_{i=1}^{N}(R_{11}(y - y_e) + R_{12}(z - z_e), \varphi_i)_\omega \varphi_i$$

is quasi-m-accretive in the space $H \times H$, but we omit the details. We shall prove that if $(y_0, z_0) \in E_\rho$ for ρ small enough, then this solution exponentially stabilizes (y_e, z_e). To this end, we substitute (y, z) by $(y + y_e, z + z_e)$ into equation (5.110) and reduce thus the problem to that of stability of the null solution to system

$$\begin{aligned}&(y_t, z_t) + \mathscr{A}(y, z) + \mathscr{A}_0(y, z) + \Phi(y, z)\\ &\quad + B\sum_{i=1}^{N}(R_{11}y + R_{12}z, \varphi_i)_\omega \varphi_i = 0, \ t > 0,\\ &(y(0), z(0)) = (y^0, z^0) \equiv (y_0 - y_e, z_0 - z_e),\end{aligned} \tag{5.111}$$

where

$$\begin{aligned}\Phi(y, z) &\equiv (\Phi^1(y, z), \Phi^2(y, z))\\ &= (f(y + y_e, z + z_e) - f(y_e, z_e), g(y + y_e, z + z_e)\\ &\quad - g(y_e, z_e)) - \mathscr{A}_0(y, z).\end{aligned}$$

By hypotheses (K_1) and (K_2), we see that

$$|\Phi(y, z)| \leq C(|y|^2 + |z|^2). \tag{5.112}$$

Multiplying (5.111) by $R(y,z)$ and using equation (5.104), we obtain after some calculation that

$$\frac{d}{dt}\langle\langle R(y,z),(y,z)\rangle\rangle+|B^*R(y,z)|^2_{H\times H}+|\mathscr{A}^{\frac{3}{4}}(y,z)|_2^2 \le 2|\langle\langle R(y,z),\Phi(y,z)\rangle\rangle|. \tag{5.113}$$

We claim that there exist $C>0$ independent of y and z, such that

$$|\langle\langle R(y,z),\Phi(y,z)\rangle\rangle| \le C(|A^{\frac{3}{4}}y|_2^2+|A^{\frac{3}{4}}z|_2^2)(|A^{\frac{1}{4}}y|_2+|A^{\frac{1}{4}}z|_2),\ \forall\,(y,z)\in D(A^{\frac{3}{4}}). \tag{5.114}$$

It is sufficient to show that

$$|(R^i(y,z),|y|^r+|z|^r)|\le C(|A^{\frac{3}{4}}y|_2^2+|A^{\frac{3}{4}}z|_2^2)(|A^{\frac{1}{4}}y|_2+|A^{\frac{1}{4}}z|_2),\ i=1,2, \tag{5.115}$$

where $r=2$ and $R^1(y,z)=R_{11}y+R_{12}z$, $R^2(y,z)=R_{12}y+R_{22}z$.

By (5.102) and by the interpolation relation $D(A^{\frac{1}{2}})=(D(A^{\frac{1}{4}}),D(A^{\frac{3}{4}}))_{\frac{1}{2}}$, we obtain that

$$|(R^i(y,z),|y|^r)|\le C\,(\|y\|+\|z\|)\,|y|^r_{2r} \le C(|A^{\frac{1}{4}}y|_2^{\frac{1}{2}}\,|A^{\frac{3}{4}}y|_2^{\frac{1}{2}}+|A^{\frac{1}{4}}z|_2^{\frac{1}{2}}\,|A^{\frac{3}{4}}z|_2^{\frac{1}{2}})|y|^r_{2r},\ i=1,2, \tag{5.116}$$

while, by Sobolev's embedding theorem (see (5.68)),

$$|y|_{2r}\le C\|y\|_\alpha,\ \text{for } 1\le r\le\frac{d}{d-2\alpha},\ d>2\alpha. \tag{5.117}$$

Then, again by the interpolation inequality, we obtain that

$$\|y\|_\alpha\le C|A^{\frac{1}{4}}y|_2^{\frac{3}{2}-\alpha}|A^{\frac{3}{4}}y|_2^{\alpha-\frac{1}{2}},$$

which together with (5.116) implies that

$$|\langle R^i(y,z),|y|^r\rangle|\le C\left(|A^{\frac{1}{4}}y|_2^{\frac{1}{2}+\left(\frac{3}{2}-\alpha\right)r}|A^{\frac{3}{4}}y|_2^{\frac{1}{2}+\left(\alpha-\frac{1}{2}\right)r} +|A^{\frac{1}{4}}z|_2^{\frac{1}{2}}|A^{\frac{1}{4}}y|_2^{\left(\frac{3}{2}-\alpha\right)r}|A^{\frac{3}{4}}z|_2^{\frac{1}{2}}|A^{\frac{3}{4}}y|_2^{\left(\alpha-\frac{1}{2}\right)r}\right). \tag{5.118}$$

Now, (5.117) holds if we choose $\alpha=\frac{5}{4}$, and so, by (5.118), we obtain that

$$|\left\langle R^i(y,z),|y|^r\right\rangle|\le C(|A^{\frac{3}{4}}y|_2^2+|A^{\frac{3}{4}}z|_2^2)(|A^{\frac{1}{4}}y|_2+|A^{\frac{1}{4}}z|_2),\ i=1,2. \tag{5.119}$$

Similarly, we may obtain that

$$|\left\langle R^i(y,z),|z|^r\right\rangle| \le C(|A^{\frac{3}{4}}y|_2^2+|A^{\frac{3}{4}}z|_2^2)(|A^{\frac{1}{4}}y|_2+|A^{\frac{1}{4}}z|_2),$$

and so, we get as above estimate (5.114), as claimed.

By (5.103), we see that, for ρ small enough, if $(y_0,z_0)\in E_\rho$, we have

$$\frac{d}{dt}\langle\langle R(y(t),z(t)),(y(t),z(t))\rangle\rangle+\frac{1}{2}|A^{\frac{3}{4}}(y(t),z(t))|_{H\times H}^2\le 0,\ \text{ a.e. } t\in(0,T^*),$$

where $(0,T^*)$ is the maximum interval of existence of the solution (y,z) with the property that

$$\langle\langle R(y(t),z(t)),(y(t),z(t))\rangle\rangle\le C_1 \tag{5.120}$$

for some constant $C_1>0$. From this, it follows that

$$\frac{d}{dt}\langle\langle R(y(t),z(t)),(y(t),z(t))\rangle\rangle+\gamma\,\langle\langle R(y(t),z(t)),(y(t),z(t))\rangle\rangle\le 0,\ \text{ a.e. } t>0,$$

which implies, as in Theorem 5.2, that

$$\int_0^\infty |A^{\frac{3}{4}}(y(t),z(t))|_{H\times H}^2dt\le 2\left\langle\left\langle R(y^0,z^0),(y^0,z^0)\right\rangle\right\rangle$$

and

$$|(y(t),z(t))|_{H\times H}^2\le\|(y(t),z(t))\|_{W\times W}^2\le C\|(y^0,z^0)\|_{W\times W}e^{-\gamma t},\ \forall t\ge 0,$$

as claimed. It should be said, however, that the existence of the interval $(0,T^*)$ with property (5.120) or, equivalently,

$$\|(y(t),z(\varepsilon))\|_{W\times W}\le C_2,\ t\in(0,T^*),$$

for some $C_2>0$, is a delicate problem which can be solved as in the proof of Theorem 5.2. Namely, one approximates (5.111) by a similar system where f,g are replaced by $f_\varepsilon,g_\varepsilon$, the Yosida approximations of f and g, and get for the corresponding solution $(y_\varepsilon,z_\varepsilon)$ the estimates (5.119)–(5.120). ■

Stabilization of the Phase-Field System

In this section, we shall use a similar method to study the stabilization of the zero equilibrium solution to the phase-field system

$$\begin{aligned}
&y_t+\ell\varphi_t-k\Delta y=mu \text{ in } Q=\mathcal{O}\times\mathbb{R}^+,\\
&\varphi_t-a\Delta\varphi-b(\varphi-\varphi^3)+dy=0,\ \text{ in } Q,\\
&y=0,\ \varphi=0 \text{ on } \Sigma=\partial\mathcal{O}\times\mathbb{R}^+,\\
&y(x,0)=y_0(x),\ \varphi(x,0)=\varphi_0(x) \text{ in } \mathcal{O}.
\end{aligned} \tag{5.121}$$

Equivalently,

$$\begin{aligned}
&y_t - k\Delta y + \ell(a\Delta\varphi + b(\varphi - \varphi^3) - dy) = mu \text{ in } Q,\\
&\varphi_t - a\Delta\varphi - b(\varphi - \varphi^3) + dy = 0 \text{ in } Q,\\
&y = 0,\ \varphi_{ij} = 0 \text{ on } \Sigma,\\
&y(x,0) = y_0(x),\ \varphi_{ij}(x,0) = \varphi_0(x) \text{ in } \mathcal{O}.
\end{aligned} \tag{5.122}$$

Here $\mathcal{O} \in \mathbb{R}^d$, $d = 1, 2, 3$, is an open and bounded subset, ℓ, k, a, b are nonnegative constants, and m is, as usually, the characteristic function of an open subset $\omega\subset\mathcal{O}$. The local controllability of steady-state solution to (5.121) with internal controllers on both equations was established in Section 3.3 via Carleman estimates for the linearized system. Here we shall assume that

(K_4) $a, b, k, \ell > 0$ and $d \neq 0$.

We notice that hypothesis (K_4) alone does not imply the stability of the zero steady-state solution to the phase-field system.

We shall denote, as above, by $|\cdot|_2$ the norm of $H = L^2(\mathcal{O})$ and, by $\langle\cdot,\cdot\rangle$ and $\langle\langle\cdot,\cdot\rangle\rangle$ the scalar products of H and $H\times H$, respectively. Throughout this section, we set $A = -\Delta$ with $D(A) = H^2(\mathcal{O}) \cap H_0^1(\mathcal{O})$ and let $\{\psi_i\}_{i=1}^\infty$ be an orthonormal basis of eigenfunctions of A. By $\{\lambda_i\}_{i=1}^\infty$ we denote the corresponding system of eigenvalues. Consider the linearized system

$$\begin{aligned}
&y_t + kAy - \ell aA\varphi + \ell b\varphi - \ell dy = mu,\ t > 0,\\
&\varphi_t + aA\varphi - b\varphi + dy = 0,\ t > 0,\\
&y(0) = y_0,\ \varphi(0) = \varphi_0.
\end{aligned} \tag{5.123}$$

Let $X_N = \text{span}\{\psi_i\}_{i=1}^N$, $P_N : L^2(\mathcal{O}) \to X_N$ be the orthonormal projection on X_N and $Q_N = I - P_N$. Applying P_N and Q_N to systems (5.123), we obtain that

$$\begin{aligned}
&\frac{d}{dt}y_N^i + (k\lambda_i - \ell d)y_N^i - (\ell a\lambda_i - \ell b)\varphi_N^i = \langle mu, \psi_i\rangle,\ t > 0,\ i = 1, \dots, N,\\
&\frac{d}{dt}\varphi_N^i + (a\lambda_i - b)\varphi_N^i + dy_N^i = 0,\ \ t > 0,\ i = 1, \dots, N,\\
&y_N^i(0) = y_{0N}^i,\ \varphi_N^i(0) = \varphi_{0N}^i
\end{aligned} \tag{5.124}$$

$$\begin{aligned}
&\frac{d}{dt}\widetilde{y}_N + kQ_NA\widetilde{y}_N - \ell aQ_NA\widetilde{\varphi}_N - \ell d\widetilde{y}_N + \ell b\widetilde{\varphi}_N = Q_N(mu),\ t > 0,\\
&\frac{d}{dt}\widetilde{\varphi}_N + aQ_NA\widetilde{\varphi}_N - b\widetilde{\varphi}_N + d\widetilde{y}_N = 0,\ t>0,\\
&\widetilde{y}_N(0) = Q_Ny_0,\ \widetilde{Q}_N(0) = Q_N\varphi_0,
\end{aligned} \tag{5.125}$$

where $y_N = P_Ny$, $\varphi_N = P_N\varphi$, $\widetilde{y}_N = Q_Ny$, $\widetilde{\varphi}_N = Q_N\varphi$, y_N^i and φ_N^i are the i-th components of y_N and φ_N, respectively.

The backward dual system of (5.124) is given by

$$\begin{aligned}&\frac{d}{dt}p^i-(k\lambda_i-\ell d)p^i-(a\lambda_i-b)q^i=0,\ i=1,\dots,N,\\&\frac{d}{dt}q^i-(\ell a\lambda_i-\ell b)q^i-dp^i=0,\ i=1,\dots,N.\end{aligned}\tag{5.126}$$

Lemma 5.7. *There are $u_j\in L^\infty(\mathbb{R}^+)$, $j=1,\dots,N$, such that for N large enough the controller*

$$u(x,t)=\sum_{i=1}^{N}u_i(t)\psi_i(x)\tag{5.127}$$

stabilizes exponentially system (5.123)*, that is,*

$$|y(t)|_2+|\varphi(t)|_2+\sum_{i=1}^{N}|u_i(t)|\le Ce^{-\gamma t}(|y_0|_2+|\varphi_0|_2),\ \ \forall t\ge 0$$

for some $\gamma>0$ and $C>0$.

Proof By (K_4) and by the same argument as that used in the proof of Lemma 5.6, it follows that system (5.124) is exactly null controllable and so there are $u_i(t)$, $i=1,\dots,N$, such that system (5.124), where $u=\sum\limits_{i=1}^{N}u_i\psi_i$, is exponentially stable with arbitrary exponent $\gamma_0>0$, that is,

$$|y_N^j(t)|+|\varphi_N^j(t)|+\sum_{i=1}^{N}|u_i(t)|\le C_{\gamma_0}e^{-\gamma_0 t}(|y_0|+|\varphi|),\ \forall t\ge 0.$$

Multiplying the first equation of (5.125) by $\widetilde{y}_N$, where u is given by (5.127) and the second equation of (5.125) by $\beta\widetilde{\varphi}_N$, where $\beta>0$ will be made precise later, we get

$$\begin{aligned}&\frac{1}{2}\frac{d}{dt}((\widetilde{y}_N^j)^2+\beta(\widetilde{\varphi}_N^j)^2)+(k\lambda_j-\ell d)(z\widetilde{y}_N^j)^2-(a\ell\lambda_j+\ell b)\widetilde{y}_N^j\widetilde{\varphi}_N^j\\&+\beta d\widetilde{y}_N^j\widetilde{\varphi}_N^j+\beta(a\lambda_j-b)(\widetilde{\varphi}_N^j)^2=\left(Q_N\left(m\sum_{i=1}^{N}u_i\psi_i\right),\psi_j\right)\widetilde{y}_N^j,\\&\qquad j=N+1,\dots\end{aligned}\tag{5.128}$$

then for β suitable chosen and N large enough we have that

$$\begin{aligned}|\widetilde{y}_N^j(t)|^2+\beta|\widetilde{\varphi}_N^j(t)|^2&\le e^{-\gamma_N t}(|\widetilde{y}_N^j(0)|^2+|\widetilde{\varphi}_N^j(0)|^2)+\int_0^t e^{-\gamma_N(t-s)}\sum_{i=1}^{N}|u_i(s)|^2ds\\&\le Ce^{-\gamma t}(|y_0|_2^2+|\varphi_0|_2^2),\ \ \forall t\ge 0,\ j=N+1,\dots,\end{aligned}$$

where $\gamma>0$. This completes the proof of Lemma 5.7.

Consider now the LQ optimal control problem

$$\Psi(y_0,\varphi_0) = \text{Min}\left\{\int_0^\infty (|A^{\frac{3}{4}}y|_2^2 + |A^{\frac{3}{4}}\varphi|_2^2 + |u|_2^2)dt;\right.$$
$$\left. u = \sum_{i=1}^N u_i(t)\psi_i(t),\ (y,\varphi,u) \text{ satisfies } (5.123)\right\}. \tag{5.129}$$

First, we claim that, $\forall\,(y_0,\varphi_0) \in D(A^{\frac{1}{4}})\times D(A^{\frac{1}{4}})$, $\Psi(y_0,z_0) < \infty$. Indeed, by Lemma 5.7, there exists a $u \in L^2(Q)$ such that the corresponding solution (y,φ) to systems (5.123) satisfies

$$|u(t)|^2 + |y(t)|_2^2 + |\varphi(t)|_2^2 \le Ce^{-\gamma t},\ \ \forall t > 0,$$

where $\gamma > 0$ and $C > 0$ are constants. We get after some calculation that

$$\begin{aligned} &\frac{1}{2}\frac{d}{dt}\left(|A^{\frac{1}{4}}y(t)|_2^2 + |A^{\frac{1}{4}}\varphi(t)|_2^2\right) + k|A^{\frac{3}{4}}y(t)|_2^2 \\ &-\ell a|A^{\frac{3}{4}}y(t)|_2\,|A^{\frac{3}{4}}\varphi(t)|_2 + a|A^{\frac{3}{4}}\varphi(t)|_2^2 \\ &\le C\left(|\varphi(t)|_2 + |y(t)|_2 + |mu(t)|_2\right)\left(|A^{\frac{3}{4}}y(t)|_2 + |A^{\frac{3}{4}}\varphi(t)|_2\right). \end{aligned} \tag{5.130}$$

Then it follows that

$$\int_0^\infty \left(|A^{\frac{3}{4}}y(t)|_2^2 + |A^{\frac{3}{4}}\varphi(t)|_2^2\right)dt < \infty,$$

as claimed. Hence

$$\Psi(y_0,\varphi_0) \le C\left(|A^{\frac{1}{4}}y_0|_2^2 + |A^{\frac{1}{4}}\varphi_0|_2^2\right).$$

Then, by the similar arguments to those in the proof of Lemma 5.6, we obtain that there exists a linear, positive, and self-adjoint operator $R : H\times H \to H\times H$, with $D(R) = D\left(A^{\frac{1}{4}}\right)\times D\left(A^{\frac{1}{4}}\right) = W\times W$, such that

$$\Psi(y_0,\varphi_0) = \langle\langle R(y_0,\varphi_0),(y_0,\varphi_0)\rangle\rangle\,.$$

We set as above

$$R = \begin{bmatrix} R_{11} & R_{12} \\ R_{12} & R_{22} \end{bmatrix}.$$

We have, therefore,

Lemma 5.8. *Let* (y^*, φ^*, u^*) *be optimal for problem* (5.129) *corresponding to* $(y_0, \varphi_0) \in W \times W$. *Then*

$$u^*(t) = -\sum_{i=1}^{N} (R_{11} y^*(t) + R_{12}\varphi^*(t), \psi_i)_\omega \psi_i, \quad \forall\, y, \varphi(t) > 0, \tag{5.131}$$

and there are constants $c_i > 0,\ i = 1, 2$, *such that*

$$c_1 \left(\|y\|_W^2 + \|\varphi\|_W^2\right) \leq \langle\langle R(y, \varphi), (y, \varphi)\rangle\rangle \leq c_2 \left(\|y\|_W^2 + \|\varphi\|_W^2\right), \quad \forall\, y, \varphi \in W. \tag{5.132}$$

The operator R *satisfies the Riccati equation*

$$\begin{aligned} &\langle kAy - \ell aA\varphi - \ell bdy + \ell b\varphi, R_{11}y + R_{12}\varphi\rangle \\ &\quad + \langle aA\varphi + dy - b\varphi, R_{12}y + R_{22}\varphi\rangle + \frac{1}{2}\sum_{i=1}^{N} |(R_{11}y + R_{12}\varphi, \psi_i)_\omega|^2 \\ &\quad = \frac{1}{2}\left(|A^{\frac{3}{4}}y|_2^2 + |A^{\frac{3}{4}}\varphi|_2^2\right), \quad \forall\, y, \varphi \in D(A). \end{aligned} \tag{5.133}$$

Now, we insert in (5.121) (equivalently, (5.122)) the feedback controller (5.131) and get the closed loop system

$$\begin{aligned} &y_t + kAy - \ell aA\varphi - \ell dy + \ell b\varphi - \ell b\varphi^3 \\ &\qquad + \sum_{i=1}^{N} m(R_{11}y + R_{12}\varphi, \psi_i)_\omega \psi_i = 0,\ t > 0, \end{aligned} \tag{5.134}$$

$$\begin{aligned} &\varphi_t + aA\varphi - b(\varphi - \varphi^3) + dy = 0,\ t > 0, \\ &y(0) = y_0,\ \varphi(0) = \varphi_0. \end{aligned} \tag{5.135}$$

By the substitution $z = y + \ell\varphi$, we rewrite system (5.134) as

$$\begin{aligned} &z_t + bAz - b\ell A\varphi + \sum_{i=1}^{N} m(R_{11}(z - \ell\varphi) + R_{12}\varphi, \psi_i)_\omega \psi_i = 0, \\ &\varphi_t + aA\varphi - b(\varphi - \varphi^3) + d(z - \ell\varphi) = 0, \\ &z(0) = z_0 - y_0 - \ell\varphi_0,\ \varphi(0) = \varphi_0. \end{aligned} \tag{5.136}$$

Equivalently,

$$\begin{aligned} &\frac{d}{dt}(z, \varphi) + \mathscr{G}(z, \varphi) = 0,\ t \geq 0, \\ &(z, \varphi)(0) = (z_0, \varphi), \end{aligned} \tag{5.137}$$

where $\mathscr{G} : H \times H$ is given by

$$\mathscr{G}(z,\varphi) = (bAz - b\ell A\varphi + \sum_{i=1}^{N} m(R_{11}(z-\ell\varphi) + R_{12}\varphi, \psi_i)_\omega \psi_i, aA\varphi - b(\varphi - \varphi^3) \\ +d(z-\ell\varphi),$$

with $D(\mathscr{G}) = (H_0^1(\mathscr{O}) \cap H^2(\mathscr{O}))^2$.

If we provide the space $H \times H$ with a suitable equivalent scalar product, we see that the operator $\mathscr{G}$ is quasi-m-accretive in $H \times H$ (see, e.g., [26], p. 237). Then, by Theorem 1.1, it follows that (see also Theorem 1.5) for all $(y_0, \varphi_0) \in H \times H$ and $T > 0$, problem (5.136) is well posed in $H \times H$, and so, system (5.134)–(5.135) has a unique solution $(y, \varphi) \in (L^2(0,T;V))^2$ such that

$$t^{\frac{1}{2}}(y,\varphi) \in (L^2(0,T;D(A)) \cap W^{1,2}([0,T];H))^2.$$

Multiplying (5.134) by $R_{11}y + R_{12}\varphi$ and (5.135) by $R_{12}y + R_{22}\varphi$, respectively, we get after some calculation that

$$\frac{d}{dt}\langle\langle R(y,\varphi),(y,\varphi)\rangle\rangle + \left(|A^{\frac{3}{4}}y|_2^2 + |A^{\frac{3}{4}}\varphi|_2^2\right) + |mR^1(y,\varphi)|_2^2 \le 2\ell b|\left\langle R^2(y,\varphi),\varphi^3\right\rangle|, \\ \forall t \in (0,T),$$

where $R^1(y,\varphi) = R_{11}y + R_{12}\varphi$, $R^2(y,\varphi) = R_{12}y + R_{22}\varphi$. We have

$$\begin{aligned} |\langle R^1(y,\varphi),\varphi^3\rangle| &\le C\,(\|y\| + \|\varphi\|)\,|\varphi|_6^3 \\ &\le C\left(|A^{\frac{1}{4}}y|_2^{\frac{1}{2}}|A^{\frac{3}{4}}y|_2^{\frac{1}{2}}|\varphi|_6^3 + |A^{\frac{1}{4}}\varphi|_2^{\frac{1}{2}}|A^{\frac{3}{4}}\varphi|_2^{\frac{1}{2}}|\varphi|_6^3\right) \\ &\le C\left(|A^{\frac{1}{4}}y|_2^2|A^{\frac{3}{4}}y|_2^2 + |A^{\frac{1}{4}}\varphi|_2^{\frac{1}{2}}|A^{\frac{1}{4}}y|_2^{\frac{3}{2}}\left(|A^{\frac{3}{4}}y|_2^2 + |A^{\frac{3}{4}}\varphi|_2^2\right)\right) \\ &\le C\left(|A^{\frac{1}{4}}y|_2^2 + |A^{\frac{1}{4}}\varphi|_2^{\frac{1}{2}}|A^{\frac{1}{4}}y|_2^{\frac{3}{2}}\right)\left(|A^{\frac{3}{4}}y|_2^2 + |A^{\frac{3}{4}}\varphi|_2^2\right), \end{aligned}$$

because $\varphi|_6 \le C|A^{\frac{3}{4}}\varphi|_2$. This yields

$$\begin{aligned} &\frac{d}{dt}\langle\langle R(y(t),\varphi(t)),(y(t),\varphi(t))\rangle\rangle + \frac{1}{2}|A^{\frac{3}{4}}y(t)|_2^2 + \frac{1}{2}|A^{\frac{3}{4}}\varphi(t)|_2^2 \\ &\quad \le C\left(|A^{\frac{1}{4}}y(t)|_2 + |A^{\frac{1}{4}}\varphi(t)|_2 - \tfrac{1}{2C}\right) \\ &\quad \le C_1(\langle\langle R(y(t),\varphi(t)),(y(t),\varphi(t))\rangle\rangle - C_2, \end{aligned} \tag{5.138}$$

where $C_1, C_2 \ge 0$.

If we replace in system (5.110) and the above computation φ^3 by its Yosida approximation $(\varphi^3)_\varepsilon$, we get, as in the proof of Theorem 5.2, that there is a maximal interval $(0, T^*)$, $T^* > 0$, where $\langle\langle R(y(t)), (y(t), \varphi(t))\rangle\rangle \le C_2$. Then, by (5.138), it follows that

$$\frac{d}{dt}\langle\langle R(y,\varphi),(y,\varphi)\rangle\rangle + \frac{1}{2}\left(|A^{\frac{3}{4}}y|_2^2 + |A^{\frac{3}{4}}\varphi|_2^2\right) \le 0,\ \forall t > 0$$

which, by virtue of (5.132), implies that

$$\|y(t)\|_W + \|\varphi(t)\|_W \le C(\|y_0\|_W + \|\varphi_0\|_W)e^{-\gamma t},\quad \forall t > 0$$

and

$$\int_0^\infty \left(|A^{\frac{3}{4}}y|_2^2 + |A^{\frac{3}{4}}\varphi|_2^2\right) dt \le C.$$

We have obtained, therefore, the following stabilization result.

Theorem 5.4. *Suppose that* (K_4) *holds. Then there exists a linear positive and self-adjoint operator* $R : H\times H \to H\times H$ *with* $V\times V\subset D(A)\subset W\times W$, *such that the feedback controller*

$$u = -\sum_{i=1}^{N}(R_{11}y + R_{12}\varphi, \psi_i)_\omega \psi_i$$

stabilizes exponentially the zero solution of the phase-field system (5.121) *for* $(y_0, \varphi_0) \in E_\rho$. *Moreover, the operator* R *is the solution to the algebraic Riccati equation* (5.133).

5.2 Boundary Stabilization of Parabolic Equations

Since in applications the boundary $\partial\mathcal{O}$ is more accessible than the interior of $\mathcal{O}$, the design of a stabilizable feedback controller supported by a boundary is an important objective in engineering control practice.

The boundary stabilization of equation (5.1) can be treated by a Riccati-based technique similar to that developed above for the internal stabilization. However, it should be said that, in spite of its theoretical simplicity and robustness, the Riccati-based approach to the boundary feedback stabilization might be impractical. One of its major drawbacks is that it involves a great computational complexity in the numerical treatment of specific problems. We shall present here a direct stabilization technique for parabolic equations which has an interest in itself and is applicable to a wider class of parabolic-like control systems and, in particular, to Navier–Stokes systems (see Chapter 6). For the sake of simplicity, we confine to the stabilization of the equilibrium solutions y_e to the parabolic equation

$$\begin{aligned}
&\frac{\partial y}{\partial t} = \Delta y + f(x, y) \text{ in } (0, \infty) \times \mathscr{O}, \\
&y(0, x) = y_0(x) \text{ in } \mathscr{O}, \\
&y = u \text{ on } (0, \infty) \times \Gamma_1, \ \frac{\partial y}{\partial n} = 0 \text{ on } (0, \infty) \times \Gamma_2,
\end{aligned} \tag{5.139}$$

where $\mathscr{O}$ is a bounded and open domain of $\mathbb{R}^d$ with a smooth boundary $\partial\mathscr{O} = \Gamma_1 \cup \Gamma_2$, Γ_1, Γ_2 being connected parts of $\partial\mathscr{O}$. Here and everywhere in the sequel, we shall denote by $\frac{\partial y}{\partial n}$ the outward normal derivative of y.

The Dirichlet controller u is applied on Γ_1 while Γ_2 is insulated. Here, $\frac{\partial}{\partial n}$ is the normal derivative and $y_e \in C^2(\overline{\mathscr{O}})$ is any solution to the equation

$$\Delta y_e + f(x, y_e) = 0 \text{ in } \mathscr{O}, \ \frac{\partial y_e}{\partial n} = 0 \text{ on } \Gamma_2.$$

Translating y_e into zero via substitution $y - y_e \to y$ we can rewrite (5.139) as

$$\begin{aligned}
&\frac{\partial y}{\partial t} = \Delta y + f(x, y + y_e) - f(x, y_e) \text{ in } (0, \infty) \times \mathscr{O}, \\
&y(0, x) = y_0^*(x) = y_0(x) - y_e(x) \text{ in } \mathscr{O}, \\
&y = u - y_e \text{ on } (0, \infty) \times \Gamma_1; \ \frac{\partial y}{\partial n} = 0 \text{ on } (0, \infty) \times \Gamma_2,
\end{aligned} \tag{5.140}$$

and the stabilization problem reduces to design a feedback controller $u = F(y)$ such that the solution to the corresponding closed loop system satisfies, for some $\gamma > 0$,

$$\int_{\mathscr{O}} |y(t, x)|^2 dx \le Ce^{-\gamma t} \int_{\mathscr{O}} |y(0, x)|^2 dx, \ \forall t \ge 0, \tag{5.141}$$

for all y_0^* in a $L^2(\mathscr{O})$–neighborhood of the origin.

The first step toward this goal is the stabilization of the linearized system associated with (5.140), that is,

$$\begin{aligned}
&\frac{\partial y}{\partial t} = \Delta y + f_y(x, y_e) y \text{ in } (0, \infty) \times \mathscr{O} \\
&y = v \text{ on } (0, \infty) \times \Gamma_1; \ \frac{\partial y}{\partial n} = 0 \text{ on } (0, \infty) \times \Gamma_2,
\end{aligned} \tag{5.142}$$

where $f_y(x, y_e) = \frac{\partial f}{\partial y}(x, y_e)$. The stabilizing feedback controller $v = F(y)$ for (5.142) will be used afterwards to stabilize locally system (5.140), and implicitly the equilibrium solution y_e.

Everywhere in the following, we shall assume that

(i) $f, f_y \in C(\overline{\mathscr{O}} \times \mathbb{R})$.

In particular, this implies that $x \to f_y(x, y_e(x))$ is continuous on $\overline{\mathscr{O}}$.

We consider the linear self-adjoint operator in $H = L^2(\mathscr{O})$,

$$Ly = \Delta y + f_y(x, y_e)y, \ \forall\, y \in \mathscr{D}(L),$$
$$\mathscr{D}(L) = \left\{ y \in H^2(\mathscr{O});\ y = 0 \text{ on } \Gamma_1,\ \frac{\partial y}{\partial n} = 0 \text{ on } \Gamma_2 \right\}.$$

The operator $-L$ has a countable set of real eigenvalues λ_j with corresponding eigenfunctions φ_j, that is, $-L\varphi_j = \lambda_j \varphi_j$, $j = 1, 2, \ldots$. Each eigenvalue λ_j is repeated here according to its multiplicity and let $N \in \mathbb{N}$ be such that $\lambda_j \le 0$ for $j = 1, \ldots, N$; $\lambda_{N+1} > 0$. (Since the resolvent of L is compact and $\langle Ly, y\rangle_2 \le -|\nabla y|_2^2 + C|y|_2^2$, it is clear that N is finite.)

Here $\frac{\partial \varphi_j}{\partial n}$ is the normal derivative of φ_j to $\partial\mathscr{O}$.

By the unique continuation property of eigenfunctions φ_j, we know that, for all j, $\frac{\partial \varphi_j}{\partial n} \not\equiv 0$ on Γ_1. In the following, we shall assume that

(ii) *The system* $\left\{ \frac{\partial \varphi_j}{\partial n},\ 1 \le j \le N \right\}$ *is linearly independent on* Γ_1.

We note that (ii) always holds if $N = 1$ and, for $d = 1$, only in this case. For $d > 1$, there are, however, significant situations where (ii) holds (see Example 5.3 below).

Consider the feedback controller

$$v = \eta \sum_{j=1}^{N} \mu_j \left\langle y, \varphi_j \right\rangle_2 \phi_j \text{ on } (0, \infty) \times \Gamma_1, \tag{5.143}$$

where $\eta, k > 0$ are parameters to be made precise later on, and

$$\mu_j = \frac{k + \lambda_j}{k + \lambda_j - \eta},\ j = 1, \ldots, N, \tag{5.144}$$

$$\phi_j = \sum_{\ell=1}^{N} a_{j\ell} \frac{\partial \varphi_\ell}{\partial n},\ j = 1, \ldots, N, \text{ on } \Gamma_1, \tag{5.145}$$

$$\sum_{\ell=1}^{N} a_{\ell j} \left\langle \frac{\partial \varphi_\ell}{\partial n}, \frac{\partial \varphi_i}{\partial n} \right\rangle_0 = \delta_{ij},\ i, j = 1, \ldots, N. \tag{5.146}$$

We note that, by assumption (ii), the Gram matrix $\left\| \left\langle \frac{\partial \varphi_\ell}{\partial n}, \frac{\partial \varphi_i}{\partial n} \right\rangle_0 \right\|_{i,\ell=1}^{N}$ is nonsingular. Hence, $a_{\ell j}$ and ϕ_j are well defined.

Theorem 5.5. *Let k and η be positive and sufficiently large such that*

$$((k + \lambda_j - \eta)\lambda_j + k\eta)(k + \lambda_j - 2\eta)^{-1} \ge \gamma_0 > 0,\ \textit{for } j = 1, \ldots, N. \tag{5.147}$$

Then, the feedback controller (5.143) *stabilizes exponentially system* (5.142). *More precisely, the solution* y *to the closed loop system*

$$\begin{aligned} &\frac{\partial y}{\partial t} = \Delta y + f_y(x, y_e)y \text{ in } (0, \infty) \times \mathscr{O}, \\ &y = \eta \sum_{j=1}^{N} \mu_j \langle y, \varphi_j \rangle_2 \phi_j \text{ in } (0, \infty) \times \Gamma_1; \\ &\frac{\partial y}{\partial n} = 0 \text{ on } (0, \infty) \times \Gamma_2, \end{aligned} \tag{5.148}$$

satisfies, for $\gamma = \inf\{\gamma_0, \lambda_{N+1}\}$, *the estimate*

$$|y(t)|_2 \leq C \exp(-\gamma t)|y(0)|_2, \ \forall t \geq 0. \tag{5.149}$$

It should be remarked that Theorem 5.5 provides a simple algorithm for the stabilization of the linear system (5.142) as well as for the nonlinear system (5.139) which, in a few words, can be described as follows. Determine first the unstable eigenvalues $\{\lambda_j\}_{j=1}^N$ and the corresponding eigenfunctions φ_j of the operator L and construct afterwards the feedback controller (5.143), where μ_j, ϕ_j are given (5.144)–(5.146) and k, η satisfy condition (5.147). In this setting, the controller u is coordinated with the measurements of boundary values of the state system and this enhances the stability of motion. Of course, in specific situations, the eigenvalues λ_j and φ_j cannot be computed exactly and so, instead of (5.143), we must consider an approximating feedback controller v_h of the form (5.143) corresponding to the approximations λ_j^h and φ_j^h of λ_j and φ_j, respectively. However, the approximating controller v_h is still stabilizing in problem (5.142) by the robustness of the stabilizer controller. Let us illustrate the method on a few examples.

Example 5.2. We consider the boundary stabilization problem

$$\begin{aligned} &y_t = y_{xx} + \lambda y, \ x \in (0, 1), \ t \geq 0, \\ &y_x(0, t) = 0, \quad y(1, t) = v(t), \ t > 0, \end{aligned} \tag{5.150}$$

with Dirichlet actuation in $x = 1$, where $\lambda > 0$ is a constant parameter. The eigenvalues of the operator $-Ly = -y'' - \lambda y$ with the domain $\mathscr{D}(L) = \{y \in H^2(0, 1);\ y'(0), y(1) = 0\}$ are $\lambda_j = \frac{(2j-1)^2\pi^2}{4} - \lambda$ with eigenfunctions $\varphi_j = \cos \frac{(2j-1)\pi}{2} x$. Then, for $\lambda < \left(\frac{3\pi}{2}\right)^2$, we have $N = 1$ and so Theorem 5.5 is applicable with the feedback controller (5.143) of the form

$$v = \eta \mu_1 \left\langle y, \cos \frac{\pi}{2} x \right\rangle_2 \phi_1, \tag{5.151}$$

where μ_1 is given by (5.144), and $\phi_1 = 1$.

By Theorem 5.5, for$(\lambda_1(k+\lambda_1-\eta)+k)(\lambda_1+k-2\eta)^{-1} > \gamma_0 > 0$, this feedback controller stabilizes (5.150) with the exponent decay

$$\gamma = \inf\left\{\gamma_0, \left(\frac{3\pi}{2}\right)^2 - \lambda\right\}.$$

In [52], a stabilizing feedback controller was first designed by the backstepping method for $\lambda < \frac{3\pi^2}{4}$ and later on, this condition was removed in [15] by a sharpening of the method. It should be said, however, that the feedback controller (5.151) is simpler than that constructed via the backstepping method for $\lambda < \left(\frac{3\pi}{2}\right)^2$.

Example 5.3. Consider the boundary stabilization of the heat equation in $\mathbb{R}^d$

$$\begin{aligned} &y_t = \Delta y + \lambda y, \ x \in (0,\pi)^d, t \geq 0, \\ &y(t,x) = v(t,x), \ x \in \partial\mathscr{O}, \ t \geq 0, \end{aligned} \tag{5.152}$$

where $x = (x_1, x_2, \dots, x_d)$ and $\mathscr{O} = (0,\pi)^d$. The eigenvalues of the operator $-Ly = -\Delta y - \lambda y$ with the domain $\{y \in H_0^1(\mathscr{O});\ \Delta y \in L^2(\mathscr{O})\}$ are given

$$\lambda_k = |k|^2 - \lambda, \ |k|^2 = k_1^2 + \cdots + k_d^2, \ (k_1, \dots, k_d) \in \mathbb{N}^d,$$

with the corresponding eigenfunctions

$$\varphi_k(x) = \sin k_1 x_1 \dots \sin k_d x_d.$$

Then, $N \in \mathbb{N}$ is determined by the condition $|k|^2 < \lambda$ and, since

$$\frac{\partial\varphi_k}{\partial n}(x) = -k_1 \sin k_2 x_2 \dots \sin k_d x_d \ \text{ for } x_1 = 0, \ (x_2, \dots, x_d) \in (0,\pi)^{d-1},$$

it turns out that assumption (ii) holds on $\partial\mathscr{O}$ (as a matter of fact on each $\Gamma_j = \{0\} \times (0,\pi)^{d-1}$). Hence, Theorem 5.5 is applicable in the present situation and so there is a feedback controller v of the form (5.143)–(5.145), which stabilizes system (5.152).

Remark 5.7. The previous equation might suggest that, for $d \geq 1$, assumption (ii) is always satisfied, but the following example shows that, in general, this is not true if $\partial\mathscr{O} = \Gamma_1 \cup \Gamma_2$, where $\Gamma_2 \neq \emptyset$. Take, for instance, equation (5.152) with $\lambda = 11$, $\mathscr{O} = (0,\pi)^2$ and boundary conditions: $y = u$ on Γ_1, $\frac{\partial y}{\partial \nu} = 0$ on Γ_2, where $\Gamma_1 = \{x_1 = 0\} \cup \{x_1 = \pi\} \times (0,\pi)$, $\Gamma_2 = (0,\pi) \times \{x_2 = 0\} \cup \{x_2 = \pi\}$. Then, $\varphi_1 = \sin x_1 \cos x_2$ is an eigenfunction for $\lambda_1 = -9$ and $\varphi_2 = \sin 3x_1 \cos x_2$ for $\lambda_2 = -1$ (both unstable eigenvalues). However, as easily seen, $\left.\frac{\partial\varphi_1}{\partial x_1}\right|_{\Gamma_1}$, $\left.\frac{\partial\varphi_2}{\partial x_1}\right|_{\Gamma_1}$ are linearly dependent.

Remark 5.8. Numerical tests for the computation of the stabilizing controller (5.143) were performed in [79] for (5.152) on $\mathscr{O} = (0, \pi) \times (0, \pi)$, and $\lambda = 3, 7$. Also, the case $\mathscr{O} = (0, 1) \times (0, 1)$, $\Gamma_1 = \{1\} \times \{0, 1\}$, $\Gamma_2 = \partial\mathscr{O} \setminus \Gamma_2$ was numerically treated.

Proof of Theorem 5.5. Consider the map $y = Dv$ defined by

$$\begin{aligned} -\Delta y - f_y(x, y_e)y + ky = 0 \text{ in } \mathscr{O}, \\ y = v \text{ on } \Gamma_1, \ \frac{\partial y}{\partial n} = 0 \text{ on } \Gamma_2 \end{aligned} \tag{5.153}$$

For k sufficiently large, the *Dirichlet map* D is well defined and $D \in L(L^2(\Gamma_1), H^{\frac{1}{2}}(\mathscr{O}))$ (see, e.g., [84]). Moreover, we have

$$\|Dv\|_{\frac{1}{2}} \le \frac{C}{k} \|v\|_{L^2(\Gamma_1)}, \ \forall\, v \in L^2(\Gamma_1). \tag{5.154}$$

Since D commutes with the operator $\frac{\partial}{\partial t}$, we may rewrite (5.142) in terms of L and D as

$$\frac{d}{dt}\, y = L(y - Dv) + kDv, \ t \ge 0.$$

Equivalently,

$$\frac{dz}{dt} = Lz - D\,\frac{dv}{dt} + kDv, \ t \ge 0, \tag{5.155}$$

where $z = y - Dv$. Moreover, for later purpose, it is convenient to express the feedback controller (5.143) in terms of z as

$$v = \eta \sum_{j=1}^{N} \langle z, \varphi_j \rangle_2\, \phi_j. \tag{5.156}$$

Indeed, taking $z = y - Dv$ in (5.143), we obtain that

$$v = \eta \sum_{j=1}^{N} \langle y, \varphi_j \rangle_2\, \phi_j - \eta \sum_{j=1}^{N} \langle v, D^*\varphi_j \rangle_0\, \phi_j \text{ on } \Gamma_1, \tag{5.157}$$

where $D^* \in L(L^2(\mathscr{O}), L^2(\Gamma_1))$ is the adjoint of D.

Now, multiplying equation (5.153), where $y = D\phi_j$, by φ_i and recalling that $L\varphi_i = -\lambda_i \varphi_i$, we obtain via Green's formula that

$$\int_{\mathscr{O}} D\phi_j \varphi_i dx = -\frac{1}{\lambda_i + k} \int_{\Gamma_1} \phi_j\, \frac{\partial \varphi_i}{\partial n}\, dx, \ i, j = 1, \ldots, N,$$

and so, by (5.144) and (5.146), we have

$$\big\langle D^*\varphi_i,\phi_j\big\rangle_0=\int_{\mathscr{O}} D\phi_j\varphi_i\,dx=-\frac{1}{\lambda_i+k}\int_{\Gamma_1}\phi_j\frac{\partial\varphi_i}{\partial n}\,dx=-\frac{\delta_{ij}}{\lambda_i+k}.$$

This yields

$$\big\langle v, D^*\varphi_i\big\rangle_0=-\frac{\eta}{k+\lambda_i-\eta}\,\langle y,\varphi_i\rangle_2\,,\ i=1,\dots,N, \tag{5.158}$$

and, substituting into (5.157), we obtain (5.156), as claimed.

Substituting (5.156) into (5.155), we obtain

$$\frac{dz}{dt}=Lz-\eta\sum_{j=1}^{N}\left\langle\frac{dz}{dt}-kz,\varphi_j\right\rangle_2 D\phi_j. \tag{5.159}$$

Let $X^1=\text{lin span}\{\varphi_j\}_{j=1}^N$, P_N the algebraic projection of $L^2(\mathscr{O})$ on X^1 and set $z^1=P_Nz$, $z^2=(I-P_N)z$. Then, we may decompose system (5.159) as follows:

$$\begin{aligned}\frac{d}{dt}z^1&=L_1z^1-\eta P_N\sum_{j=1}^{N}\left\langle\frac{dz^1}{dt}-kz^1,\varphi_j\right\rangle_2 D\phi_j,\\ \frac{d}{dt}z^2&=L_2z^2-\eta(I-P_N)\sum_{j=1}^{N}\left\langle\frac{dz^1}{dt}-kz^1,\varphi_j\right\rangle_2 D\phi_j,\end{aligned} \tag{5.160}$$

where $L_1=P_NL$, $L_2=(I-P_N)L$ and $z=z^1+z^2$.

If we represent z^1 as $\sum_{j=1}^{N} z_j\varphi_j$, we see by (5.158) that

$$z_j'+\lambda_jz_j=\frac{\eta}{k+\lambda_j-\eta}\,(z_j'-kz_j),$$

or, equivalently,

$$z_j'+\frac{(k+\lambda_j-\eta)\lambda_j+k\eta}{k+\lambda_j-2\eta}\,z_j=0,\ j=1,\dots,N, \tag{5.161}$$

and, by condition (5.147), we have that

$$|z_j'(t)|+|z_j(t)|\le Ce^{-\gamma_0t}|z_j(0)|,\ j=1,\dots,N. \tag{5.162}$$

On the other hand, taking into account that the spectrum of $L_2,\sigma(L_2)=\{\lambda_j\}_{j=N+1}^{\infty}$, we have

$$|e^{tL_2}z_0^2|_2\le Ce^{-\lambda_{N+1}t}|z_0^2|_2,\ \forall t\ge 0,\ z_0^2\in X^2,$$

because the C_0-semigroup e^{tL_2} generated by the operator L_2 is analytic in $L^2(\mathcal{O})$. Taking into account the second equation in (5.160) and (5.162), we obtain that

$$|z(t)|_2 \leq Ce^{-\gamma t}|z(0)|_2, \ \forall t \geq 0, \tag{5.163}$$

where $\gamma = \inf\{\gamma_0, \lambda_{N+1}\}$. Keeping in mind that $y = z + Du$, by (5.163) and (5.156) we see that (5.149) holds. ■

Remark 5.9. The above design of a stabilizable feedback controller applies as well to equation (5.139) with homogeneous Dirichlet condition on Γ_2 and Dirichlet actuation on Γ_1, that is, $y = u$ on Γ_1; $y = 0$ on Γ_2, or to the Neumann boundary control $\frac{\partial y}{\partial n} = u$ on Γ_1; $\frac{\partial y}{\partial n} = 0$ on Γ_2, but we omit the details.

We also note that Theorem 5.5 and the above stabilization construction extend word by word to the controlled parabolic linear equation

$$\begin{aligned} &\frac{\partial y}{\partial t} - \operatorname{div}(a(x)\nabla y) + b(x)y = 0 \text{ in } (0, T) \times \mathcal{O}, \\ &y = v \text{ on } (0, \infty) \times \Gamma_1, \ a\nabla y \cdot n = 0 \text{ on } (0, T) \times \Gamma_2, \end{aligned}$$

where $a, b \in C(\overline{\mathcal{O}})$, $a(x) > 0$, $\forall x \in \overline{\mathcal{O}}$.

5.3 Stabilization of Semilinear Equations

We strengthen assumption (i) to

$$|f_y(x, y)| \leq C(|y|^m + 1), \ \forall x \in \overline{\mathcal{O}}, \ y \in \mathbb{R}, \tag{5.164}$$

where $0 < m < \infty$ for $d = 1, 2$, $m = \frac{d}{d-2}$ for $d \geq 3$.

If $y_e \in C^2(\overline{\mathcal{O}})$ is an equilibrium solution to (5.139), we consider the feedback controller

$$u = F(y) = \eta \sum_{j=1}^{N} \mu_j \langle y - y_e, \varphi_j \rangle_2 \phi_j + y_e \text{ on } \Gamma_1. \tag{5.165}$$

where μ_j, ϕ_j are given by (5.144)–(5.146).

Theorem 5.6. *Let* $1 \leq d \leq 3$. *Then, under assumptions (i), (ii),* (5.164), *and* (5.147), *the feedback controller* (5.165) *stabilizes exponentially the solutions* y_e *to system* (5.139). *More precisely, the solution* y *to the closed loop system*

$$\begin{aligned}
&\frac{\partial y}{\partial t} = \Delta y + f(x, y) \textit{ in } (0, \infty) \times \mathcal{O}, \\
&y = F(y) \textit{ on } (0, \infty) \times \Gamma_1; \\
&\frac{\partial y}{\partial n} = 0 \textit{ on } (0, \infty) \times \Gamma_2, \\
&y(0, x) = y_0(x), \;\; x \in \mathcal{O},
\end{aligned} \tag{5.166}$$

satisfies for $|y_0 - y_e|_2 \le \rho$ *and* ρ *sufficiently small, the estimate*

$$|y(t) - y_e|_2 \le C \exp(-\gamma t)|y_0 - y_e|_2, \; \forall t \ge 0, \tag{5.167}$$

where $\gamma > 0$.

Proof. By substitution $y - y_e \to y$, we reduce the problem to the stability of the null solution to system (5.140) with the boundary controller (5.143), that is,

$$\begin{aligned}
&\frac{\partial y}{\partial t} = Ly + g(x, y) \text{ in } (0, \infty) \times \mathcal{O}, \\
&y = G(y) \text{ on } (0, \infty) \times \Gamma_1; \\
&\frac{\partial y}{\partial n} = 0 \text{ on } (0, \infty) \times \Gamma_2,
\end{aligned} \tag{5.168}$$

where

$$g(x, y) = f(x, y + y_e) - f(x, y_e) - f_y(x, y_e)y,$$

$$G(y) = \eta \sum_{j=1}^{N} \mu_j \langle y, \varphi_j \rangle_2 \phi_j.$$

Arguing as in the previous case, we write (5.168) as

$$\frac{dy}{dt} = L(y - DG(y)) + g(x, y) + kDG(y), \; t \ge 0, \tag{5.169}$$

and setting $z = y - DG(y)$, we obtain that

$$\frac{dz}{dt} = Lz + g\left(x, (I - DG)^{-1}z\right) - \eta \sum_{j=1}^{N} \left\langle \frac{dz}{dt} - kz, \varphi_j \right\rangle_2 D\phi_j. \tag{5.170}$$

We set as above

$$z = z^1 + z^2, z^1 \in X^1, z^2 \in X^2 and z = \sum_{j=1}^{N} z_j \varphi_j.$$

This yields

$$z_j' + \frac{(k+\lambda_j-\eta)\lambda_j + k\eta}{k+\lambda_j-2\eta}\, z_j + \frac{k+\lambda_j}{k+\lambda_j-2\eta}\left\langle g(x,(I-DG)^{-1}z,\varphi_j)\right\rangle_2$$

and

$$\begin{aligned} z_j' - kz_j &= K_j(z) \\ &= -\frac{1}{k+\lambda_j-2\eta}((k+\lambda_j)^2 z_j + (k+\lambda_j))\left\langle g(x,(I-DG)^{-1},\varphi_j)\right\rangle_2 . \end{aligned}$$

We get

$$\frac{dz^1}{dt} = \widetilde{L}_1 z^1 + \sum_{j=1}^{N} \frac{z+\lambda_j}{k+\lambda_j-2\eta}\left\langle g(x,(I-DG)^{-1}z),\varphi_j\right\rangle_2 \varphi_j = 0 \tag{5.171}$$

$$\frac{dz^2}{dt} = L_2 z^2 + (I-P_N)g(x,(I-DG)^{-1}z) - \eta(I-P_N)\sum_{j=1}^{N} K_j(z)D\phi_j, \tag{5.172}$$

$$\widetilde{L}_1 z^1 = \sum_{j=1}^{N} \frac{(k+\lambda_j-\eta)\lambda_j + k\eta}{k+\lambda_j-2\eta}\, z_j\varphi_j. \tag{5.173}$$

By virtue of (5.163), both operators $\widetilde{L}_1$ and L_2 are exponentially stable on X^1, respectively X^2, and therefore so is the operator $z \to Bz = \widetilde{L}_1 z^1 + L_2 z_2$, $z = z_1 + z_2$ on the space $L^2(\mathcal{O})$. We set

$$\begin{aligned} \mathcal{G}(z) = &-\sum_{j=1}^{N} \frac{k+\lambda_j}{k+\lambda_j+\eta}\left\langle g(x,(I-DG)^{-1}z),\varphi_j\right\rangle_2 \varphi_j \\ &-(I-P_N)g(x,(I-DG)^{-1}z)) - \eta(I-P_N)\sum_{j=1}^{N} K_j(z)D_j\phi, \end{aligned}$$

and rewrite (5.171), (5.172) as

$$\frac{dz}{dt} = Bz + \mathcal{G}(z),\ t \geq 0;\ \ z(0) = z_0 = y_0 - DGy_0.$$

Equivalently,

$$z(t) = e^{tB}z_0 + \int_0^t e^{(t-s)B}\mathcal{G}(z(s))ds,\ t \geq 0. \tag{5.174}$$

We are going to show that, for $|z_0|_2 \le \rho$ sufficiently small, equation (5.174) has a unique solution $z \in L^{m+1}(0,\infty; H^1(\mathscr{O}))$. To this end, we proceed as in [15]. Namely, consider the map $\Lambda : L^{m+1}(0,\infty; H^1(\mathscr{O})) \to L^{m+1}(0,\infty; H^1(\mathscr{O}))$ defined by

$$\Lambda z(t) = e^{tB} z_0 + \int_0^t e^{(t-s)B} \mathscr{G}(z(s))ds \tag{5.175}$$

and show that, for r sufficiently small, it maps the ball

$$\{z \in L^{m+1}(0,\infty; H^1(\mathscr{O})); \|z\|_{L^{m+1}(0,\infty;H^1(\mathscr{O}))} \le r\} = S(0,r)$$

into itself and is a contraction on $S(0,r)$ for $|z_0|_2$ sufficiently small and r suitable chosen. Indeed, by assumption (5.164), we have

$$\begin{aligned}&|g(x,(I-DG)^{-1}z) - g(x,(I-DG)^{-1}\bar{z})|\\ &\quad\le C|(I-DG)^{-1}z - (I-DG)^{-1}\bar{z}|(|(I-DG)^{-1}z|^m + |(I-DG)^{-1}\bar{z}|^m + 1).\end{aligned}$$

Taking into account Sobolev's embedding theorem (recall that $1 \le d \le 3$) this yields

$$\begin{aligned}&|g(x,(I-DG)^{-1}z) - g(x,(I-DG)^{-1}\bar{z})|_2\\ &\quad\le C\|(I-DG)^{-1}z-(I-DG)^{-1}\bar{z}\|_{L^4(\mathscr{O})}\\ &\qquad(\|(I-DG)^{-1}z\|^m_{L^{2m}(\mathscr{O})}+\|(I-DG)^{-1}\bar{z}\|^m_{L^{2m}(\mathscr{O})})\\ &\quad\le C_1\|(I-DG)^{-1}z-(I-DG)^{-1}\bar{z}\|_{H^1(\mathscr{O})}\\ &\qquad(\|(I-DG)^{-1}z\|^m_{H^1(\mathscr{O})}+\|(I-DG)^{-1}\bar{z}\|^m_{H^1(\mathscr{O})}+1)\\ &\quad\le C_2\|z-\bar{z}\|_{H^1(\mathscr{O})}(\|z\|^m_{H^1(\mathscr{O})} + \|\bar{z}\|^m_{H^1(\mathscr{O})}),\ \forall z,\bar{z} \in H^1(\mathscr{O}).\end{aligned}$$

Hence,

$$|\mathscr{G}(z) - \mathscr{G}(\bar{z})|_2 \le C_2\|z-\bar{z}\|_{H^1(\mathscr{O})}(\|z\|^m_{H^1(\mathscr{O})} + \|\bar{z}\|^m_{H^1(\mathscr{O})}), \tag{5.176}$$

while

$$|\mathscr{G}(z)|_2 \le C_3\|z\|^{m+1}_{H^1(\mathscr{O})}. \tag{5.177}$$

Taking into account that, by (5.171)–(5.172),

$$\|e^{tB}z_0\|_{H^1(\mathscr{O})} \le Ce^{-\gamma t}|z_0|_2,\ \forall t \ge 0,$$

we see by (5.175) that Λ maps $S(0,r)$ into itself for r suitable chosen and $|z_0|_2$ sufficiently small. Moreover, by (5.176), it follows that Λ is a contraction on $S(0,r)$.

Hence, equation (5.174) has, for $|y_0|_2 \le \rho$ sufficiently small, a unique solution $z \in L^{m+1}(0,\infty; H^1(\mathcal{O}))$. By a standard argument (see, e.g., Proposition 5.9 in [28]), this implies also that $|z(t)|_2 \le Ce^{-\gamma t}|z(0)|_2$, $\forall t \ge 0$, and the latter extends to the solution y to (5.168). Then (5.167) follows.

Example 5.4. Consider the classical Fitzhugh–Nagumo equation which models the dynamics of electrical impulses across a cell membrane of an excited neuron. Namely, in its simplified version, it has the form

$$\begin{aligned} y_t &= y_{xx} + y(1-y)(y-a),\ 0 < x < \ell,\ t \ge 0,\\ y_x(t,\ell) &= 0,\ y(t,0) = u(t),\ t \ge 0, \end{aligned} \tag{5.178}$$

where $0 < a < \frac{1}{2} \cdot$ This equation has the unstable equilibrium solution $y = a$.

The linearized operator

$$-Ly = -(y'' + a(1-a)y),$$

$$D(L) = \{y \in H^2(0,1);\, y(0) = 0,\ y'(\ell) = 0\}$$

has the eigenvalues $\lambda_j = \frac{(2j-1)^2\pi^2}{4\ell^2} - a(1-a)$ with the eigenfunctions $\varphi_j(x) = \sin \frac{(2j-1)\pi}{2\ell} x$, $j = 1, 2, \ldots$ Then, for $\frac{\ell}{\pi}\sqrt{a(1-a)} < \frac{3}{2}$, we have $N = 1$ and so, by Theorem 5.6, the controller

$$u(t) = \eta\mu_1 \left\langle y(t) - a, \sin\frac{\pi}{2\ell} x\right\rangle_2 \phi_1 + a, \tag{5.179}$$

where $\mu_1 = \frac{k+\lambda_1}{k+\lambda_1-\eta}$, stabilizes exponentially the equilibrium solution $y = a$ for η chosen in such a way that $(\lambda_1(k+\lambda_1-\eta)+k)(\lambda_1+k-2\eta)^{-1} > 0$.

5.4 Internal Stabilization of Stochastic Parabolic Equations

Consider the stochastic nonlinear controlled parabolic equation

$$\begin{aligned} &dX(t) - \Delta X(t)dt + a(t,\xi)X(t)dt + b(t,\xi)\cdot\nabla_\xi X(t)dt\\ &\qquad + f(X(t))dt = X(t)dW(t) + \mathbf{1}_{\mathcal{O}_0}u(t)dt \text{ in } (0,\infty)\times\mathcal{O},\\ &X = 0 \text{ on } (0,\infty)\times\partial\mathcal{O},\quad X(0) = x \text{ in } \mathcal{O}. \end{aligned} \tag{5.180}$$

Here, $\mathcal{O}$ is a bounded and open domain of $\mathbb{R}^d$, $d \ge 1$, with smooth boundary $\partial\mathcal{O}$ and $W(t)$ is a Wiener cylindrical process of the form

$$W(t) = \sum_{k=1}^{\infty} \mu_k e_k(\xi)\beta_k(t),\quad t \ge 0,\ \xi \in \mathcal{O}, \tag{5.181}$$

where μ_k are real numbers, $\{e_k\} \subset C^2(\overline{\mathcal{O}})$ is an orthonormal system in $L^2(\mathcal{O})$ and $\{\beta_k\}_{k=1}^\infty$ are independent Brownian motions in a stochastic basis $\{\Omega, \mathcal{F}, \mathcal{F}_t, \mathbb{P}\}$.

We assume that

$$\sum_{k=1}^{\infty} \mu_k^2 |e_k|_\infty^2 < \infty, \tag{5.182}$$

where $|\cdot|_\infty$ denotes the $L^\infty(\mathcal{O})$-norm.

The functions $a : [0, \infty) \times \mathcal{O} \to \mathbb{R}$, $b : [0, \infty) \times \mathcal{O} \to \mathbb{R}^d$ and $f : \mathbb{R} \to \mathbb{R}$ are assumed to satisfy

$$a \in C([0, \infty) \times \overline{\mathcal{O}}), \; b \in C^1([0, \infty) \times \overline{\mathcal{O}}) \tag{5.183}$$

$$\sup\{|a(t)|_\infty + |\nabla b(t)|_\infty; \; t \geq 0\} < \infty \tag{5.184}$$

$$f \in \mathrm{Lip}(\mathbb{R}), \; f(0) = 0. \tag{5.185}$$

Finally, $\mathcal{O}_0$ is an open subdomain of $\mathcal{O}$ with smooth boundary, $\mathbf{1}_{\mathcal{O}_0}$ is its characteristic function, and $u = u(t, \xi)$ is an adapted controller with respect to the natural filtration $\{\mathcal{F}_t\}$.

The main problem we address here is the design of a feedback controller $u = F(X)$ such that the corresponding closed loop system (5.180) is asymptotically stable in probability, that is,

$$\lim_{t \to \infty} X(t) = 0, \; \mathbb{P}\text{-a.s.}$$

(As seen in Section 3.6, a stronger property, the exact null controllability of (5.180) in finite time, is in general an open problem.)

Given an $\mathcal{F}_t$-adapted process $u \in L^2(0, T; L^2(\mathcal{O}, L^2(\mathcal{O})))$, a continuous $\mathcal{F}_t$-adapted process $X : [0, T] \to L^2(\mathcal{O})$ is said to be a solution to (5.180) if

$$X \in L^2(\mathcal{O}; L^\infty(0, T; L^2(\mathcal{O}))) \cap C([0, T]; L^2(\mathcal{O}; L^2(\mathcal{O}))) \tag{5.186}$$

and

$$\begin{aligned} X(t, \xi) = &\int_0^t (\Delta X(s, \xi) - a(t, \xi) X(s, \xi) - b(s, \xi) \cdot \nabla_\xi X(s, \xi) \\ &- f(X(s, \xi))) ds + \int_0^t \mathbf{1}_{\mathcal{O}_0} u(s, \xi) ds + \int_0^t X(s, \xi) dW(s), \\ &\qquad \xi \in \mathcal{O}, \; t \in (0, T), \; \mathbb{P}\text{-a.s.} \end{aligned} \tag{5.187}$$

Taking into account assumptions (5.182)–(5.185), we may conclude (see, e.g., [64], p. 208) that (5.180) has a unique solution X satisfying (5.186), (5.187).

Let $\mathscr{O}_0$ be an open subset of $\mathscr{O}$. We set $\mathscr{O}_1 = \mathscr{O} \setminus \overline{\mathscr{O}}_0$ and denote by $A_1 : D(A_1) \subset L^2(\mathscr{O}_1) \to L^2(\mathscr{O}_1)$ defined by

$$A_1 y = -\Delta y, \ y \in D(A_1) = H_0^1(\mathscr{O}_1) \cap H^2(\mathscr{O}_1), \tag{5.188}$$

or, equivalently,

$$\langle A_1 y, z \rangle_1 = \int_{\mathscr{O}_1} \nabla y \cdot \nabla z \, d\xi, \ \forall\, y, z \in H_0^1(\mathscr{O}_1), \tag{5.189}$$

where $\langle \cdot, \cdot \rangle_1$ is the duality on $H_0^1(\mathscr{O}_1) \times H^{-1}(\mathscr{O}_1)$ induced by $L^2(\mathscr{O}_1)$ as pivot space.

Denote by $\lambda_1^*(\mathscr{O}_1)$ the first eigenvalue of the operator A_1, that is,

$$\lambda_1^*(\mathscr{O}_1) = \inf \left\{ \int_{\mathscr{O}_1} |\nabla y|^2 d\xi; \ y \in H_0^1(\mathscr{O}_1), \ \int_{\mathscr{O}_1} y^2 d\xi = 1 \right\}. \tag{5.190}$$

Consider in (5.180) the feedback controller

$$u = -\eta X, \ \eta \in \mathbb{R}^+, \tag{5.191}$$

and the corresponding closed loop system

$$\begin{aligned} &dX - Xdt + aXdt + b \cdot \nabla X dt + f(X)dt = Xdt - \eta \mathbf{1}_{\mathscr{O}_0} X dt \ \text{ in } (0, \infty) \times \mathscr{O}, \\ &X(0) = x \text{ in } \mathscr{O}, \ \ X = 0 \text{ on } (0, \infty) \times \partial \mathscr{O}. \end{aligned} \tag{5.192}$$

Theorem 5.7 is the main result.

Theorem 5.7. *Assume that*

$$\begin{aligned} \lambda_1^*(\mathscr{O}_1) &- \frac{1}{2} \sum_{j=1}^{\infty} \mu_j^2 |e_j|_\infty^2 - \|f\|_{\mathrm{Lip}} \\ &- \sup \left\{ -a(t, \xi) + \frac{1}{2} \operatorname{div}_\xi b(t, \xi); \ (t, \xi) \in \mathbb{R}^+ \times \mathscr{O} \right\} > 0. \end{aligned} \tag{5.193}$$

Then, for each $x \in L^2(\mathscr{O})$ and for η sufficiently large (independent of x), the feedback controller (5.191) *exponentially stabilizes in probability equation* (5.180). *More precisely, there is $\gamma > 0$ such that the solution X to* (5.192) *satisfies*

$$\lim_{t \to \infty} e^{\gamma t} |X(t)|_2^2 = 0, \ \mathbb{P}\text{-a.s.} \tag{5.194}$$

$$e^{\gamma t} \mathbb{E} |X(t)|_2^2 + \mathbb{E} \int_0^\infty e^{\gamma t} |X(t)|_2^2 dt \le C |x|_2^2. \tag{5.195}$$

We recall that, by the classical Rayleigh–Faber–Krahn perimetric inequality in dimension $d \geq 2$, we have

$$\lambda_1^*(\mathcal{O}_1) \geq \left(\frac{\omega_d}{|\mathcal{O}_1|}\right)^{\frac{2}{d}} J_{\frac{d}{2}-1,1}, \tag{5.196}$$

where $|\mathcal{O}_1| = \mathrm{Vol}(\mathcal{O}_1)$, $\omega_d = \pi^{\frac{d}{2}} / \left(\Gamma\left(\frac{d}{2}+1\right)\right)$, and $J_{m,1}$ is the first positive zero of the Bessel function $I_m(r)$. In particular, by Theorem 5.7, we conclude that, if $|\mathcal{O}_1|$ is sufficiently small, then the feedback controller (5.191) is exponentially stabilizable. More precisely, we have

Corollary 5.1. *Assume under hypotheses* (5.182)–(5.185) *that*

$$|\mathcal{O}_1| \leq \omega_d J_{\frac{d}{2}-1,1}^{\frac{d}{2}} \left(\frac{1}{2}\sum_{Lip}^{\infty} \mu_j^2 |e_j|_\infty^2 + \sup_{\mathbb{R}^+\times\mathcal{O}} \left\{-a + \frac{1}{2}\,\mathrm{div}_\xi b\right\} + \|f\|_{\mathrm{Lip}}\right)^{-\frac{d}{2}}. \tag{5.197}$$

Then, for each $x \in L^2(\mathcal{O})$, the feedback controller (5.191) *exponentially stabilizes system* (5.180) *in sense of* (5.194), (5.195).

An Example

The stochastic equation

$$\begin{aligned} &dX - X_{\xi\xi}dt + (aX + bX_\xi)dt = \mu X d\beta + V dt,\ 0 < \xi < 1, \\ &X(t,0) = X(t,1) = 0,\ t \geq 0, \end{aligned}$$

where β is a Brownian motion and $\mu \in \mathbb{R}$, $a \in C([0,T]\times\mathbb{R})$, $b \in C^1([0,1]\times\mathbb{R})$, is exponentially stabilizable in probability by any feedback controller $V = -\eta \mathbf{1}_{[a_1,a_2]} X$, where $\eta > 0$ is sufficiently large and $0 < a_1 < a_2 < 1$ are such that

$$\pi \inf\left\{\frac{1}{a_1}, \frac{1}{1-a_2}\right\} > \frac{\mu^2}{2} + \sup_{(t,\xi)\in\mathbb{R}^+\times(0,1)} \left\{-a(t,\xi) + \frac{1}{2} b_\xi(t,\xi)\right\}.$$

Proof of Theorem 5.7. The main ingredient of the proof is the following lemma.

Lemma 5.9. *For each $\varepsilon > 0$ there is $\eta_0 = \eta_0(\varepsilon)$ such that*

$$\begin{aligned} \int_{\mathcal{O}} |\nabla y(\xi)|^2 d\xi + \eta \int_{\mathcal{O}_0} y^2(\xi) d\xi \geq (\lambda_1^*(\mathcal{O}_1) - \varepsilon)|y|_2^2, \\ \forall\, y \in H_0^1(\mathcal{O}),\ \eta \geq \eta_0. \end{aligned} \tag{5.198}$$

Proof. Denote by ν_1 the first eigenvalue of the self-adjoint operator

$$A^\eta y = Ay + \eta \mathbf{1}_{\mathcal{O}_0} y,\ \forall\, y \in D(A^\eta) = H_0^1(\mathcal{O}) \cap H^2(\mathcal{O}),$$

where $A = -\Delta$, $D(A) = H_0^1(\mathcal{O}) \cap H^2(\mathcal{O})$, and $\eta \in \mathbb{R}^+$.

We have by the Rayleigh formula

$$\nu_1^\eta = \inf\left\{\int_{\mathscr{O}} |\nabla y|^2 d\xi + \eta \int_{\mathscr{O}_0} |y|^2 d\xi;\ |y|_2 = 1\right\} \le \lambda_1^*(\mathscr{O}_1) \tag{5.199}$$

because any function $y \in H_0^1(\mathscr{O}_1)$ can be extended by zero to $H_0^1(\mathscr{O})$ across the smooth boundary $\partial\mathscr{O}_1 = \partial\mathscr{O}_0$. Let $\varphi_1^\eta \in H_0^1(\mathscr{O}) \cap H^2(\mathscr{O})$ be such that

$$A^\eta \varphi_1^\eta = \nu_1^\eta \varphi_1^\eta,\ \ |\varphi_1^\eta|_2 = 1.$$

We have by (5.199) that

$$\int_{\mathscr{O}} |\nabla \varphi_1^\eta|^2 d\xi + \eta \int_{\mathscr{O}_0} |\varphi_1^\eta|^2 d\xi = \nu_1^\eta \le \lambda_1^*(\mathscr{O}),\ \forall \eta > 0.$$

Then, on a subsequence, again denoted η, we have for $\eta \to \infty$, that $\nu_1^\eta \to \nu^*$ and

$$\begin{aligned} \varphi_1^\eta &\longrightarrow \varphi_1 \ \text{ weakly in } H_0^1(\mathscr{O}),\ \text{ strongly in } L^2(\mathscr{O}) \\ \eta \int_{\mathscr{O}_0} |\varphi_1^\eta|^2 d\xi &\longrightarrow 0. \end{aligned}$$

We have, therefore, $\varphi_1 \in H_0^1(\mathscr{O}_1)$, $|\varphi_1|_2 = 0$ and, since $A^\eta \varphi_1^\eta|_{\mathscr{O}_1} = A_1 \varphi_1^\eta$, we have also that $A_1\varphi_1 = \nu^*\varphi_1$. Moreover, by (5.199) we see that $\nu^* \le \lambda_1^*(\mathscr{O}_1)$. Since $\lambda_1^*(\mathscr{O}_1)$ is the first eigenvalue of A_1, we have that $\nu^* = \lambda_1^*(\mathscr{O}_1)$ and

$$\lim_{\eta\to\infty}\ \inf\left\{\int_{\mathscr{O}} |\nabla y|^2 d\xi + \eta \int_{\mathscr{O}_0} y^2 d\xi;\ |y|_1 = 1\right\} = \lambda_1^*(\mathscr{O}_1).$$

This yields (5.198), as claimed.

Proof of Theorem 5.7 (Continued). By applying Itô's formula in (5.192), which by virtue of (5.186) is possible, we obtain that

$$\begin{aligned} &\frac{1}{2}\, d(e^{\gamma t}|X(t)|_2^2) + \int_{\mathscr{O}} e^{\gamma t} |\nabla X(t,\xi)|^2 d\xi + \int_{\mathscr{O}} (a(t,\xi) - \gamma \\ &\quad -\frac{1}{2}\,\mathrm{div}_\xi b(t,\xi)) e^{\gamma t} X^2(t,\xi) d\xi = \frac{1}{2}\int_{\mathscr{O}} e^{\gamma t} \sum_{k=1}^\infty |(Xe_k)(t,\xi)|^2 d\xi \\ &\quad -\int_{\mathscr{O}} e^{\gamma t} f(X(t,\xi)) X(t,\xi) d\xi \\ &\quad -\eta \int_{\mathscr{O}_0} e^{\gamma t} |X(t,\xi)|^2 d\xi + \int_{\mathscr{O}} e^{\gamma t} \sum_{k=1}^\infty (Xe_k)(t,\xi) X(t,\xi) d\beta_k(t),\ \mathbb{P}\text{-a.s.},\ t \ge 0. \end{aligned}$$

Equivalently,

$$\frac{1}{2}\, e^{\gamma t} |X(t)|_2^2 + \int_0^t K(s) ds = \frac{1}{2}\, |x|_2^2 + M(t),\ t \ge 0,\ \mathbb{P}\text{-a.s.}, \tag{5.200}$$

where

$$K(t) = \int_{\mathscr{O}} e^{\gamma t}(|\nabla X(t,\xi)|^2 + (a(t,\xi) - \gamma - \frac{1}{2}\,\mathrm{div}_\xi b(t,\xi))|X(t,\xi)|^2$$
$$-\frac{1}{2}\sum_{k=1}^{\infty} X^2(t,\xi)e_k^2(\xi))d\xi + \eta\int_{\mathscr{O}_0} e^{\gamma t}|X(t,\xi)|^2 d\xi + \int_{\mathscr{O}} e^{\gamma t} f(X(t,\xi))X(t,\xi)d\xi, \tag{5.201}$$

$$M(t) = \int_0^t \int_{\mathscr{O}} \sum_{k=1}^{\infty} e^{\gamma s} X^2(s,\xi)e_k(s,\xi)d\beta_k(s),\ t \geq 0. \tag{5.202}$$

By (5.199) and (5.193), we see that, for $\eta \geq \eta_0$ sufficiently large and $0 < \gamma \leq \gamma_0$ sufficiently small, we have

$$K(t) \geq \varepsilon_0 \int_{\mathscr{O}} e^{\gamma t}|X(t,\xi)|^2 d\xi,\ \forall t > 0,\ \mathbb{P}\text{-a.s.}, \tag{5.203}$$

where $\varepsilon_0 > 0$. Taking expectation into (5.200), we obtain that

$$\frac{1}{2}\,\mathbb{E}[e^{\gamma t}|X(t)|_2^2] + \varepsilon_0 \int_0^t e^{\gamma s}\mathbb{E}|X(s)|_2^2 ds \leq \frac{1}{2}\,|x|_2^2,\ \forall t \geq 0. \tag{5.204}$$

Since $t \to \int_0^t K(s)$ is an a.s. nondecreasing stochastic process and $t \to M(t)$ is a continuous local martingale, we infer by (5.200), (5.204) that there exist

$$\lim_{t\to\infty}(e^{\gamma t}|X(t)|_2^2) < \infty,\ K(\infty) < \infty,$$

which imply (5.194), (5.195), as claimed. ■

Remark 5.10. By the proof, it is clear that Theorem 5.7 extends to more general nonlinear functions $f = f(t,\xi,x)$, as well as to smooth functions $\mu_k = \mu_k(t,\xi)$. Also, the Lipschitz condition (5.185) can be weakened to f continuous, monotonically increasing and with polynomial growth. Moreover, Δ can be replaced by any strongly elliptic linear operator in $\mathscr{O}$. The details are omitted.

5.5 Stabilization of Navier–Stokes Equations Driven by Linear Multiplicative Noise

We consider here the stochastic Navier–Stokes equation

$$\begin{aligned} &dX(t) - \nu_0\Delta X(t)dt + (a(t)\cdot\nabla)X(t)dt \\ &\qquad + (X(t)\cdot\nabla)b(t)dt + (X(t)\cdot\nabla)X(t)dt \\ &\qquad = X(t)dW(t) + \nabla p(t)dt + \mathbf{1}_{\mathscr{O}_0}u(t)dt \text{ in } (0,\infty)\times\mathscr{O}, \\ &\nabla\cdot X(t) = 0 \quad \text{in } (0,\infty)\times\mathscr{O}, \\ &X(t) = 0 \quad \text{on } (0,\infty)\times\partial\mathscr{O},\ X(0) = x \quad \text{in } \mathscr{O}, \end{aligned} \tag{5.205}$$

where $\nu > 0$, $a, b \in (C^1((0,\infty) \times \overline{\mathcal{O}}))^2$, $\nabla \cdot a = \nabla \cdot b = 0$, $a \cdot \mathbf{n} = b \cdot \mathbf{n} = 0$ on $\partial\mathcal{O}$. Here $\mathcal{O}$ is a bounded and open domain of $\mathbb{R}^2$ and $\mathcal{O}_0$ is an open subset of $\mathcal{O}$. The boundaries $\partial\mathcal{O}$ and $\partial\mathcal{O}_0$ are assumed to be smooth. We set

$$H = \{y \in (L^2(\mathcal{O}))^2;\ \nabla \cdot y = 0,\ y \cdot \mathbf{n} = 0 \text{ on } \partial\mathcal{O}\},$$

where $\mathbf{n}$ is the normal to $\partial\mathcal{O}$. We denote by $\langle \cdot, \cdot \rangle_H$ the scalar product of H and by $|\cdot|_H$ the norm. The Wiener process $W(t)$ is of the form (5.181), where $\{e_k\}$ is the orthonormal system in $L^2(\mathcal{O})$ given by $-\Delta e_k = \lambda_k e_k$ in $\mathcal{O}$, $e_k = 0$ on $\partial\mathcal{O}$, and $\mu_k \in \mathbb{R}$. As in the previous case, the main objective here is the design of a stabilizable feedback controller u for equation (5.205).

We use the standard notations (see Section 1.4)

$$\begin{aligned} H &= \{y \in (L^2(\mathcal{O}))^d;\ \nabla \cdot y = 0 \text{ in } \mathcal{O},\ y \cdot \mathbf{n} = 0 \text{ on } \partial\mathcal{O}\}, \\ V &= \{y \in (H_0^1(\mathcal{O}))^d;\ \nabla \cdot y = 0 \text{ in } \mathcal{O}\}, \\ A &= -\nu_0 \Pi \Delta,\ D(A) = (H^2(\mathcal{O}))^d \cap V, \end{aligned}$$

where Π is the Leray projector on H.

Consider the Stokes operator A_1 on $\mathcal{O}_1 = \mathcal{O} \setminus \mathcal{O}_0$, that is,

$$\langle A_1 y, \varphi \rangle = \nu_0 \sum_{i=1}^{d} \int_{\mathcal{O}_1} \nabla y_i \cdot \nabla \varphi_i d\xi,\ \forall \varphi \in V_1,$$

where $V_1 = \{y \in (H_0^1(\mathcal{O}_1))^d;\ \nabla \cdot y = 0 \text{ in } \mathcal{O}_1\}$. Denote again by $\lambda_1^*(\mathcal{O}_1)$ the first eigenvalue of A_1, that is,

$$\lambda_1^*(\mathcal{O}_1) = \inf\left\{ \nu_0 \sum_{i=1}^{d} \int_{\mathcal{O}} |\nabla \varphi_i|^2 d\xi,\ \varphi \in V_1,\ \int_{\mathcal{O}_1} |\varphi|^2 d\xi = 1 \right\}. \tag{5.206}$$

Also, in this case, we have (see Lemma 1 in [35]), for $\eta \geq \eta_0(\varepsilon)$ and $\varepsilon > 0$,

$$\langle Ay, y \rangle_H + \eta \big\langle \Pi(\mathbf{1}_{\mathcal{O}_0} y), y \big\rangle_H \geq (\lambda_1^*(\mathcal{O}_1) - \varepsilon)|y|_H^2,\ \forall y \in V. \tag{5.207}$$

We consider in system (5.205) the linear feedback controller

$$u = -\eta X,\ \eta > 0. \tag{5.208}$$

We set $\gamma^*(t) = \sup\left\{ \int_{\mathcal{O}} |y_i D_i b_j y_j d\xi|;\ |y|_H = 1 \right\} < \infty$, where $b = \{b_1, b_2\}$.

The closed loop system (5.205) with the feedback controller (5.208) has a unique strong solution in the sense of (5.186), (5.187). (See, e.g., [64], p. 281.) We have

Theorem 5.8. *Assume that*

$$\lambda_1^*(\mathcal{O}_1) > \frac{1}{2} \sum_{j=1}^{\infty} \mu_j^2 |e_j|_\infty^2 + \sup_{t \in \mathbb{R}^+} \gamma^*(t). \tag{5.209}$$

Then, for each $x \in H$ and η sufficiently large independent of x, the solution X to the closed loop system (5.205) *with the feedback controller* (5.208) *satisfies*

$$\mathbb{E}[e^{\gamma t}|X(t)|_H^2] + \int_0^\infty e^{\gamma t}\mathbb{E}|X(t)|_H^2 dt < C|x|_H^2, \tag{5.210}$$

$$\lim_{t\to\infty} e^{\gamma t}|X(t)|_H^2 = 0, \ \mathbb{P}\text{-a.s.}, \tag{5.211}$$

for some $\gamma > 0$.

The proof is essentially the same as that of Theorem 5.7, and so it will be sketched only. Taking into account that

$$\langle (X\cdot\nabla)X, X\rangle_H + \langle (a(t)\cdot\nabla)X, X\rangle_H = 0, \ t > 0, \mathbb{P}\text{-a.s.},$$

we obtain by (5.205), (5.208), via Itô's formula, that

$$\begin{aligned}
&\frac{1}{2}\, e^{\gamma t}|X(t)|_H^2 + \int_0^t e^{\gamma s}\Big(\langle AX(s), X(s)\rangle_H + \langle X(s)\cdot\nabla b(s), X(s)\rangle_H \\
&\quad -\frac{1}{2}\sum_{j=1}^\infty |X(s)e_j|_H^2 + \eta\left\langle \mathbf{1}_{\mathcal{O}_0} X(s), X(s)\right\rangle_H\Big)ds \\
&= \frac{1}{2}\,|x|_H^2 + \int_0^t e^{\gamma s}\sum_{j=1}^\infty \left\langle X(s)e_j, X(s)\right\rangle_H d\beta_j(s), \ t \ge 0.
\end{aligned} \tag{5.212}$$

Then, by virtue of (5.207) and (5.209), we have, by (5.212), that

$$\frac{1}{2}\, e^{\gamma t}|X(t)|_H^2 + I(t) = \frac{1}{2}\,|x|_H^2 + M^*(t), \ t \ge 0, \ \mathbb{P}\text{-a.s.},$$

where $I(t)$ is a nondecreasing process, which satisfies

$$\mathbb{E}[I(t)] \ge \varepsilon_0 \int_0^t e^{\gamma s}\mathbb{E}|X(s)|_H^2 ds, \ \forall t \ge 0,$$

for η sufficiently large, and

$$M^*(t) = \int_0^t e^{\gamma s}\sum_{j=1}^\infty \left\langle X(s)e_j, X(s)\right\rangle_H d\beta_j(s)$$

is a continuous local martingale.

As in the previous case, this implies that $\lim\limits_{t\to\infty} e^{\gamma t}|X(t)|_H^2$ exists $\mathbb{P}$-a.s. and, therefore, (5.210) and (5.211) hold.

Remark 5.11. By (5.206), we see that $\lambda_1^*(\mathcal{O}_1) > \nu\mu_1$, where μ_1 is the first eigenvalue on $\mathcal{O}$ of the Stokes operator $-P\Delta$ and we have also that

$$\lambda_1^*(\mathcal{O}_1) \geq C\nu \left(\sup_{x \in \mathcal{O}_1} \operatorname{dist}(x, \partial\mathcal{O}) \right)^{-2}.$$

5.6 Notes on Chapter 5

Section 5.1 is based on the stabilization method developed in [37] (see also [28, 39, 44]) for stabilization of Navier–Stokes equations. Earlier results were given in [36]. In fact, the stabilization of a linearized parabolic system closely follows the corresponding treatment in [37] for the Oseen–Stokes equation. The spectral decomposition method was first used for the stabilization of parabolic equations by R. Triggiani in [115] and it is confined to parabolic-like systems that is to infinite-dimensional systems with compact resolvents and with spectrum determined growth property [114]. For other resent contributions in this direction, we mention the work of H. Badra and T. Takahashi [13].

It should be said that this method leaves out the stabilization of nonsteady-state trajectories of time-dependent systems. However, the latter case can be also treated by the method developed in [47] for the Navier–Stokes equations. A different approach to the stabilization of time-dependent parabolic equations based on the cost of approximate controllability was developed by C. Lefter [86].

In the works [49, 92] there are designed – by the above spectral decomposition method combined with a fixed point technique on suitable functions space – stabilizable feedback controllers for Cahn–Hillard type systems, while in [53] is studied, by similar methods, the internal stabilization of a FitzHugh–Nagumo model with diffusion.

Section 5.2 is based on the author work [30]. An extension of these results to more general systems are given by I. Munteanu in [98, 99]. A different stabilization approach, called *backstepping*, was developed by A. Balough et al. in [15, 16] (see also [111]).

The results of Section 5.4 are taken from the author work [32]. Earlier results for the diffusion equation were given in [5, 6].

Another internal and boundary stabilization technique not developed here, but which was treated in some detail in [28], is the stabilization by Gaussian noises. For system (5.1), such a stabilizing stochastic controller u is of the form

$$u(t) = \sum_{i=1}^{N} (y(t), \varphi_i^*)_2 \phi_i \dot{\beta}_i(t),$$

where $\{\varphi_i\}_{i=1}$, $\{\phi_i\}_{i=1}^N \subset L^2(\mathcal{O})$ are suitable chosen and $\{\beta_i\}$ is a system of independent Brownian motions on a probability space. The corresponding closed loop system is a stochastic parabolic equation with linear multiplicative noise.

As mentioned earlier, a general procedure to stabilize the nonlinear systems, also called the *linearization technique*, is to use the stabilizable feedback controllers for the linearized equations. The use of this linear feedback law in the nonlinear system is advantageous for its simplicity and elegance, but it might be not efficient enough to stabilize systems with high order nonlinearity. One might suspect that, choosing in Theorem 5.2 a feedback controller u of the form

$$u = \psi\left(\sum_{i=1}^{2k}(G(y - y_e), \psi_i)_\omega \psi_i\right),$$

the stabilization effect is effective for functions f of higher order. However, this remains to be done.

Chapter 6
Boundary Stabilization of Navier–Stokes Equations

The stabilization of fluid flows and, in particular, of Navier–Stokes equations was extensively studied via the Riccati-based approach in the last decade and the main references are the works [24, 28, 44, 45, 74, 107]. Here, following [29], we shall briefly present a few results concerning the stabilization with oblique feedback controllers and, more precisely, with *almost normal* boundary feedback controllers.

6.1 The Main Stabilization Results

It is well known that the steady-state solutions to the Navier–Stokes equation might be unstable for a large Reynold number (that is, for small viscosity ν_0). The objective here is to design a stabilizable feedback controller defined on the boundary of the domain guaranteeing the stability of the fluid flow for large Reynolds numbers.

The Navier–Stokes system considered here is of the form (see (1.48))

$$\begin{aligned} &\frac{\partial y}{\partial t} - \nu_0 \Delta y + (y \cdot \nabla) y = \nabla p + f_e && \text{in } (0,\infty) \times \mathcal{O}, \\ &\nabla \cdot y = 0 && \text{in } (0,\infty) \times \mathcal{O}, \\ &y = v, \ y \cdot \mathbf{n} = 0 && \text{on } (0,\infty) \times \partial\mathcal{O}, \\ &y(0) = y_0 && \text{in } \mathcal{O}, \end{aligned} \tag{6.1}$$

in a bounded open domain $\mathcal{O} \subset \mathbb{R}^d$, $d = 2, 3$, with the boundary $\partial\mathcal{O}$ which is assumed to be a finite union of $d-1$ dimensional C^2– connected manifolds. Here $\nu_0 > 0$, f_e is a given smooth function and v is a boundary input. If y_e is an equilibrium (steady-state) solution to (6.1), then (6.1) can be, equivalently, written as

V. Barbu, *Controllability and Stabilization of Parabolic Equations*,
Progress in Nonlinear Differential Equations and Their Applications 90,
https://doi.org/10.1007/978-3-319-76666-9_6

$$\begin{aligned}
&\frac{\partial y}{\partial t} - \nu_0 \Delta y + (y_e \cdot \nabla) y + (y \cdot \nabla) y_e + (y \cdot \nabla) y = \nabla p && \text{in } (0,\infty) \times \mathscr{O}, \\
&\nabla \cdot y = 0 && \text{in } (0,\infty) \times \mathscr{O}, \\
&y = u, \ y \cdot \mathbf{n} = 0 && \text{on } (0,\infty) \times \partial\mathscr{O}, \\
&y(0) = y_0 - y_e && \text{in } \mathscr{O}.
\end{aligned} \tag{6.2}$$

As seen in Section 1.5 (Remark 1.1), in $3-D$, equation (6.2) ha a local solution which is global if $\nu_0^{-1}\|y_0 - y_e\|^2$ is sufficiently small. However, in general, the null solution to (6.2) is not asymptotically stable. As seen below, this can be achieved, however, by choosing a suitable feedback controller supported by the boundary $\partial\mathscr{O}$.

Let Γ be a connected component of $\partial\mathscr{O}$. Our main concern here is the design of an oblique boundary feedback controller with support in Γ which stabilizes exponentially the equilibrium state y_e, or, equivalently, the zero solution to (6.2). The main step toward this end is the stabilization of the linear system corresponding to (6.2) or, more generally, of the Oseen–Stokes system

$$\begin{aligned}
&\frac{\partial y}{\partial t} - \nu_0 \Delta y + (y \cdot \nabla) a_1 + (a_2 \cdot \nabla) y = \nabla p && \text{in } (0,\infty) \times \mathscr{O}, \\
&\nabla \cdot y = 0 && \text{in } (0,\infty) \times \mathscr{O}, \\
&y = u, \ y \cdot \mathbf{n} = 0 && \text{on } (0,\infty) \times \partial\mathscr{O},
\end{aligned} \tag{6.3}$$

where $a_1, a_2 \in (C^2(\overline{\mathscr{O}}))^d$, $\nabla \cdot a_1 = \nabla \cdot a_2 = 0$ in $\mathscr{O}$. Besides its significance as first order linear approximation of (6.2), this system models the dynamics of a Stokes flow with inclusion of a convection acceleration $(a_2 \cdot \nabla) y$ and also the disturbance flow induced by a moving body in a Stokes fluid flow.

In its complex form, the main result Theorem 6.1 below, amounts to saying that, if the unstable eigenvalues of system (6.3) are semi-simple and a certain unique continuation type property for eigenfunctions of the dual linearized system holds, then there is a boundary feedback controller of the form

$$\begin{aligned}
u(t,x) = \eta \mathbf{1}_\Gamma \sum_{j=1}^{N} \mu_j \left(\int_{\mathscr{O}} y(t,x) \overline{\varphi}_j^*(x) dx \right) (\phi_j(x) + \alpha(x)\mathbf{n}(x)), \\
t \ge 0, \ x \in \partial\mathscr{O},
\end{aligned} \tag{6.4}$$

which stabilizes exponentially system (6.1). Here, $\mathbf{1}_\Gamma$ is the characteristic function of Γ as subset of $\partial\mathscr{O}$, $\phi_j \subset (C^2(\Gamma))^d$ are suitably chosen functions and $\{\varphi_j^*\}_{j=1}^N$ is an eigenfunction system for the adjoint $\mathscr{L}^*$ of the Stokes–Oseen operator

$$\begin{aligned}
\mathscr{L}\varphi &= -\nu_0 \Delta \varphi + (a \cdot \nabla)\varphi + (\varphi \cdot \nabla) b, \ \varphi \in D(\mathscr{L}), \\
D(\mathscr{L}) &= \{\varphi \in (H^2(\mathscr{O}))^d \cap (H_0^1(\mathscr{O}))^d; \ \nabla \cdot \varphi = 0 \ \text{in } \mathscr{O}\},
\end{aligned} \tag{6.5}$$

that is,

$$(\mathscr{L}^* \psi)_j = -\nu_0 \Delta \psi_j - \sum_{i=1}^{N} (D_i(a_i \psi_j) - \psi_i D_j b_i), \ \ j = 1,..,d. \tag{6.6}$$

It turns out (see Theorem 6.2) that this feedback controller also stabilizes the Navier–Stokes system (6.2) in a neighborhood of the origin.

In (6.4), N is the number of the eigenvalue λ_j of $\mathscr{L}$ with $\operatorname{Re}\lambda_j \leq 0$ and $\alpha \in C^2(\partial\mathscr{O})$ is an arbitrary function with zero circulation on Γ, that is,

$$\int_\Gamma \alpha(x)dx = 0. \tag{6.7}$$

If α is identically zero, then the controller (6.4) is tangential but in general it is oblique to domain $\mathscr{O}$ which makes it more effective for control actuation.

As a matter of fact, we shall see below (see Corollary 6.1) that, with exception of a set of Lebesgue measure arbitrarily small, the controller u given by (6.4) can be chosen in a direction close to $\mathbf{n}$, that is, *almost normal*. It should be mentioned that, in literature, only in a few situations normal stabilizing controllers for equation (6.1) were designed, and this happened mostly for periodic flows in $2-D$ channels only (see, e.g., [1, 16, 25, 117, 118]). However, there is a large body of results on boundary stabilization of system (6.1) by tangential or not made precise boundary feedback controllers and here the works [12, 74, 75, 106, 107] should be cited. The Riccati-based approach used in these works is essentially the same, as that developed in Chapter 5 for the internal stabilization of parabolic equations and can be described in a few words as follows: one decomposes system (6.1) in a finite-dimensional unstable part which is exactly controllable and an infinite-dimensional part which is exponentially stable and proves so its stabilization by an open loop boundary controller with finite-dimensional structure. Then, one designs in a standard way a stabilizing feedback controller via the algebraic Riccati equation associated with an infinite horizon quadratic optimal control problem. As mentioned earlier, a major drawback of this method is that it involves a large amount of computation. Here we shall construct an explicit feedback stabilizing controller using the same method as in Section 5.3.

Everywhere in the following, $\mathscr{O}$ is a bounded and open domain of $\mathbb{R}^d$, $d = 2, 3$, its boundary $\partial\mathscr{O}$ is a finite union of $d-1$ dimensional C^2– connected manifolds and Γ is a connected component of $\partial\mathscr{O}$.

We set $H = \{y \in (L^2(\mathscr{O}))^d;\ \nabla\cdot y = 0 \text{ in } \mathscr{O},\ \ y\cdot\mathbf{n} = 0 \text{ on } \partial\mathscr{O}\}$ and denote by $\Pi : (L^2(\mathscr{O}))^d \to H$ the Leray projector on H. We consider the operator $A : D(A)\subset H \to H$, $\mathscr{A} : D(\mathscr{A})\subset H \to H$,

$$Ay = -\nu_0\Pi(\Delta y),\ \forall\, y \in D(A) = (H_0^1(\mathscr{O}))^d \cap (H^2(\mathscr{O}))^d \cap H, \tag{6.8}$$

$$\begin{aligned}\mathscr{A}y &= \Pi(-\nu_0\Delta y + (y\cdot\nabla)a + (b\cdot\nabla)y) \\ &= Ay + \Pi((y\cdot\nabla)a + (b\cdot\nabla)y),\ \forall\, y \in D(\mathscr{A}) = D(A).\end{aligned} \tag{6.9}$$

We denote, as usually, by $\widetilde{H}$ the complexified space $\widetilde{H} = H + iH$ and consider the extension $\widetilde{\mathscr{A}}$ of $\mathscr{A}$ to $\widetilde{H}$, that is, $\widetilde{\mathscr{A}}(y + iz) = \mathscr{A}y + i\mathscr{A}z$ for all $y, z \in D(\mathscr{A})$.

The scalar product of H and of $\widetilde{H}$ are denoted by $\langle\cdot,\cdot\rangle$ and $\langle\cdot,\cdot\rangle_{\widetilde{H}}$, respectively. The corresponding norms are denoted by $|\cdot|_H$ and $|\cdot|_{\widetilde{H}}$, respectively.

For simplicity, we denote in the following again by $\mathscr{A}$ the operator $\widetilde{\mathscr{A}}$ and the difference will be clear from the content. The operator $\mathscr{A}$ has a compact resolvent $(\lambda I-\mathscr{A})^{-1}$. Consequently, $\mathscr{A}$ has a countable number of eigenvalues $\{\lambda_j\}_{j=1}^{\infty}$ with corresponding eigenfunctions φ_j each with finite algebraic multiplicity m_j. In the following, each eigenvalue λ_j is repeated according to its algebraic multiplicity ℓ_j.

Note also that there is a finite number of eigenvalues $\{\lambda_j\}_{j=1}^{N}$ with $\operatorname{Re}\lambda_j\le 0$ and that the spaces $X_u=\text{lin span}\{\varphi_j\}_{j=1}^{N}=P_N\widetilde{H}$, $X_s=(I-P_N)\widetilde{H}$ are invariant with respect to $\mathscr{A}$. Here, P_N is the algebraic projection of $\widetilde{H}$ on X_u and is defined by

$$P_N=\frac{1}{2\pi i}\int_{\Gamma_0}(\lambda I-\mathscr{A})^{-1}d\lambda,$$

where Γ_0 is a closed curve which contains in interior the eigenvalues $\{\lambda_j\}_{j=1}^{N}$.

If we set $\mathscr{A}_u=\mathscr{A}|_{X_u}$, $\mathscr{A}_s=\mathscr{A}|_{X_s}$, then we have

$$\sigma(\mathscr{A}_u)=\{\lambda_j:\operatorname{Re}\lambda_j\le 0\},\quad \sigma(\mathscr{A}_s)=\{\lambda_j:\operatorname{Re}\lambda_j>0\}. \tag{6.10}$$

We recall that the eigenvalue λ_j is called semi-simple if its algebraic multiplicity m_j coincides with its geometric multiplicity m_j^g. In particular, this happens if λ_j is simple and it turns out that the property of the eigenvalues λ_j to be all simple is generic (see [28], p. 164). The dual operator $\mathscr{A}^*$ has the eigenvalues $\overline{\lambda}_j$ with the eigenfunctions φ_j^*, $j=1,\dots$.

For the time being, the following hypotheses will be assumed.

(L_1) *The eigenvalues* λ_j, $j=1,\dots,N$, *are semi-simple.*

This implies that

$$\mathscr{A}\varphi_j=\lambda_j\varphi_j,\quad \mathscr{A}^*\varphi_j^*=\overline{\lambda}_j\varphi_j^*,\quad j=1,\dots,N, \tag{6.11}$$

or, equivalently,

$$\mathscr{L}\varphi_j=\lambda_j\varphi_j+\nabla p_j,\quad \mathscr{L}^*\varphi_j^*=\overline{\lambda}_j\varphi_j^*+\nabla p_j^*,\quad j=1,\dots,N, \tag{6.12}$$

and so we can choose systems $\{\varphi_j\}$, $\{\varphi_j^*\}$ in such a way that

$$\langle\varphi_j,\varphi_k^*\rangle_{\widetilde{H}}=\delta_{jk},\quad j,k=1,\dots,N. \tag{6.13}$$

The next hypothesis is a unique continuation type assumption on the normal derivatives $\dfrac{\partial\varphi_j^*}{\partial n}$, $j=1,\dots,N$.

(L_2) *The system* $\left\{\dfrac{\partial\varphi_j^*}{\partial n}\right\}_{j=1}^{N}$ *is linearly independent on* Γ.

In the special case where the unstable spectrum $\mathscr{A}$ has only one distinct eigenvalue λ_1 (eventually multivalued), hypothesis (L_2) is implied by the following weaker assumption

$(L_2)'$ $\frac{\partial \varphi^*}{\partial n}$ *is not identically zero on* Γ,

where φ^* is any eigenfunction corresponding to the unstable eigenvalue $\bar{\lambda}_1$.

Since any linear combination of this system of eigenfunctions is again an eigenfunction corresponding to $\bar{\lambda}_1$, it is clear that in this case (L_2) is implied by $(L_2)'$.

It is not known whether $(L_2)'$ is always satisfied, but likewise hypothesis (L_1), if $\Gamma = \partial\mathscr{O}$, it holds, however, for "almost all a, b" in the generic sense (see [28, 39]). In Section 6.5, it is presented, however, a significant example, where $(L_2)'$ holds.

As regards hypothesis (L_1), it is necessary here for the existence of the biorthogonal system $\{\varphi_j\}$, $\{\varphi_j^*\}$, whose existence simplifies the construction of the stabilizable controller.

Consider the feedback boundary controller

$$u = \eta \mathbf{1}_\Gamma \sum_{j=1}^{N} \mu_j \left\langle P_N y, \varphi_j^* \right\rangle_{\widetilde{H}} (\phi_j + \alpha \mathbf{n}), \tag{6.14}$$

where

$$\mu_j = \frac{k + \lambda_j}{k + \lambda_j - \nu_0 \eta} \quad j = 1, \ldots, N, \tag{6.15}$$

$$\phi_j = \sum_{\ell=1}^{N} a_{j\ell} \frac{\partial \varphi_\ell^*}{\partial n}, \quad j = 1, \ldots, N, \tag{6.16}$$

and $a_{\ell j} \in \mathbb{C}$ are chosen in such a way that

$$\sum_{\ell=1}^{N} a_{\ell j} \int_\Gamma \frac{\partial \varphi_\ell^*}{\partial n} \frac{\partial \overline{\varphi}_i^*}{\partial n}\, dx = \delta_{ij} - \frac{1}{\nu_0} \langle \alpha, p_i^* \rangle_0 \text{ for } i, j = 1, \ldots, N. \tag{6.17}$$

By virtue of hypothesis (L_2), such a system $\{a_{\ell j}\}_{\ell,j=1}^N$ exists because the Gram matrix

$$\left\| \int_\Gamma \frac{\partial \varphi_\ell^*}{\partial n} \frac{\partial \overline{\varphi}_i^*}{\partial n} \right\|_{i,\ell=1}^N = Z_0,$$

is not singular.

By (6.13), (6.17), we have

$$\int_\Gamma \phi_j \frac{\partial \overline{\varphi}_i^*}{\partial n}\, dx = \delta_{ij} - \frac{1}{\nu_0} \langle \alpha, p_i^* \rangle_0, \quad i, j = 1, \ldots, N. \tag{6.18}$$

Here $\{p_i^*\}$ are given by (6.12) and $\langle \cdot, \cdot \rangle_0$ is the scalar product in $L^2(\Gamma)$.

The first stabilization result refers to the Oseen-Stokes system (6.3).

Theorem 6.1. *Assume that* $d = 2, 3$, (L_1), (L_2), (6.7) *hold, and that* $\operatorname{Re}\lambda_j \le 0$ *for* $j = 1, \dots, N$, $\operatorname{Re}\lambda_j > 0$ *for* $j > N$. *Let* $k > 0$ *sufficiently large and* $\eta > 0$ *be such that*

$$\operatorname{Re}\lambda_j + \eta\nu_0 + \eta^2\nu_0^2(\operatorname{Re}\lambda_j + k - \eta\nu_0)|k + \lambda_j - \eta|^{-2} > 0 \text{ for } j = 1, \dots, N. \tag{6.19}$$

Then the feedback controller (6.14) *stabilizes exponentially system* (6.3), *that is, the solution* y *to the closed loop system*

$$\begin{aligned} &\frac{\partial y}{\partial t} - \nu_0 \Delta y + (y \cdot \nabla)a + (b \cdot \nabla)y = \nabla p && \text{in } (0, \infty) \times \mathcal{O}, \\ &\nabla \cdot y = 0 && \text{in } (0, \infty) \times \mathcal{O}, \\ &y = \eta \mathbf{1}_\Gamma \sum_{j=1}^N \mu_j \left\langle P_N y, \varphi_j^* \right\rangle_{\widetilde{H}} (\phi_j + \alpha \mathbf{n}) && \text{on } (0, \infty) \times \partial\mathcal{O}, \end{aligned} \tag{6.20}$$

satisfies for some $\gamma > 0$ *the estimate*

$$|y(t)|_{\widetilde{H}} \le Ce^{-\gamma t}|y(0)|_{\widetilde{H}}, \quad \forall t \ge 0. \tag{6.21}$$

It is easily seen that (6.19) holds for k sufficiently large and $\eta > 0$ such that $\operatorname{Re}\lambda_j + \eta\nu_0 > 0$, $\forall j = 1, \dots, N$.

If λ_j are complex, then the controller (6.14) is complex valued too and plugged into system (6.3) leads to a real closed loop system in the state variables $(\operatorname{Re} y, \operatorname{Im} y)$. In order to circumvent such a situation, we shall construct in Section 6.4 a real stabilizing feedback controller of the form (6.14) which has a similar stabilization effect. (See Theorem 6.3.)

A problem of major interest is whether the controller u can be chosen *almost* normal, that is, its normal component $u_{\mathbf{n}}$ is close to the normal $\mathbf{n}$. We have

Corollary 6.1. *There is a stabilizing controller* u *of the form* (6.14), (6.16), *with* $\|a_{ij}\|_{i,j=1}^N = Z_0^{-1}$ *and* $\alpha = \lambda\alpha^*$, *where* $\lambda \in \mathbb{C}$ *is arbitrary and* $\alpha^* \in C^2(\Gamma)$ *satisfies* (6.7) *and*

$$|\alpha^*(x)| \ne 0, \quad a.e.\ x \in \Gamma. \tag{6.22}$$

The exact significance of this result is that, for each $\varepsilon > 0$, there is a Lebesgue measurable subset Γ_ε such that $m(\Gamma \setminus \Gamma_\varepsilon) \le \varepsilon$ and, on Γ_ε, the normal component $\lambda\alpha^*\mathbf{n}$ of the controller u is $\ne 0$ and arbitrarily large with respect to the tangential component represented by ϕ_j. (Here m is the Lebesgue measure on Γ.)

Proof of Corollary 6.1. We set

$$X = \left\{ \psi \in L^2(\Gamma);\ \int_\Gamma \psi\, dx = 0 \right\},$$

$$Y = \left\{ \sum_{j=1}^N \gamma_j \left(p_j^* - \frac{1}{m(\Gamma)} \int_\Gamma p_j^* dx \right);\ \gamma_j \in \mathbb{C} \right\}, Y_1 = X \cap Y^\perp.$$

Then $Y_1 = \{\psi \in X;\ \left\langle X, p_j^* \right\rangle_0 = 0,\ \forall j = 1, \dots, N\}$, $L^2(\Gamma) = Y \oplus Y_1 \oplus \mathbb{C}$, and so, any $\psi \in C^2(\Gamma)$ can be written as

$$\psi = \widetilde{\alpha} + \sum_{j=1}^{N} \gamma_j \left(p_j^* - \frac{1}{m(\Gamma)} \int_{\Gamma} p_j^* dx \right) + \gamma_0, \tag{6.23}$$

for some $\widetilde{\alpha} \in Y_1$ and $\gamma_j \in \mathbb{C}$, $j = 0, 1, \dots, N$. We note that there are $\psi^* \in C^2(\Gamma)$ and $\gamma_j^* \in \mathbb{C}$ such that

$$|\psi^*(x) - \sum_{j=1}^{N} \gamma_j^* \left(p_j^*(x) - \frac{1}{m(\Gamma)} \int_{\Gamma} p_j^* dx \right) - \gamma_0^*| > 0, \text{ a.e. } x \in \Gamma. \tag{6.24}$$

Otherwise, for each $\widetilde{\psi} \in C^2(\Gamma)$ and $\{\widetilde{\gamma}_j\}_{j=0}^N \subset \mathbb{C}$, there is a Lebesgue measurable subset $\widetilde{\Gamma} \subset \Gamma$ such that $m(\widetilde{\Gamma}) > 0$ and

$$\widetilde{\psi} - \sum_{j=1}^{N} \widetilde{\gamma}_j \left(p_j^* - \frac{1}{m(\Gamma)} \int_{\Gamma} p_j^* dx \right) - \widetilde{\gamma}_0 = 0, \text{ in } \widetilde{\Gamma},$$

which, by virtue of arbitrarity of $\widetilde{\psi}$ and $\widetilde{\gamma}_j$, is absurd. Indeed, it suffices to fix $\widetilde{\gamma}_j \in \mathbb{C}$, $j = 0, 1, \dots, N$, and take $\widetilde{\psi} \in C^2(\Gamma)$ in such a way that

$$\inf_{\Gamma} |\widetilde{\psi}| > \sup_{\Gamma} \left| \sum_{j=1}^{N} \widetilde{\gamma}_j \left(p_j^* - \frac{1}{m(\Gamma)} \int_{\Gamma} p_j^* dx \right) - \widetilde{\gamma}_0 \right|$$

to arrive to a contradiction. Then, for the corresponding $\widetilde{\alpha}$ in (6.23), denoted α^*, we have (6.22). Now, we see that, for each $\lambda \in \mathbb{C}$, $\alpha = \lambda\alpha^*$, we have $\left\langle \alpha, p_i^* \right\rangle_0 = 0$, $\forall i = 1, \dots, N$, and so, by (6.17) we have $\|a_{ij}\|_{i,j=1}^N = Z_0^{-1}$, as claimed. ∎

Remark 6.1. The idea of the proof already used in the previous works mentioned above is to decompose system (6.20) or the space $\widetilde{H}$ in a finite differential system corresponding to unstable eigenvalue $\{\lambda_j\}_{j=1}^N$ and an infinite and stable differential system. For this, Hypothesis (L_1) is not absolutely necessary (see [97]) and it was assumed only for convenience in order to get a simple diagonal form for the finite-dimensional unstable system and implicitly for the stabilizing feedback. As regards (L_2), one might suspect that it can be replaced by the weaker assumption $(\mathrm{L}_2)'$ eventually modifying the form of stabilizing controller.

Remark 6.2. As easily follows from the proof in (6.14), the function α can be replaced by a system of functions $\{\alpha_j\}_{j=1}^N$ satisfying (6.7) with the corresponding modification of (6.17).

The above construction works also in more general case where Γ is a smooth part (not necessarily connected) of $\partial\mathcal{O}$ but in this case $\mathbf{1}_\Gamma$ should be replaced by a C^2– function on $\partial\mathcal{O}$ with compact support in Γ and in condition (6.7) α should be replaced by $\mathbf{1}_\Gamma\alpha$.

In the boundary stabilization of Navier–Stokes equation (6.3) with finite dimensional controllers due to compatibility of the boundary trace of state y with the boundary control there are two feasible regularity levels for the solution y, namely $(H^{\frac{1}{2}-\varepsilon}(\mathcal{O}))^d$ for $d=2$ and $(H^{\frac{1}{2}+\varepsilon}(\mathcal{O}))^d$ for $d=3$. (See [28, 84].) However, in $3-D$ the high topological level $(H^{\frac{1}{2}+\varepsilon}(\mathcal{O}))^d$ is not appropriate for some technical reasons related to properties of the inertial term $(y\cdot\nabla)y$ and so, contrary to the linear case, the treatment should be confined to $d=2$.

Consider the Sobolev spaces $W=(H^{\frac{1}{2}-\varepsilon}(\mathcal{O}))^2\cap\widetilde{H}$, $Z=(H^{\frac{3}{2}-\varepsilon}(\mathcal{O}))^2\cap\widetilde{H})$, where $0<\varepsilon<\frac{1}{2}$, with the norms denoted by $\|\cdot\|_W, \|\cdot\|_Z$. The main stabilization result for the Navier–Stokes system (6.1) is Theorem 6.2 below.

Theorem 6.2. *Let $d=2$ and $a=b=y_e$. Then, under the assumptions of Theorem* 6.1, *the feedback boundary controller* (6.14) *stabilizes exponentially system* (6.2) *in a neighborhood $\mathcal{W}=\{y_0\in W;\ \|y_0\|_W<\rho\}$. More precisely, the solution $y\in C([0,\infty);W)\cap L^2(0,\infty;Z)$ to the closed loop system*

$$\begin{aligned}&\frac{\partial y}{\partial t}-\nu_0\Delta y+(y\cdot\nabla)y_e+(y_e\cdot\nabla)y+(y\cdot\nabla)y=\nabla p \text{ in } (0,\infty)\times\mathcal{O},\\&y=\eta\mathbf{1}_\Gamma\sum_{j=1}^N\mu_j\left\langle P_Ny,\varphi_j^*\right\rangle_{\widetilde{H}}(\phi_j+\alpha\mathbf{n}) \text{ on } (0,\infty)\times\partial\mathcal{O},\end{aligned}\tag{6.25}$$

satisfies for $y(0)\in\mathcal{W}$ and ρ sufficiently small

$$\|y(t)\|_W\le Ce^{-\gamma t}\|y(0)\|_W,\ \ \forall t\ge 0,\tag{6.26}$$

for some $\gamma>0$.

In particular, it follows that the boundary feedback controller

$$u=\eta\sum_{j=1}^N\mu_j\left\langle P_N(y-y_e),\varphi_j^*\right\rangle_{\widetilde{H}}(\phi_j+\alpha\mathbf{n})\tag{6.27}$$

stabilizes exponentially the equilibrium solution y_e to (6.1) in a neighborhood $\{y_0\in W;\ \|y_0-y_e\|_W<\rho\}$.

6.2 Proof of Theorem 6.1

We set

$$U^0=\left\{u\in(L^2(\partial\mathcal{O}))^d;\ \int_{\partial\mathcal{O}}u(x)\cdot\mathbf{n}(x)dx=0\right\}.$$

Then, for $k > 0$ sufficiently large, there is a unique solution $y \in (H^{\frac{1}{2}}(\mathscr{O}))^d$ to the equation

$$-\nu_0 \Delta y + (y \cdot \nabla)a + (b \cdot \nabla)y + ky = \nabla p \text{ in } \mathscr{O},$$
$$\nabla \cdot y = 0 \text{ in } \mathscr{O}, \quad y = u \text{ on } \partial\mathscr{O}.$$

We set $y = Du$ and note that

$$D \in L((H^s(\partial\mathscr{O}))^d \cap U^0; (H^{s+\frac{1}{2}})\mathscr{O}))^d), \text{ for } s \geq -\frac{1}{2} \cdot$$

In terms of $\mathscr{A}$ (see (6.9)) and of the Dirichlet map D, system (6.3) can be written as

$$\begin{aligned} &\Pi \frac{d}{dt} y(t) + \mathscr{A}(y(t) - Du(t)) = k\Pi Du, \quad t \geq 0, \\ &y(0) = y_0. \end{aligned} \tag{6.28}$$

Equivalently,

$$\begin{aligned} &\frac{d}{dt} z(t) + \mathscr{A} z(t) = -\Pi \left(D \frac{du}{dt}(t) - kDu(t) \right), \quad t \geq 0, \\ &z(0) = y_0 - Du(0), \end{aligned} \tag{6.29}$$

$$z(t) = y(t) - Du(t), \quad t \geq 0. \tag{6.30}$$

In the following, we fix $k > 0$ sufficiently large and $\eta > 0$ such that (6.19) holds. In particular, for this choice of k and η, we also have

$$\lambda_i + k - \nu_0 \eta \neq 0 \text{ for } i = 1, 2, \ldots, N. \tag{6.31}$$

We note first that in terms of z the controller (6.14) can be, equivalently, expressed as

$$u(t) = \eta \mathbf{1}_\Gamma \sum_{j=1}^{N} \left\langle P_N z(t), \varphi_j^* \right\rangle_{\widetilde{H}} (\phi_j + \alpha \mathbf{n}). \tag{6.32}$$

Indeed, by (6.30) and (6.32), we have

$$\begin{aligned} u(t) = {} & \eta \mathbf{1}_\Gamma \sum_{j=1}^{N} \left\langle P_N y(t), \varphi_j^* \right\rangle_{\widetilde{H}} (\phi_j + \alpha \mathbf{n}) \\ & - \eta \mathbf{1}_\Gamma \sum_{j=1}^{N} \left\langle u(t), D^* \varphi_j^* \right\rangle_0 (\phi_j + \alpha \mathbf{n}), \end{aligned} \tag{6.33}$$

where D^* is the adjoint of D.

On the other hand, if we set $\psi = D\mathbf{1}_\Gamma(\phi_j + \alpha\mathbf{n})$ and recall that

$$\begin{aligned}&\mathscr{L}^*\varphi_i^* - \overline{\lambda}_i\varphi_i^* = \nabla p_i^* \text{ in } \mathscr{O}, \quad \varphi_i^* = 0 \text{ on } \partial\mathscr{O}, \quad \nabla\cdot\varphi_j^* = 0,\\ &\mathscr{L}\psi + k\psi = \nabla\widetilde{p} \text{ in } \mathscr{O}, \quad \psi = \mathbf{1}_\Gamma(\phi_j + \alpha\mathbf{n}) \text{ on } \partial\mathscr{O}, \quad \nabla\cdot\psi = 0,\end{aligned}$$

we get by (6.18) via Green's formula

$$\begin{aligned}\langle\phi_j+\alpha\mathbf{n}, D^*\varphi_i^*\rangle_0 &= \int_{\mathscr{O}}\psi\cdot\overline{\varphi}_i^*dx = -\frac{\nu_0}{\lambda_i+k}\int_\Gamma(\phi_j+\alpha\mathbf{n})\cdot\frac{\partial\overline{\varphi}_i^*}{\partial n}\,dx\\ &-\frac{1}{k+\lambda_i}\langle\alpha, p_i^*\rangle_0 = -\frac{\nu_0}{\lambda_i+k}\,\delta_{ij},\ \forall\, i,j=1,\dots,N,\end{aligned} \tag{6.34}$$

because $\mathbf{n}\cdot\dfrac{\partial\varphi_i^*}{\partial n} = 0$, a.e. on $\partial\mathscr{O}$ (see [44]. Then, by (6.33), (6.34), we see that

$$\langle u(t), D^*\varphi_i^*\rangle_0 = \frac{-\eta\nu_0}{k+\lambda_i-\nu_0\eta}\,\langle P_N y, \varphi_i^*\rangle_{\widetilde{H}} \tag{6.35}$$

and, substituting into (6.33), we get (6.14) as claimed.

Now, by (6.32) and (6.29), we obtain that

$$\begin{aligned}&\frac{dz}{dt}+\mathscr{A}z = -\eta\sum_{j=1}^N\left\langle P_N\left(\frac{d}{dt}z(t)-kz(t)\right), \varphi_j^*\right\rangle_{\widetilde{H}}\Pi D(\mathbf{1}_\Gamma(\phi_j+\alpha\mathbf{n})),\\ &z(0) = z_0 = y_0 - Du(0).\end{aligned} \tag{6.36}$$

It is convenient to decompose system (6.36) into a finite dimensional part corresponding to the unstable spectrum $\{\lambda_j, j = 1,\dots N\}$ of $\mathscr{A}$ and an infinite dimensional one which corresponds to the stable spectrum $\{\lambda_j, j > N\}$. Namely, we write (6.36) as

$$\frac{dz_u}{dt} + \mathscr{A}_u z_u = -\eta P_N\sum_{j=1}^N\left\langle P_N\left(\frac{dz}{dt} - kz\right), \varphi_j^*\right\rangle_{\widetilde{H}}\Pi D(\mathbf{1}_\Gamma(\phi_j + \alpha\mathbf{n})), \tag{6.37}$$

$$\frac{dz_s}{dt} + \mathscr{A}_s z_s = -\eta(I - P_N)\sum_{j=1}^N\left\langle P_N\left(\frac{dz}{dt} - kz\right), \varphi_j^*\right\rangle_{\widetilde{H}}\Pi D(\mathbf{1}_\Gamma(\phi_j + \alpha\mathbf{n})), \tag{6.38}$$

where $z = z_u + z_s$, $z_u \in X_u$, $z_s \in X_s$ and P_N is the algebraic projection on X_u defined in Section 2.1. If we represent z_u as

$$z_u = \sum_{j=1}^N z_j\varphi_j,$$

and recall (6.35), we can rewrite (6.37) as

$$z_j' + \lambda_j z_j = \frac{\eta\nu_0}{k+\lambda_j}\,(z_j' - kz_j),\ \ t \geq 0. \tag{6.39}$$

Equivalently,

$$z_j' + \frac{(k+\lambda_j)\lambda_j + k\eta\nu_0}{k+\lambda_j - \eta\nu_0}\, z_j = 0,\ \ j = 1,\dots,N. \tag{6.40}$$

By (6.19) we have

$$\operatorname{Re}\frac{(k+\lambda_j)\lambda_j + k\eta\nu_0}{k+\lambda_j - \eta\nu_0} > 0 \ \text{ for } j = 1,\dots,N.$$

Then, by (6.39) there is $\gamma_0 > 0$ such that

$$|z_j(t)| \leq e^{-\gamma_0 t}|z_j(0)|,\ \ j = 1,\dots,N. \tag{6.41}$$

On the other hand, by (6.39) we have

$$\frac{dz_s}{dt} + \mathscr{A}_s z_s = -\eta(I - P_N)\sum_{j=1}^{N}(z_j' - kz_j)\Pi D(\mathbf{1}_G(\phi_j + \alpha\mathbf{n})), \tag{6.42}$$

and since

$$\|e^{-\mathscr{A}_s t}\|_{L(\widetilde{H},\widetilde{H})} \leq Ce^{-\gamma_1 t},\ \ \forall t \geq 0,$$

for some $\gamma_1 > 0$, we see by (6.40), (6.42) that

$$|z_s(t)|_{\widetilde{H}} \leq C\exp(-\gamma_0 t)|z_s(0)|_{\widetilde{H}},\ \ \forall t \geq 0, \tag{6.43}$$

which together with (6.41) yields

$$|z(t)|_{\widetilde{H}} \leq C\exp(-\gamma_0 t)|z(0)|_{\widetilde{H}},\ \ \forall t \geq 0.$$

Now, recalling (6.30) and (6.32), we obtain the estimate (6.21), thereby completing the proof.

6.3 Proof of Theorem 6.2

System (6.2) with the feedback controller

$$u = Fy = \eta\mathbf{1}_\Gamma\sum_{j=1}^{N}\mu_j\left\langle P_N y, \varphi_j^*\right\rangle_{\widetilde{H}}(\phi_j + \alpha\mathbf{n})$$

can be written as (see(6.28))

$$\Pi \frac{dy}{dt} + \mathscr{A}(y - DFy) + By = k\Pi\, DFy,\ t > 0,$$
$$y(0) = y_0,$$

where $By = \Pi((y \cdot \nabla)y)$. Setting $z = y - DFy$ we rewrite it as (see (6.36))

$$\frac{dz}{dt} + \mathscr{A}z + B((I - DF)^{-1}z)$$
$$= -\eta \sum_{j=1}^{N} \left\langle P_N \left(\frac{d}{dt} z(t) - kz(t) \right), \varphi_j^* \right\rangle_{\widetilde{H}} \Pi D(\mathbf{1}_\Gamma(\phi_j + \alpha\mathbf{n})).$$

We set, as in previous case,

$$z = z_u + z_s,\ z_u \in X_u,\ z_s \in X_s \text{ and } z_u = \sum_{j=1}^{N} z_j \varphi_j.$$

Recalling (6.39), (6.40), and (6.42), we get for $j = 1, \ldots, N$

$$z_j' + \frac{(k + \lambda_j)\lambda_j + k\eta\nu_0}{k + \lambda_j - \eta\nu_0} z_j + \frac{k + \lambda_j}{k + \lambda_j - \eta\nu_0} \left\langle B((I - DF)^{-1}(z), \varphi_j^*) \right\rangle_{\widetilde{H}} = 0$$

and $z_j' - kz_j = K_j(z)$, where

$$K_j(z) = -\frac{1}{k + \lambda_j - \eta\nu_0} \left((k + \lambda_j)^2 z_j + (k + \lambda_j) \left\langle B((I - DF)^{-1}z, \varphi_j^*) \right\rangle_{\widetilde{H}} \right).$$

Then, we may write the above system as

$$\frac{dz_u}{dt} + \widetilde{\mathscr{A}}_u z_u + \sum_{j=1}^{N} \frac{k + \lambda_j}{k + \lambda_j - \eta\nu_0} \left\langle B((I - DF)^{-1}(z), \varphi_j) \right\rangle_{\widetilde{H}} P_N\varphi_j = 0, \tag{6.44}$$

$$\frac{dz_s}{dt} + \mathscr{A}_s z_s + (I - P_N)B((I - DF)^{-1})z$$
$$= -\eta(I - P_N) \sum_{j=1}^{N} K_j(z)\Pi D(\mathbf{1}_\Gamma(\phi_j + \alpha\mathbf{n})). \tag{6.45}$$

Here $\widetilde{\mathscr{A}_u} \in L(X_u, X_u)$ is the operator defined by

$$\widetilde{\mathscr{A}_u} z_u = \sum_{j=1}^{N} \frac{(k + \lambda_j)\lambda_j + k\eta\nu_0}{k + \lambda_j - \eta\nu_0} z_j\varphi_j.$$

By virtue of (6.41), (6.43), both operators $\widetilde{\mathscr{A}_u}$, $\mathscr{A}_s$ are exponentially stable on spaces X_u, respectively X_s and, therefore, so is the operator

$$\mathscr{C}(z) = \widetilde{\mathscr{A}_u} z_u + \mathscr{A}_s z_s \, , z = z_u + z_s$$

on the space $\widetilde{H}$. We set

$$\mathscr{B}(z) = \sum_{j=1}^{N} \frac{k+\lambda_j}{k+\lambda_j+\eta\nu_0} B((I-DF)^{-1}z), \varphi_j)_{\widetilde{H}} P_N \varphi_j$$

$$+(I-P_N)B((I-DF)^{-1}z) + \eta(I-P_N)\sum_{j=1}^{N} K_j(z)\Pi D(\mathbf{1}_\Gamma(\phi_j+\alpha\mathbf{n}))$$

and rewrite (6.44), (6.45) as

$$\frac{dz}{dt} + \mathscr{C}z + \mathscr{B}(z) = 0, \ t \geq 0, \ z(0) = z_0 = y_0 - DFy_0.$$

Equivalently,

$$z(t) = e^{-t\mathscr{C}} z_0 - \int_0^t e^{-(t-s)\mathscr{C}} \mathscr{B}(z(s))ds, t \geq 0. \tag{6.46}$$

We recall that (see Section 1.3)

$$\|B(z_1) - B(z_2)\|_W \leq C(\|z_1\|_Z + \|z_2\|_Z)\|z_1 - z_2\|_Z, \forall z_1, z_2 \in Z. \tag{6.47}$$

On the other hand, we have

$$|Dz|_{s+\frac{1}{2}} \leq \frac{C}{k-c} |z|_s, \ \ \forall z \in (H^s(\partial\mathscr{O}))^d, \ s \geq -\frac{1}{2},$$

where c, C are independent of k, and this yields, for $s = 1 - \varepsilon$,

$$\|DFy\|_Z \leq \frac{C\eta}{k-c} \|y\|_Z, \ \ \forall y \in Z.$$

This implies that, for k large enough and η as in condition (6.19), the operator $(I - DF)^{-1}$ is Lipschitz on the space Z and this implies that the local Lipschitz property (6.47) extends to the operator $\mathscr{B}$. Then arguing exactly as in the proof of Theorem 5.1 in [28] (see, also, [44] and [45]) we conclude that the integral equation (6.46) has for $||z_0||_W \leq \rho$ sufficiently small, a unique solution $z \in C([0,\infty); W) \cap L^2(0,\infty; Z)$ which has the exponential decay

$$||z(t)||_W \leq Me^{-\gamma t}||z_0||_W, \forall t > 0$$

which completes the proof.

6.4 Real Stabilizing Feedback Controllers

We shall construct here a real stabilizing feedback controller of the form (6.14). To do this we replace the system of functions $\{\varphi_j\}$ by that obtained taking the real and imaginary parts of this one. Namely, we consider the following system of functions in the space H:

$$\psi_{2j-1} = \operatorname{Re}\varphi_j, \quad \psi_{2j} = \operatorname{Im}\varphi_j,$$

and similarly for the adjoint system

$$\psi^*_{2j-1} = \operatorname{Re}\varphi^*_j, \quad \psi^*_{2j} = \operatorname{Im}\varphi^*_j.$$

For the sake of simplicity, we assume that all unstable eigenvalues λ_j, $j = 1, \dots, N$ are simple mentioning, however, that the general case can be treated completely similar.

We set $\widetilde{X}_u = \text{lin span}\{\psi_j; j = 1., ..N\}$. It should be mentioned that the dimension of this space is still N and denote again by P_N the algebraic projection of H on $\tilde{X}_u$. Then we decompose the space as $H = \widetilde{X}_u \oplus \widetilde{X}_s$ and note that the real operator $\mathscr{A}$ leaves invariant both spaces $\widetilde{X}_s$ and $\widetilde{X}_u$ and since $\widetilde{X}_s + i\widetilde{X}_s = X_s$ we infer that the operator $\widetilde{\mathscr{A}^*_s} = \mathscr{A}|_{\widetilde{X}_s}$ generates an exponential stable semigroup on $\widetilde{X}_s \subset H$. We set also $\widetilde{\mathscr{A}^*_u} = \mathscr{A}|_{\widetilde{X}_u}$.

We have

$$\begin{aligned} \mathscr{A}\psi_{2j-1} &= \operatorname{Re}\lambda_{2j-1}\,\psi_{2j-1} - \operatorname{Im}\lambda_{2j-1}\,\psi_{2j}, \\ \mathscr{A}\psi_{2j} &= \operatorname{Im}\lambda_{2j-1}\,\psi_{2j-1} + \operatorname{Re}\lambda_{2j-1}\,\psi_{2j}, \end{aligned} \tag{6.48}$$

and, similarly for ψ^*_j, i.e.,

$$\begin{aligned} \mathscr{A}^*\psi^*_{2j-1} &= \operatorname{Re}\lambda_{2j-1}\,\psi^*_{2j-1} - \operatorname{Im}\lambda_{2j-1}\,\psi^*_{2j}, \\ \mathscr{A}^*\psi^*_{2j} &= \operatorname{Im}\lambda_{2j-1}\,\psi^*_{2j-1} + \operatorname{Re}\lambda_{2j-1}\,\psi^*_{2j}. \end{aligned} \tag{6.49}$$

Equivalently,

$$\begin{aligned} \mathscr{L}^*\psi^*_{2j-1} &= \operatorname{Re}\lambda_{2j-1}\,\psi^*_{2j-1} - \operatorname{Im}\lambda_{2j-1}\,\psi^*_{2j} + \nabla p^*_{2j-1}, \\ \mathscr{L}^*\psi^*_{2j} &= \operatorname{Im}\lambda_{2j-1}\,\psi^*_{2j-1} + \operatorname{Re}\lambda_{2j-1}\,\psi^*_{2j} + \nabla p^*_{2j}. \end{aligned} \tag{6.50}$$

Under assumption (L_2), the following real version of this hypothesis holds.

$(L_2)^*$ *The system* $\{\frac{\partial\psi^*_j}{\partial n}, \ , j = 1, .., N\}$ *is linearly independent on* Γ.

Then, following (6.14) consider the real feedback controller

$$u^* = \eta \mathbf{1}_\Gamma \sum_{j=1}^{N} \left(\left\langle P_N y, \psi_j^* \right\rangle - \sum_{\ell=1}^{N} K_{j\ell} \left\langle P_N y, \psi_\ell^* \right\rangle \right) (\phi_j^* + \alpha \mathbf{n}), \tag{6.51}$$

where $K_{j\ell}$ are made precise later on, ϕ_j^* is of the form

$$\phi_j^* = \sum_{i=1}^{N} a_{ij}^* \frac{\partial \psi_i^*}{\partial n}, \quad j = 1, \dots, N, \tag{6.52}$$

and a_{ij}^* are chosen in a such a way that (see (6.18)),

$$\left\langle \frac{\partial \psi_i^*}{\partial n}, \phi_j^* \right\rangle_0 = -\frac{1}{\nu_0} \langle p_i^*, \alpha \rangle_0 + \delta_{ij}, \; i, j = 1, \dots, N. \tag{6.53}$$

(As seen earlier, this choice is possible by virtue of $(L_2)^*$.)

Now, proceeding as in Section 5.1, we show that for $K_{j\ell}$ suitably chosen the feedback controller (6.51) can be put in the form

$$u = \eta \mathbf{1}_\Gamma \sum_{j=1}^{N} \left\langle P_N z, \psi_j^* \right\rangle (\phi_j^* + \alpha \mathbf{n}), \tag{6.54}$$

where z is given by (6.30). Indeed, in terms of y, (6.54) can be written as (see (6.33))

$$u = \eta \mathbf{1}_\Gamma \sum_{j=1}^{N} \left\langle P_N y, \psi_j^* \right\rangle (\phi_j^* + \alpha \mathbf{n}) - \eta \mathbf{1}_\Gamma \sum_{j=1}^{N} \left\langle u, D^* \psi_j^* \right\rangle_0 (\phi_j^* + \alpha \mathbf{n}). \tag{6.55}$$

This yields

$$\begin{aligned} \langle u, D^* \psi_i \rangle_0 &= \eta \sum_{j=1}^{N} \left\langle P_N y, \psi_j^* \right\rangle \left\langle \phi_j^* + \alpha \mathbf{n}, D^* \psi_i^* \right\rangle_0 \\ &- \eta \sum_{j=1}^{N} \left\langle u, D^* \psi_j^* \right\rangle_0 \left\langle \phi_j^* + \alpha \mathbf{n}, D^* \psi_i^* \right\rangle_0, \; i = 1, \dots, N. \end{aligned} \tag{6.56}$$

On the other hand, by (6.50), (6.54), we see that

$$
\begin{aligned}
&(\operatorname{Re}\lambda_{2i-1}+k)\left\langle\psi^*_{2i-1},\,D(\mathbf{1}_\Gamma(\phi^*_j+\alpha\mathbf{n}))\right\rangle\\
&\quad-\operatorname{Im}\lambda_{2i-1}\left\langle\psi^*_{2i},\,D(\mathbf{1}_\Gamma(\phi^*_j+\alpha\mathbf{n}))\right\rangle\\
&\quad=-\nu_0\left\langle\frac{\partial\psi^*_{2i-1}}{\partial n},\phi^*_j+\alpha\mathbf{n}\right\rangle_0-\langle p^*_{2i-1},\alpha\rangle_0=-\nu_0\delta_{2i-1j},\\
&(\operatorname{Re}\lambda_{2i-1}+k)\left\langle\psi^*_{2i},\,D(\mathbf{1}_\Gamma(\phi^*_j+\alpha_k\mathbf{n}))\right\rangle\\
&\quad-\operatorname{Im}\lambda_{2i-1}\left\langle\psi^*_{2i-1},\,D(\mathbf{1}_\Gamma(\phi^*_j+\alpha\mathbf{n}))\right\rangle\\
&\quad=-\nu_0\left\langle\frac{\partial\psi^*_{2i}}{\partial n},\phi^*_j+\alpha\mathbf{n}\right\rangle_0-\langle p^*_{2i},\alpha\rangle_0=-\nu_0\delta_{2ij}.
\end{aligned}
\tag{6.57}
$$

We set

$$
\begin{aligned}
&\left\langle\psi^*_i,\,D(\mathbf{1}_\Gamma(\phi^*_j+\alpha\mathbf{n}))\right\rangle=\eta_{ij},\\
&\gamma_i=((\operatorname{Re}\lambda_i+k)^2+(\operatorname{Im}\lambda_i)^2)^{-1},\ i,j=1,\dots,N.
\end{aligned}
\tag{6.58}
$$

This yields

$$
\begin{aligned}
(\operatorname{Re}\lambda_{2i-1}+k)\eta_{2i-1j}+\operatorname{Im}\lambda_{2i-1}\,\eta_{2ij}&=-\nu_0\delta_{2i-1j}\\
(\operatorname{Re}\lambda_{2i-1}+k)\eta_{2ij}-\operatorname{Im}\lambda_{2i-1}\,\eta_{2i-1j}&=-\nu_0\delta_{2ij}.
\end{aligned}
\tag{6.59}
$$

Then, by (6.56), (6.58), we obtain

$$
\langle u,D^*\psi^*_i\rangle_0=\eta\sum_{j=1}^N\left\langle P_Ny,\psi^*_j\right\rangle\eta_{ij}-\eta\sum_{j=1}^N\left\langle u,D^*\psi^*_j\right\rangle_0\eta_{ij}.
$$

Equivalently,

$$
\sum_{j=1}^N(\delta_{ij}+\eta\eta_{ij})\langle u,D^*\psi_j\rangle_0=\eta\sum_{j=1}^N\left\langle P_Ny,\psi^*_j\right\rangle\eta_{ij}.
\tag{6.60}
$$

By (6.58), (6.59) we have

$$
\begin{aligned}
\eta_{2i-1j}&=-\nu_0\gamma_{2i-1}(\operatorname{Re}\lambda_{2i-1}+k)\delta_{2i-1j}-\operatorname{Im}\lambda_{2i-1}\delta_{2ij})\\
\eta_{2ij}&=-\nu_0\gamma_{2i-1}(\operatorname{Im}\lambda_{2i-1}\delta_{2i-1j}+(\operatorname{Re}\lambda_{2i-1}+k)\delta_{2i}).
\end{aligned}
\tag{6.61}
$$

We set $K=\|K_{j\ell}\|^N_{j,\ell=1}$, where

$$
K=\eta(\|\delta_{ij}+\nu_0\eta\eta_{ij}\|^N_{i,j=1})^{-1}\times\|\eta_{ij}\|.
\tag{6.62}
$$

By (6.61) we see that, for k sufficiently large, K is well defined. By (6.60), we have

$$\langle u, D^*\psi_j\rangle_0 = \sum_{\ell=1}^{N} K_{j\ell}\langle P_N y, \psi_\ell^*\rangle, \ j = 1, \dots, N.$$

Substituting the latter into (6.56), we see that u is of the form (6.51), where $K_{j\ell}$ is given by (6.62).

Now, we rewrite system (6.28) as (see (6.36), (6.54))

$$\frac{dz}{dt} + \mathscr{A}z = -\eta \sum_{j=1}^{N} \left\langle P_N\left(\frac{dz}{dt} - kz\right), \psi_j^*\right\rangle \Pi D(\mathbf{1}_\Gamma(\phi_j^* + \alpha\mathbf{n})) \tag{6.63}$$

with the corresponding projection on X_u^* (see (6.37))

$$\frac{dz_u}{dt} + \widetilde{\mathscr{A}}_u^* z_u = -\eta P_N \sum_{j=1}^{N} \left\langle P_N\left(\frac{dz}{dt} - kz\right), \psi_j^*\right\rangle \Pi D(\mathbf{1}_\Gamma(\phi_j^* + \alpha\mathbf{n})). \tag{6.64}$$

We set $z_u = \sum_{i=1}^{N} z_i\psi_i$ and so, we may write (6.58) as

$$\sum_{i=1}^{N}(b_{i\ell}z_i' + a_{i\ell}z_i) = -\eta\sum_{j=1}^{N}\sum_{i=1}^{N} b_{ij}(z_i' - kz_i)\eta_{\ell j}, \tag{6.65}$$

where $b_{i\ell} = \langle\psi_i, \psi_\ell^*\rangle$, $a_{i\ell} = \langle\mathscr{A}\psi_i, \psi_\ell^*\rangle$ and $\eta_{\ell j}$ are given by (6.61).

We set $B = \|b_{i\ell}\|_{i,\ell=1}^N$, $A_0 = \|a_{i\ell}\|_{i,\ell=1}^N$, $E = \|\eta_{\ell j}\|_{\ell,j=1}^N$ and rewrite (6.64) as

$$Bz' + A_0z + \eta EB(z' - kz) = 0, \ t \geq 0, \tag{6.66}$$

where $z = \{z_i\}_{j=1}^N$.

To study the stability of system (6.66) it is convenient to consider the limit case $k = \infty$. Taking into account (6.61), we see that for $k \to \infty$ $\eta_{ij}k \to \delta_{ij}$ and so system (6.66) reduces to

$$Bz' + A_0z + \eta Bz = 0,$$

which is exponentially stable if $\eta > 0$ is sufficiently large because

$$|Bz(t)| \leq e^{-\eta t}|Bz(0)| + \int_0^t e^{-\eta(t-s)}|A_0z(s)|ds$$

and B is invertible as consequence of the independence of the systems $\{\psi_j\}_{j=1}^N$ and $\{\psi_j^*\}_{j=1}^N$. This implies that system (6.66) and, consequently, (6.63) is exponentially stable for k and η sufficiently large.

Then, we have the following real version of Theorem 6.1.

Theorem 6.3. *Under assumptions* (L_1), (L_2) *and* (6.7) *for k and η sufficiently large, there is a boundary feedback controller u^* of the form* (6.62) *which stabilizes exponentially system* (6.3).

6.5 An Example to Stabilization of a Periodic Flow in a 2D Channel

The previous results remain true for the Navier–Stokes system in a $2D$ channel $\mathcal{O} = \{(x, y) \in \mathbb{R} \times (0, 1)\}$ with periodic condition in direction x. We illustrate this on the standard problem of laminar flows in a two-dimensional channel with the walls located at $y = 0, 1$. (See, e.g., [1, 16, 25, 28].) We assume that the velocity field $(u(t, x, y), v(t, x, y))$ and the pressure $p(t, x, y)$ are 2π periodic in x. Then, the dynamic of flow is governed by the system

$$\begin{aligned}
&u_t - \nu_0 \Delta u + u u_x + v u_y = p_x, \quad x \in \mathbb{R},\ y \in (0, 1),\\
&v_t - \nu_0 \Delta v + u v_x + v v_y = p_y, \quad x \in \mathbb{R},\ y \in (0, 1),\\
&u_x + v_y = 0,\\
&u(t, x + 2\pi, y) \equiv u(t, x, y),\ \ v(t, x + 2\pi, y) \equiv v(t, x, y),\ \ y \in (0, 1).
\end{aligned} \tag{6.67}$$

Consider a steady-state flow with zero vertical velocity component, that is, $(U(x, y), 0)$. We have $U(x, y) = U(y) = c(y^2 - y)$, $\forall\, y \in (0, 1)$, and take $c = -\frac{a}{2\nu_0}$, $a \in \mathbb{R}^+$. The linearization of (6.67) around the steady-state flow $(U(y), 0)$ leads to the following system

$$\begin{aligned}
&u_t - \nu_0 \Delta u + u_x U + v U' = p_x, \ \ y \in (0, 1),\ x, t \in \mathbb{R},\\
&v_t - \nu_0 \Delta v + v_x U = p_y, \ \ u_x + v_y = 0,\\
&u(t, x + 2\pi, y) \equiv u(t, x, y),\ \ v(t, x + 2\pi, y) \equiv v(t, x, y).
\end{aligned} \tag{6.68}$$

A convenient way to treat this system is to represent u, v as Fourier series. Let us briefly recall this standard procedure. Denote by $L^2_\pi(Q)$, $Q = (0, 2\pi) \times (0, 1)$ the space of all the functions $u \in L^2_{\text{loc}}(R \times (0, 1))$ which are 2π-periodic in x. These functions are characterized by their Fourier series

$$u(x, y) = a_0(y) + \sum_{k \neq 0} a_k(y) e^{ikx},\ \ a_k = \bar{a}_{-k},\ \sum_{k \in \mathbb{Z}} \int_0^1 |a_k|^2 dy < \infty.$$

Similarly, there are defined the Sobolev spaces $H^1_\pi(Q)$, $H^2_\pi(Q)$.

We set $H = \{(u, v) \in (L^2_\pi(Q))^2;\ u_x + v_y = 0,\ v(x, 0) = v(x, 1) = 0\}$.

If $u_x + v_y = 0$, then the trace of v at $y = 0, 1$ is well defined as an element of $H^{-1}(0, 2\pi) \times H^{-1}(0, 2\pi)$. We also set

$$V = \{(u, v) \in H \cap H^1_\pi(Q);\ u(x, 0) = u(x, 1) = v(x, 0) = v(x, 1) = 0\}.$$

The space H can be defined equally as

$$H = \Big\{u = \sum_{k\in\mathbb{Z}} u_k(y)e^{ikx},\ v = \sum_{k\in\mathbb{Z}} v_k(y)e^{ikx},\ v_k(0) = v_k(1) = 0,$$
$$\sum_{k\in\mathbb{Z}} \int_0^1 (|u_k|^2 + |v_k|^2)dy < \infty,\ iku_k(y) + v_k'(y) = 0,\ \text{a.e. } y \in (0, 1),\ k \in \mathbb{Z}\Big\}.$$

Let $\Pi : L^2_\pi(Q) \to H$ be the Leray projector and $\mathscr{A} : D(\mathscr{A}) \subset H \to H'$ the operator

$$\begin{aligned}\mathscr{A}(u, v) &= \Pi\{-\nu_0\Delta u + u_xU + vU',\ -\nu_0\Delta v + v_xU\},\\ &\forall (u, v) \in D(\mathscr{A}) = (H^2((0, 2\pi) \times (0, 1)) \cap V.\end{aligned} \tag{6.69}$$

We associate with (6.69) the boundary value conditions

$$\begin{aligned}u(t, x, 0) &= u^0(t, x),\ u(t, x, 1) = u^1(t, x),\ t \ge 0,\ x \in \mathbb{R},\\ v(t, x, 0) &= v^0(t, x),\ v(t, x, 1) = v^1(t, x),\ t \ge 0,\ x \in \mathbb{R},\end{aligned} \tag{6.70}$$

and, for $k^* > 0$ sufficiently large, we consider the Dirichlet map $D : X \to L^2_\pi(Q)$ defined by $D(u^*, v^*) = (\widetilde{u}, \widetilde{v})$,

$$\begin{aligned}&-\nu_0\Delta\widetilde{u} + \widetilde{u}_xU + \widetilde{v}U' + k^*\widetilde{u} = p_x,\ x \in \mathbb{R},\ y \in (0, 1),\\ &-\nu_0\Delta\widetilde{v} + \widetilde{v}_xU + k^*\widetilde{v} = p_y,\ x \in \mathbb{R},\ y \in (0, 1),\\ &\widetilde{u}_x + \widetilde{v}_y = 0,\ \widetilde{u}(x + 2\pi, y) = \widetilde{u}(x, y),\ \widetilde{v}(x + 2\pi, y) = \widetilde{v}(x, y),\\ &\widetilde{u}(x, y) = u^*(x, y),\ \widetilde{v}(x, y) = v^*(x, y),\ y = 0, 1.\end{aligned} \tag{6.71}$$

Here $X = \big\{(u^*, v^*) \in L^2((0, 2\pi) \times \partial(0, 1));\ u^*(x + 2\pi, y) = u^*(x, y),\ v^*(x + 2\pi, y) = v^*(x, y),\ \int_0^{2\pi} v^*(x, 0)dx = \int_0^{2\pi} v^*(x, 1)dx\big\}$. Then system (6.69), (6.70) can be written as

$$\begin{aligned}&\Pi \frac{d}{dt} y(t) + \mathscr{A}(y(t) - DU^*(t)) = k^*\Pi\, DU^*(t),\ t \ge 0,\\ &y(0) = (u_0, v_0),\end{aligned} \tag{6.72}$$

where $y = (u, v)$, $U^* = (u^*, v^*)$.We denote again by $\mathscr{A}$ the extension of $\mathscr{A}$ on the complexified space $\widetilde{H}$ and by λ_j, φ_j the eigenvalues and corresponding eigenvectors of the operator $\mathscr{A}$. By φ_j^*, we denote the eigenvector to the dual operator $\mathscr{A}^*$. Written into this form, which is exactly (6.28), it is clear that one can apply Theorem 6.1 provided hypotheses (L_1), (L_2) are satisfied for $\mathscr{A}$. We check below the unique continuation hypothesis $(L_2)'$ which has also an interest in itself.

Lemma 6.1. *Assume that all eigenvalues* λ_j, $j = 1, 2, \dots, N$, *are semi-simple. Then we have*

$$\frac{\partial \varphi_j}{\partial n}(x, y) \not\equiv 0, \quad x \in (0, 2\pi), \quad y = 0, 1, \tag{6.73}$$

$$\frac{\partial \varphi_j^*}{\partial n}(x, y) \not\equiv 0, \quad x \in (0, 2\pi), \quad y = 0, 1. \tag{6.74}$$

Proof. If we represent $\varphi_j = (u^j, v^j)$, then (6.73) reduces to

$$\left| \frac{\partial}{\partial y} v^j(x, y) \right| + \left| \frac{\partial}{\partial y} u^j(x, y) \right| > 0, \; x \in (0, 2\pi), \; y = 0, 1. \tag{6.75}$$

We set $\lambda = \lambda_j$ and $\varphi_j = (u, v)$. This means that, if λ is semisimple, then

$$\begin{aligned} &-\nu_0 \Delta u + u_x U + v U' = \lambda u + p_x, \; x \in \mathbb{R}, \; y \in (0, 1), \\ &-\nu_0 \Delta v + v_x U = \lambda v + p_y, \; x \in \mathbb{R}, \; y \in (0, 1), \\ &u_x + v_y = 0, \; u(x + 2\pi, y) = u(x, y), \; v(x + 2\pi, y) = v(x, y). \end{aligned} \tag{6.76}$$

If we represent u, v, p as Fourier series with coefficients u_k, v_k, p_k we reduce (6.76) to the system

$$\begin{aligned} &-\nu_0 u_k'' + (\nu_0 k^2 + ikU) u_k + U' v_k = ikp_k + \lambda u_k, \; y \in (0, 1), \\ &-\nu_0 v_k'' + (\nu_0 k^2 + ikU) v_k = p_k' + \lambda v_k, \; iku_k + v_k' = 0 \text{ in } (0, 1), \\ &u_k(0) = u_k(1) = 0, \;\; v_k(0) = v_k(1) = 0. \end{aligned}$$

Equivalently,

$$\begin{aligned} &-\nu_0 v_k^{\mathrm{iv}} + (2\nu_0 k^2 + ikU) v_k'' - k(\nu_0 k^3 + ik^2 U + iU'') v_k - \lambda(v_k'' - k^2 v_k) = 0, \\ &\hspace{25em} y \in (0, 1), \\ &v_k(0) = v_k(1) = 0, \; v_k'(0) = v_k'(1) = 0, \; \forall k \neq 0. \end{aligned} \tag{6.77}$$

Now, let us check (6.74) or, equivalently, (6.76). We have

$$\frac{\partial}{\partial n} u(x, y) = -i \sum_{k \neq 0} \frac{e^{ikx}}{k} v_k''(y), \quad \forall x, y \in 0, 1,$$

and so (6.76) reduces to

$$|v_k''(0)| + |v_k''(1)| > 0 \text{ for all } k. \tag{6.78}$$

Assume that $v_k''(0) = v_k''(1) = 0$ for all k and lead from this to a contradiction. To this end we set $W_k = v_k'' - k^2 v_k$ and rewrite (6.78) as

$$\begin{aligned} -\nu_0 W_k'' + (\nu_0 k^2 + ikU - \lambda)W_k &= ikU'' v_k \text{ in } (0,1), \\ W_k(0) = W_k(1) &= 0. \end{aligned} \tag{6.79}$$

If we multiply (6.79) by $\overline{W}_k$, integrate on $(0, 1)$, and take the real part, we obtain that

$$\int_0^1 (\nu_0 |W_k'|^2 + (\nu_0 k^2 - \operatorname{Re}\lambda)|W_k|^2) dy = 0, \ \forall k$$

and since $\operatorname{Re}\lambda = \operatorname{Re}\lambda_j \leq 0$ for all $j = 1, \ldots, N$, we get $W_k = 0$, and so $v_k \equiv 0$. The contradiction we arrived at proves (6.78) and (6.73). To prove (6.74) one proceeds similarly with dual system of eigenfunction but since the proof is more delicate we refer to [97].

We suspect that the same argument applies to prove hypothesis (L_2), that is, $\left\{\frac{\partial \varphi_j^*}{\partial n}\right\}_{j=1}^N$ is linearly independent on $\partial\mathscr{O}$ but this remains to be done. Of course, if all φ_j^*, $j = 1, \ldots, N$, are eigenvectors corresponding to the same eigenvalue, the independence follows by Lemma 6.1. Then, following the general case (6.14), we can design a feedback controller $(\widetilde{u}, \widetilde{v})$ for system (6.68). (In terms of (6.70), $\widetilde{u} = (u^0, u^1)$, $\widetilde{v} = (v^0, v^1)$.) We set $\varphi_j^* = (u_j^*, v_j^*)$, $j = 1, \ldots, N$, where φ_j^* are eigenvectors of the dual operator $\mathscr{A}^*$ with corresponding eigenvalues $\overline{\lambda}_j$ and $\operatorname{Re}\lambda_j < 0$ for $j = 1, \ldots, N$. (Recall that the eigenvalues λ_j are repeated according to their multiplicity.)

We consider the boundary feedback controller

$$\begin{aligned} \widetilde{u}(t,x,y) &= \eta \sum_{j=1}^N \mu_j v_j(t) \phi_j^1(x,y), \ x \in \mathbb{R}, \ y = 0, 1, \\ \widetilde{v}(t,x,y) &= \eta \sum_{j=1}^N \mu_j v_j(t)(\phi_j^2(x,y) + aH(y)), \ x \in \mathbb{R}, \ y = 0, 1, \\ v_j(t) &= \int_0^{2\pi} (u(t,x,y)\bar{u}_j^*(x,y) + v(t,x,y)\bar{v}_j^*(x,y)) dx\, dy \\ &= \sum_{k \neq 0} (u_k(t,y)(\bar{u}_j^*)_k(y) + v_k(t,y)(\bar{v}_k^*)_k(y)). \end{aligned} \tag{6.80}$$

Here a is an arbitrary constant, H is a smooth function such that $H(0) = -1$, $H(1) = 1$, μ_j are defined as in (6.15), ϕ^i_j, $i = 1, 2$, are of the form (see (6.16)) $\phi^1_j = \sum_{j=1}^{N} a_{ij}\chi^1_i$, $\phi^2_j = \sum_{i=1}^{N} a_{ij}\chi^2_i$, where $(\chi^1_i, \chi^2_i) = \chi_i$, $\chi_i(x, 0) = -\frac{\partial \varphi^*_i}{\partial y}(x, 0)$, $\chi_i(x, 1) = \frac{\partial \varphi^*_i}{\partial y}(x, 1)$ and a_{ij} are chosen as in (6.17). By Theorem 6.1, we have

Corollary 6.2. *If there is at most one unstable semi-simple eigenvalue for* (6.76) *(eventually multiple), then, for each $a \in \mathbb{R}$ and $\eta > 0$ suitably chosen, the feedback boundary controller* (6.80) *stabilizes exponentially system* (6.68).

We note also that, by Theorem 6.2, the feedback controller (6.80) stabilizes the Navier–Stokes equation (6.67).

6.6 Notes on Chapter 6

This chapter is based on the author work [29]. The method developed here was extended by I. Munteanu [100] to the boundary stabilization of 2-D periodic MHD equations. It should be said that the design of a normal stabilizable feedback controller for general domains is of crucial importance in fluid flow control and the results presented here only partially respond to this demanding objective.

References

1. Aamo, O.M., Krstic, M., Bewley, T.R.: Control of mixing by boundary feedback in $2D$-channel. Automatica **39**, 1597–1606 (2003)
2. Adams, D.: Sobolev Spaces. Academic, New York (1975)
3. Ammar Khodja, F., Benabdallah, A., Dupaix, C., Kostin, L.: Controllability to the trajectories of two-phase field models by one control force. SIAMJ. Control Optim. **42**(5), 1661–1680 (2003)
4. Andreu, F., Vaselles, V., Díaz, J.I., Mazón, J.M.: Some qualitative properties for the total variation flow. J. Funct. Anal. **188**(2), 516–547 (2002)
5. Aniculăesei, Gh., Aniţa, S.: Stabilization of the heat equation via internal feedback. Nonlinear Funct. Anal. Appl. **6**(2), 271–280 (2001)
6. Aniţa, S.: Internal stabilization of diffusion equation. Nonlinear Stud. **8**, 193–213 (2001)
7. Aniţa, S., Barbu, V.: Null controllability of nonlinear convective heat equations. ESAIM: Control Optim. Calc. Var. **5**, 157–173 (2000)
8. Aniţa, S., Barbu, V.: Local exact controllability of a reaction diffusion system. Differ. Integr. Equ. **14**(5), 577–587 (2001)
9. Aniţa, S., Tataru, D.: Null controllability of dissipative semilinear heat equation. Appl. Math. Optim. **46**(2), 97–105 (2002)
10. Araruna, F.D., Boldrini, J.L., Calsavara, B.M.R.: Optimal control and controllability of a phase field system with one control force. Appl. Math. Optim. **70**, 539–563 (2014)
11. Arris, R.: On stability criteria of chemical reaction engineering. Chem. Eng. Sci. **24**, 149–169 (1969)
12. Badra, M.: Feedback stabilization of the 2-D and 3-D Navier–Stokes equations based on an extended system. ESAIM: Control Optim. Calc. Var. **15**, 934–968 (2009)
13. Badra, M., Takahashi, T.: Stabilization of parabolic nonlinear systems with finite dimensional feedback or dynamical controllers: applications to the Navier-Stokes systems. SIAM J. Control Opitm. **49**(2), 420–463 (2011)
14. Bak, P., Tang, C., Wiesenfeld, K.: Self-organized criticality: an explanation of the $1/f$ noise. Phys. Rev. Lett. **59**, 381–394 (1987)
15. Balogh, A., Krstić, M.: Infinite dimensional backstepping-style feedback transformation for the heat equation with an arbitrary level of instability. Eur. J. Control **8**, 165–172 (2002)
16. Balogh, A., Liu, W.-L., Krstic, M.: Stability enhancement by boundary control in $2D$ channel flow. IEEE Trans. Autom. Control **11**, 1696–1711 (2001)
17. Barbu, V.: Nonlinear Semigroups and Differential Equations in Banach Spaces. Noordhoff, Leyden (1976)

V. Barbu, *Controllability and Stabilization of Parabolic Equations*,
Progress in Nonlinear Differential Equations and Their Applications 90,
https://doi.org/10.1007/978-3-319-76666-9

18. Barbu, V.: Analysis and Control of Nonlinear Infinite Dimensional Systems. Academic, Boston (1993)
19. Barbu, V.: Mathematical Methods in Optimization of Differential System. Kluwer, Dordrecht (1994)
20. Barbu, V.: Exact controllability of the superlinear heat equations. Appl. Math. Optim. **42**, 73–89 (2000)
21. Barbu, V.: The Carleman inequality for linear parabolic equations in L^q norm. Differ. Integr. Equ. **15**, 363–372 (2001)
22. Barbu, V.: Controllability of parabolic and Navier–Stokes equations. Sci. Math. Japonicae **56**, 143–210 (2002)
23. Barbu, V.: Local controllability of the phase field system. Nonlinear Anal. Theory Methods Appl. **50**, 363–372 (2002)
24. Barbu, V.: Feedback stabilization of Navier–Stokes equations. ESAIM: Control Optim. Calc. Var. **9**, 197–206 (2003)
25. Barbu, V.: Stabilization of a plane channel flow by wall normal controllers. Nonlinear Anal. Theory. Methods Appl. **56**, 145–168 (2007)
26. Barbu, V.: Nonlinear Differential Equations of Monotone Type in Banach Spaces. Springer, New York (2010)
27. Barbu, V.: Self-organized criticality and convergence to equilibrium of solutions to nonlinear diffusion problems. Annu. Rev. Control. **340**, 52–61 (2010)
28. Barbu, V.: Stabilization of Navier-Stokes Flows. Springer, Berlin (2011)
29. Barbu, V.: Stabilization of Navier-Stokes equations by oblique boundary feedback controllers. SIAM J. Control Optim. **50**, 2288–2307 (2012)
30. Barbu, V.: Boundary stabilization of equilibrium solutions to parabolic equations. IEEE Trans. Autom. Control **58**(9), 2416–2420 (2013)
31. Barbu, V.: Exact null internal controllability for the heat equation on unbounded convex domains. ESAIM: Control Optim. Calc. Var. **6**(1), 222–235 (2013)
32. Barbu, V.: Note on the internal stabilization of stochastic parabolic equations with linearly multiplicative Gaussian noise. ESAIM: Control Optim. Calc. Var. **19**(4), 1055–1063 (2013)
33. Barbu, V.: Self-organized criticality of cellular automata model; absorbtion in finite-time of supercritical region into a critical one. Math. Methods Appl. Sci. 1–8 (2013)
34. Barbu, V., Iannelli, M.: Controllability of the heat equation with memory. Differ. Integr. Equ. **13**, 1393–1412 (2000)
35. Barbu, V., Lefter, C.: Internal stabilizability of Navier–Stokes Equations. Syst. Control Lett. **48**, 161–167 (2003)
36. Barbu, V., Wang, G.: Internal stabilization of semilinear parabolic systems. J. Math. Anal. Appl. **285**, 387–407 (2003)
37. Barbu, V., Triggiani, R.: Internal stabilization of Navier–Stokes equations with finite dimensional controllers. Indiana Univ. Math. J. **53**, 1443–1494 (2004)
38. Barbu, V., Da Prato, G.: The generator of the transition semigroup corresponding to a stochastic partial differential equation. Commun. Part. Differ. Equ. **33**, 1318–1330 (2008)
39. Barbu, V., Lasiecka, I.: The unique continuation property of eigenfunctions of Stokes–Oseen operator is generic with respect to coefficients. Nonlinear Anal. **75**, 3964–3972 (2012)
40. Barbu, V., Röckner, M.: Stochastic porous media equation and self-organized criticality: convergence to the critical state in all dimensions. Commun. Math. Phys. **311**, 539–555 (2012)
41. Barbu, V., Röckner, M.: Stochastic variational inequalities and applications to the total variation flow perturbed by linear multiplicative noise. Arch. Ration. Mech. Anal. **209**(3), 797–834 (2013)
42. Barbu, V., Tubaro, L.: Exact controllability of stochastic differential equations with multiplicative noise (submitted)
43. Barbu, V., Rascanu, A., Tessitore, G.: Carleman estimates and controllability of stochastic heat equation with multiplicative noise. Appl. Math. Optim. **47**, 97–120 (2003)

44. Barbu, V., Lasiecka, I., Triggiani, R.: Boundary stabilization of Navier–Stokes equations. Am. Math. Soc. **852**, 1–145 (2006)
45. Barbu, V., Lasiecka, I., Triggiani, R.: Abstract setting for tangential boundary stabilization of Navier–Stokes equations by high and low-gain feedback controllers. Nonlinear Anal. **64**, 2704–2746 (2006)
46. Barbu, V., Da Prato, G., Röckner, M.: Stochastic porous media equation and self-organized criticality. Commun. Math. Phys. **285**, 901–923 (2009)
47. Barbu, V., Rodrigues, S., Shiriakyan, A.: Internal exponential stabilization to a nonstationary solution for 3D Navier–Stokes equations. SIAM J. Control Optim. **49**, 1454–1478 (2011)
48. Barbu, V., Da Prato, G., Röckner, M.: Stochastic Porous Media Equations. Springer, Berlin (2016)
49. Barbu, V., Colli, J.P., Gillardi, G., Marinoschi, G.: Feedback stabilization of the Cahn-Hilliard type system for phase separation. J. Differ. Equ. **262**, 2286–2334 (2017)
50. Beceanu, M.: Local exact controllability of nonlinear diffusion equation in 1-*D*. Abstr. Appl. Anal. **14**, 793–811 (2003)
51. Bensoussan, A., Da Prato, G., Delfour, M.C., Milter, S.K.: Representation and Control of Infinite Dimensional Systems. Systems & Control Foundations & Applications, vol. I. Birkhäuser, Boston (2005)
52. Bošković, D.M., Krstić, M., Liu, W.: Boundary control of an unstable heat equation. IEEE Trans. Autom. Control **46**, 2028–2038 (2001)
53. Breiten, T., Kunisch, K.: Riccati based feedback control of the monodomain equations with the FitzHugh-Nagumo model. SIAM J. Control Optim. **52**, 4057–4081 (2014)
54. Brezis, H.: Opérateurs Maximaux Monotones et semigroupes de contractions dans les espaces de Hilbert. North Holland, Amsterdam (1973)
55. Brezis, H.: Functional Analysis, Sobolev Spaces and Partial Differential Equations. Springer, Berlin (2011)
56. Cafiero, R., Loreto, V., Pietronero, A., Zapperi, Z.: Local rigidity and self-organized criticality for avalanches. Europhys. Lett. **29**, 111–116 (1995)
57. Caginalp, C.: An analysis of a phase field model of a free boundary. Arch. Ration. Mech. Anal. **92**, 205–243 (1986)
58. Cannarsa, P., Fragnelli, G., Rocchetti, D.: Controllability results for a class of one-dimensional degenerate parabolic problems in nondivergence form. J. Evol. Equ. **8**, 583–616 (2008)
59. Cannarsa, P., Martinez, P., Vancostenoble, J.: Global Carleman estimates for degenerate parabolic operators with applications. Mem. Am. Math. Soc. **239** (2016)
60. Coron, J.M.: Control and Nonlinearity. Mathematical Survey and Monographs, vol. 136. American Mathematical Society, Providence (2007)
61. Coron, J.M., Guilleron, J.Ph.: Control of three heat equations coupled with two cubic nonlinearities. SIAM J. Control Optim. **55**, 989–1019 (2017)
62. Coron, J.M., Guerrero, S., Rossier, L.: Null controllability of a parabolic system with a cubic coupling term. SIAM J. Control Optim. **48**, 5629–5635 (2010)
63. Coron, J.-M., Díaz, J.I., Drici, A., Mingazzinbi, T.: Global null controllability of the 1-dimensional nonlinear slow diffusion equation. Chin. Ann. Math. B **34**(3), 333–344 (2013)
64. Da Prato, G., Zabczyk, J.: Stochastic Equations in Infinite Dimension. Cambridge University Press, Cambridge (2013)
65. Deimling, K.: Nonlinear Analysis. Springer, Berlin (1990)
66. Diaz, J.I., Henry, J., Ramos, A.M.: On the approximate controllability of some semilinear parabolic boundary-value problems. Appl. Math. Optim. **37**, 71–97 (1998)
67. Dubova, A., Osses, A., Puel, J.P.: Exact controllability to trajectories for semilinear heat equations with discontinuous coefficients. ESAIM: Control Optim. Calc. Var. **8**, 621–661 (2002)
68. Dubova, A., Fernandez-Cara, E., Gonzales-Burgos, M., Zuazua, E.: On the controllability of parabolic systems with a nonlinear term involving the state and the gradient. SIAM J. Control Optim. **41**, 718–819 (2002)

69. Fattorini, H.D., Russell, D.L.: Exact controllability theorems for linear parabolic equations in one space dimension. Arch. Ration. Mech. Anal. **43**, 272–292 (1971)
70. Fernandez, L.A., Zuazua, E., Approximate controllability for the semilinear heat equation involving gradient terms. J. Optim. Theory Appl. **101**, 307–328 (1999)
71. Fernandez-Cara, E.: Null controllability of the semilinear heat equation. ESAIM: Control Optim. Calc. Var. **2**, 87–107 (1997)
72. Fernandez-Cara, E., Zuazua, E.: Null and approximate controllability for weakly blowing up semilinear heat equations. Ann. IHP. Anal. Nonlinéaire **17**(5), 583–616 (2000)
73. Fernandez-Cara, E., Guerrero, S.: Global Carleman inequalities for parabolic systems and applications to controllability. SIAM J. Control Optim. **45**, 1305–1446 (2006)
74. Fursikov, A.V.: Real processes of the 3-D Navier–Stokes systems and its feedback stabilization from the boundary. In: Agranovic, M.S., Shubin, M.A. (eds.) AMS Translations. Partial Differential Equations, M. Vishnik Seminar, pp. 95–123 (2002)
75. Fursikov, A.V.: Stabilization for the $3-D$ Navier–Stokes systems by feedback boundary control. Discr. Contin. Dyn. Syst. **10**, 289–314 (2004)
76. Fursikov, A., Imanuvilov, O.Yu.: Controllability of Evolution Equations, Lecture Notes, vol. 34. Seoul National University, Seoul (1996)
77. Gonzales-Burgos, M., Perez-Garcia, R.: Controllability to the trajectories of phase-field models by one control force. Asymptot. Anal. **46**, 123–162 (2006)
78. Guerrero, S., Imanuvilov, O.Yu.: Remarks on noncontrollability of the heat equation with memory. ESAIM: Control Optim. Calc. Var. **19**, 288–300 (2013)
79. Halanay, A., Murea, C., Safta, C.: Numerical experiment for stabilization of the heat equation by Dirichlet boundary control. Numer. Funct. Anal. Optim. **34**(12), 1317–1327 (2013)
80. Henry, D.: Geometric Theory of Semiliner Parabolic Equations, Lecture Notes Mathematics, vol. 840. Springer, Berlin (1981)
81. Imanuvilov, O.Yu., Yamamoto, M.: Carleman inequalities for parabolic equations in Sobolev spaces of negative order and exact controllability for semiliner parabolic equations. Publ. Res. Inst. Math. Sci. **39**, 227–274 (2003)
82. Kato, T.: Perturbation Theory of Linear Operators. Springer, Berlin (1966)
83. Ladyzenskaia, O.A., Solonnikov, V.A., Ural'ceva, N.N.: Linear and Quasilinear Equations of Parabolic Type. Translations Mathematical Monographs, vol. 23. American Mathematical Society, Providence (1968)
84. Lasiecka, I., Triggiani, R.: Control Theory for Partial Differential Equations Continuous and Approximation Theory. Encyclopedia of Mathematics and Its Applications. Cambridge University Press, Cambridge (2000)
85. Lebeau, G., Robbiano, L.: Contrôle exact de l'équation de la chaleur. Commun. Part. Differ. Equ. **30**, 335–357 (1995)
86. Lefter, C.: Internal feedback stabilization of nonstationary solutions to semilinear parabolic equations. J. Optim. Theory Appl. **170**(3), 960–976 (2016)
87. Le Rousseau, J., Lebeau, G.: On Carleman estimates for elliptic and parabolic operators. Applications to unique continuation and control of parabolic equations. Control Optim. Calc. Var. **18**, 712–747 (2012)
88. Le Rousseau, J., Robbiano, L.: Local and global Carleman estimates for parabolic operators with coefficients with jumps at interfaces. Invent. Math. **183**, 245–336 (2011)
89. Lions, J.L.: Quelques Méthodes de resolution des problèmes aux limites Non Linéaires. Dunod, Gauthier-Villars (1969)
90. Liu, Q.: Some results on the controllability of forward stochastic heat equations with control on drift. J. Funct. Anal. **260**, 832–851 (2011)
91. Korevaar, N.J., Lewis, J.L.: Convex solutions of certain elliptic equations have constant rank Hessians. Arch. Ration. Mech. Anal. **97**(1), 19–32 (1987)
92. Marinoschi, G.: Internal feedback stabilization of Cahn–Hilliard system with viscosity effect. Pure Appl. Funct. Anal. **3**, 107–135 (2018)
93. Micu, S., Zuazua, E.: On the lack of null controllability of the heat equation on the half-line. Trans. Am. Math. Soc. **353**, 1635–1659 (2000)

94. Micu, S., Zuazua, E.: On the lack of null controllability of the heat equation on the half-space. Part. Math. **58**, 1–24 (2001)
95. Miller, L.: Unique continuation estimates for the Laplacian and the heat equation on non-compact manifolds. Math. Res. Lett. **12**, 37–47 (2005)
96. Mizel, V.J., Seidman, T.: Observation and prediction for the heat equation. J. Math. Anal. Appl. **28**, 303–312 (1969)
97. Munteanu, I.: Normal feedback stabilization of periodic flows in a 2-D channel. J. Optim. Theory Appl. **152**, 413–438 (2012)
98. Munteanu, I.: Boundary stabilization of the phase field system by finite dimensional feedback controllers. J. Math. Anal. Appl. **412**, 964–975 (2014)
99. Munteanu, I.: Stabilization of parabolic semilinear equations. Int. J. Control **90**, 1063–1076 (2017)
100. Munteanu, I.: Boundary stabilization of a 2-D periodic MHD channel flow, by proportional feedback. ESAIM: Control Optim. Calc. Var. **23**, 1253–1266 (2017)
101. Murray, J.: Mathematical Biology, Spatial Models and Biomedical Applications. Springer, Berlin (2003)
102. Pandolfi, L.: The controllability of the Gurtin–Pipkin equation: a cosine operator approach. Appl. Math. Optim. **52**, 143–165 (2005)
103. Pandolfi, L.: Riesz systems and controllability of heat equation with memory. Integr. Equ. Oper. Theory **64**, 429–453 (2009)
104. Pazy, A.: Semigroups of Linear Operators and Applications to Partial Differential Equations. Springer, New York (1983)
105. Qin, S., Wang, G.: Controllability of impulse controlled systems of heat equations coupled by constant matrices. J. Differ. Equ. **263**(10), 6456–6493 (2017)
106. Raymond, J.P.: Feedback boundary stabilization of the two dimensional Navier–Stokes equations. SIAM J. Control Optim. **45**, 790–828 (2006)
107. Raymond, J.P.: Feedback boundary stabilization of the three dimensional Navier–Stokes equations. J. Math. Pures Appl. **87**, 627–669 (2007)
108. Russell, D.L.: A unified boundary controllability theory for hyperbolic and parabolic partial differential equations. Stud. Appl. Math. **52**, 189–212 (1973)
109. Russell, D.: Controllability and stabilizability theory for linear partial differential equations: recent progress and open questions. SIAM Rev. **20**(4), 639–739 (1978)
110. Sirbu, M.: Feedback null controllability of the semilinear heat equation. Differ. Integr. Equ. **15**, 115–128 (2002)
111. Smyshlyaev, A., Krstic, M.: Closed ferm boundary state feedbacks for a class of $1D$ partial differential equations. IEEE Trans. Autom. Control **49**, 2185–2202 (2004)
112. Tang, S., Zhang, X.: Null controllability for forward and backward stochastic parabolic equations. SIAM J. Control Optim. **48**, 2191–2216 (2009)
113. Temam, R.: Navier-Stokes Equations and Nonlinear Functional Analysis. SIAM, Philadelphia (1983)
114. Triggiani, R.: On the stabilizability problem in Banach spaces. J. Math. Anal. Appl. **52**(3), 383–403 (1975)
115. Triggiani, R.: Boundary feedback stabilization of parabolic equations. Appl. Math. Optim. **6**, 2185–2202 (1980)
116. Tucsnak, M., Weiss, G.: Observations and Control for Operator Semigroups. Birkhäuser, Basel (2009)
117. Vazquez, R., Krstic, M.: A closed-form feedback controller for stabilization of linearized Navier–Stokes equations: the $2D$ Poisseuille flow. IEEE Trans. Autom. (2005). https://doi.org/10.1109/CDC.2005.1583349
118. Vazquez, R., Tvelat, E., Coron, J.M.: Control for fast and stable Laminar-to-High-Reynolds-Number transfer in a $2D$ channel flow. Discr. Contin. Dyn. Syst. Ser. B **10**(4), 925–956 (2008)
119. Wang, G.: L^{∞}-controllability for the heat equation and its consequence for the time optimal control problems. SIAM J. Control Optim. **47**, 1701–1720 (2008)

120. Zhang, X.: A unified controllability and observability theory for some stochastic and deterministic partial differential equations. In: Proceedings of the International Congress of Mathematicians, Hyderabad, pp. 3008–3033 (2010)
121. Zuazua, E.: Finite dimensional null controllability of the semilinear heat equations. J. Math. Pures Appl. **76**, 237–264 (1997)
122. Zuazua, E.: Approximate controllability for semilinear heat equation with globally Lipschitz nonlinearities. Control. Cybern. **28**, 665–683 (1999)

Index

V. Barbu, *Controllability and Stabilization of Parabolic Equations*,
Progress in Nonlinear Differential Equations and Their Applications 90,
https://doi.org/10.1007/978-3-319-76666-9

Printed in the United States
By Bookmasters